T0231106

Water

The Forgotten Biological Molecule

Edited by

Denis Le Bihan

NeuroSpin, France

Hidenao Fukuyama

Kyoto University, Japan

Routledge
Taylor & Francis Group

LONDON AND NEW YORK

WATER: THE FORGOTTEN BIOLOGICAL MOLECULE

First published 2011 by Pan Stanford Publishing Pte. Ltd.

Published 2018 by Routledge
2 Park Square, Milton Park, Abingdon, Oxon OX14 4RN
52 Vanderbilt Avenue, New York, NY 10017

Routledge is an imprint of the Taylor & Francis Group, an informa business

Copyright © 2011 by Taylor & Francis.

British Library Cataloguing-in-Publication Data
A catalogue record for this book is available from the British Library.

ISBN 13: 978-981-4267-52-6 (hbk)

Dropwise

Oceanwise

Cloudwise

Rainwise

Riverwise

Lakewise

Being water

Being stream

Being waterfall

Being wave

Being dew

Being trickle

Being steam

Being ice

The being of water is

A guest homing on water

—Dilip Chitre, *Shesha: Selected Marathi Poems (1954–2008)*

Contents

Another Book on Water?

At the onset of the 21st century, humankind is focusing its attention on a very small molecule made of three atoms, one carbon atom and two oxygen atoms. Controlling CO_2 in the atmosphere is becoming a major goal economically, socially, and politically. Yet, there is another small molecule, also made of three atoms, one oxygen atom and two hydrogen atoms, that is going to play a similarly, if not more, prominent role in the near future. While an excessive concentration of CO_2 might be harmful to life on earth as we know it, H_2O, especially in its liquid form, the "Blue Gold" (hence the color of the book cover), is just indispensable to our lives. Over the past centuries, lack of access to water has triggered death, through drought, famine, and even wars. Most capitals have been established near large rivers. The preservation of the quality and abundance of drinking water will become a major challenge for nations during this century. This is no surprise. Water makes 60% to 70% of the human body weight and is crucial to the working of the biological machinery. So far humankind has more or less imprinted the importance of water for its survival in its collective unconsciousness, and it is noticeable that water has a central position in many religions on earth and is sometimes even considered as a sacred element.

Although the concept that water is the molecule of life is today evident, the exact mechanisms underlying the contribution of water to life remain largely unclear. Worse, the physics of this tiny, extremely common, and abundant molecule is still not fully understood. Water is a very atypical molecule in the first place. Why is it in the liquid state at the earth's temperature and pressure, while other molecules of similar size, again CO_2, are in the gaseous state? Why is ice, the solid state of water, lighter than liquid water, so that it can float, like icebergs? Why is water in the liquid form in clouds, while the temperatures there are well below the freezing level? All these "abnormal" features make water as we know it on earth and give it a form that is compatible with life. Water is abundant in the universe, but we have not yet found another place where similar atmospheric conditions exist. Hence, one should not be surprised that the physics of water is still an object of intense research, with reports appearing in the highest-ranking scientific journals.

The abnormal properties of water are not so easy to explain, but most of them lie in the geometry of its molecule, especially the angle between the hydrogen and oxygen atoms, which is about 105°. Yes, life on earth has been made possible because of this peculiar angle. Had this angle been a little larger (109°), water molecules would have condensed themselves as crystals, like carbon atoms in diamonds, and we would not be here. Beside the physics of single water molecules resides the "sociology" of water molecules: how water molecules are organized in space and time as a dynamic network because of their electrical charges and their location around the molecule, thereby allowing "hydrogen bounding," which is especially important to understand liquid water.

Life has been made possible because the water molecules can accommodate other atoms or molecules, smaller or bigger, within their network in a particular way, whether water "likes" or "dislikes" them. It should be emphasized that water is never an "inert" compound, but a rather reactive molecule, which will interact, negatively or positively, with the other moieties present within its network. Reciprocally, those moieties interact with the water network and might induce changes in the water network structure. Those interactions explain why water is so important in almost all chemical reactions taking place in cells, photosynthesis, enzyme reactions, metabolism, etc., without a need for the cells to provide energy. The shape of those very complex biological molecules, such as enzymes or cell receptors, is crucial for them to interact with their intended targets. Water molecules play a direct role in giving those molecules their shape. Water may also play a more direct role in the reactions, for instance by dissociating molecules in ions. How could such interactions take place in living systems is the object of this book.

Progress has been made in our understanding of protein hydration and protein dynamics, on the interaction of water with hydrophobic or hydrophilic interfaces, such as membranes, but many questions remain largely unanswered and sometimes even subject of big controversies. There is no reason to think that water molecules in cells would be different than water molecules in a glass. But life has capitalized on water. It might well be that cell compounds that are extremely numerous and dense are interacting so strongly with the water network so as to affect the mobility of the water molecules, or perhaps modulate the structure of the entire network. Those effects could be very local, encompassing a layer of one or two water molecules at the contact of the proteins, or propagate themselves to hundred layers through electrostatic interactions within the water network. Some researchers consider that water in cells has the structure of a gel, whereas others do not. Those different views are presented in this book.

It should be pointed out, however, that results of experiments are extremely dependent on the methods used to obtain them. It is not at all the same to "observe" the behavior of single water molecules at a time resolution of picoseconds and to "integrate" the movement of billions of water molecules in a dynamic network over time intervals of milliseconds. Scaling in space and time is clearly an issue that must be kept in mind when comparing results obtained with different techniques, as it could contribute to the discrepancies. What is true at a molecular scale might not hold at a network scale. Interactions between scales are complex and not necessarily linear, especially in the crowed environment present in living cells. Furthermore, it is also quite different to observe the effects of water on complex molecules *in vitro* in a carefully controlled environment, and *in vivo* in intact cell systems. Unfortunately, there is no perfect technique at our disposal.

Researchers have so far used mainly neutron scattering methodology and nuclear magnetic resonance to study the interactions of biological molecules with water. Both approaches have their benefits and limitations, but they do not address at all the same scales in time and space. Hence, caution is required when extrapolating results obtained by one method to the other. Both provide extremely valuable information and should rather be considered as complementary. Another powerful approach that has been introduced more recently with the availability of powerful computing systems is the simulation of molecular dynamics of water molecules. With reasonable hypotheses on the local forces involved in molecular interactions it is possible to simulate the individual movement of a large number of water molecules in the vicinity of complex molecules, such as proteins or receptors, or interfaces, such as membranes. Predictions can then be made on the interactions, which can be tested against experimental evidence.

There is still a higher order of interaction and integration to consider when dealing with cells, tissues, or even organisms as a whole. The implication of water in molecular biology and basic cellular mechanisms, such as those defining the cells' shape or volume or regulating interaction between cells, cannot be ignored. A lot of work remains to be done in order to understand the mechanisms in detail, but recent studies of water mobility, whether diffusion or flow, in biological structures have pointed out its importance to cellular physiology. Different organisms have adopted different strategies in the way they get the most out of water, depending on their environment, and water contributes to the biodiversity. Faulty mechanisms in the use of water by tissues may lead to severe diseases or death.

Clearly, water deserves to be seen as the prime "biological molecule," and its importance should not be forgotten and taken for granted. How such a tiny molecule with its 105° "magic" angle could have been at the origin of life remains clearly an amazing question, which this book modestly attempts to address. This is motivation that led us to cover a large variety of timely issues regarding the importance of water for biological systems, from molecular biology to whole organisms. Life has led to intelligence, and recent studies have suggested that water (9 out of 10 molecules in the brain are water molecules) may also actively contribute to the mechanisms underlying brain function. Could the "molecule of life" also be the "molecule of the mind"?

Advice to readers

The book covers a wide variety of topics. Depending on the advancement of knowledge for each of those topics and the methods used to investigate them, the technical level of the chapters greatly varies. Indeed, the book was rather conceived as a collection of independent chapters dealing with those issues, and not written for a continuous reading. Readers are invited to browse the book freely, as chapters could easily be omitted in a first approach and read in any order, without altering its intended meaning.

Denis Le Bihan
Kyoto, February 2010

Acknowledgments

This book has been made possible thanks to the individual contribution of some of the best experts in the world, and we would like to gratefully acknowledge them. We met with some of them during a workshop organized in November 2008 in Kyoto to discuss various issues about the role of water in biological processes. Finally, it is a pleasure to thank Stanford Chong, Sarabjeet Garcha, Jenny Rompas, and those members of the staff of Pan Stanford Publishing Pte. Ltd. who have been concerned with the production of this book for their kindness and consideration they have shown to us.

Denis Le Bihan and Hidenao Fukuyama

Part I

THE STRUCTURE OF THE WATER MOLECULE

Chapter 1

THE WATER MOLECULE, LIQUID WATER, HYDROGEN BONDS, AND WATER NETWORKS

Martin Chaplin

Department of Applied Science, London South Bank University, Borough Road, London SE1 0AA, UK

martin.chaplin@lsbu.ac.uk

1.1 INTRODUCTION

Water (H_2O) is often perceived to be a perfectly ordinary substance because it is transparent, odorless, tasteless, and ubiquitous. We interact with it all the time in our everyday lives, and it is, not surprisingly, the most studied material on earth. Unfortunately, we often erroneously think that the properties of liquids are those that we find for liquid water and the changes we see as water changes to steam or ice as those commonly found among other liquids. These assumptions are much mistaken as the properties of liquid water are unique, because it is considerably different from other liquids. Liquid water is the most extraordinary substance, and life on earth and its probability elsewhere in the universe cannot have evolved or continue without it. Organisms consist mostly of liquid water, and this water performs many functions. Therefore it cannot be considered simply as an inert diluent. It transports, lubricates, reacts, enables, stabilizes, signals, structures, and partitions. The living world should be thought of as an equal partnership between the biological molecules and water where each is required and structured by the other.

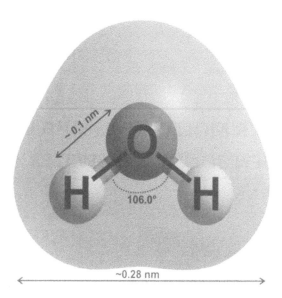

Figure 1.1 The water molecule. Water molecules have the molecular formula H_2O. The "average" water molecule has two mirror planes (in the plane of the picture and vertically through the oxygen atom at a right angle) and a twofold axis of symmetry (vertically through the oxygen atom). As experimentally determined, a gaseous water molecule has an O–H length of 0.095718 nm and H–O–H angle of 104.474°, somewhat less than the tetrahedral angle (109.47°), because of the lower repulsion between the hydrogen atoms than between them and the unbonded electrons. Water molecules in the liquid state have the dimension s as shown. The mean van der Waals diameter of water has been reported to be identical with that of isoelectronic neon (0.282 nm). Molecular model values and intermediate peak radial distribution data indicate, however, that it is somewhat greater than this (~0.32 nm). The molecule is clearly not perfectly spherical, however, with about a ±5% variation in van der Waals diameter dependent on the axis chosen—approximately tetrahedrally placed slight indentations being apparent opposite the four (putative) electron pairs.

The water molecule is the simplest compound of the two most common reactive elements in the universe, consisting of just two hydrogen atoms attached to a single oxygen atom. Indeed, very few molecules are smaller or lighter (Fig. 1.1). This small size belies the complexity of its actions and its singular capabilities. Unusually for molecules in the liquid state, water molecules are unstable as the three atoms only stay together momentarily in spite of 80% of the molecule's electrons being concerned with bonding. The electronic structure of water is given in Fig. 1.2.

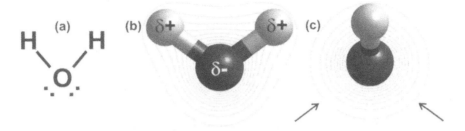

Figure 1.2 The water molecule is often described as having four, approximately tetrahedrally arranged, sp³-hybridized electron pairs, two of which are associated with hydrogen atoms leaving the two remaining lone pairs (a). In a perfect tetrahedral arrangement the bond–bond, bond–lone pair, and lone pair–lone pair angles would all be 109.47° and such tetrahedral bonding patterns are found (on average) in condensed phases such as hexagonal ice (i.e., normally found ice). *Ab initio* calculations on isolated molecules, however, do not confirm the presence of significant directed electron density where lone pairs are expected (b, shown by the arrows, in c), but rather an sp²-hybridized structure with a pz orbital is indicated. The negative charge is more evenly smeared out along the line between where these lone pairs would have been expected and lies closer to the center of the O atom than the centers of the positive charge on the hydrogen atoms. Thus, the oxygen atom possesses a partial negative charge, and the hydrogen atoms are somewhat positively charged.

The hydrogen atoms in water may possess parallel (paramagnetic *ortho*-H_2O, magnetic moment = 1) or antiparallel (nonmagnetic *para*-H_2O, magnetic moment = 0) nuclear spins. The equilibrium ratio of these nuclear spin states in H_2O is all *para* at 0 K or when absorbed at surfaces, where the molecules have no rotational spin in their ground state, shifting to 3:1 *ortho:para* at less cold temperatures (>50 K) in the bulk. This means that liquid H_2O effectively consists of a mixture of slowly interconverting nonidentical molecules. Although the properties of pure liquid *ortho*-H_2O or *para*-H_2O are unknown, the difference in energy between these two forms of water is small, and it is questionable, although uncertain, whether their properties will differ significantly from those of "normal" liquid water.

The water molecule consists of two light atoms (H) and a relatively heavy atom (O). The approximately 16-fold difference in mass gives rise to its ease of molecular rotation and the substantial relative movements of the hydrogen nuclei, which significantly affect their interaction with other molecules (Fig. 1.3).

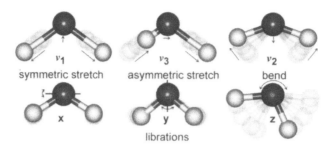

Figure 1.3 The vibrations and librations of a water molecule in liquid water. The vibrations in liquid water give broad peaks due to the range of interactions centered on about 3400 cm^{-1} for v_1 and v_3, about 1650 cm^{-1} for v_2, and about 750 cm^{-1} for the librations.

Although not often perceived as such, water is a very reactive molecule available at a high concentration. Pure liquid water is 55.345 M at 25°C, containing over 110 M hydroxyl groups. This reactivity, however, is greatly moderated at ambient temperatures due to the extensive hydrogen bonding (see below) but is highly significant in supercritical water (>374°C, >22 MPa). Water molecules each possess a strongly nucleophilic oxygen atom that enables many of life's reactions as well as ionizing to produce reactive hydrogen ions and hydroxide ions (Fig. 1.4). Dissociation is a rare event, occurring only twice a day for any particular water molecules; that is, only once for every 10^{16} times the hydrogen bond breaks.

Figure 1.4 In liquid water, water molecules can ionize to give H_3O^+ and OH^- ions. In neutral water, the concentration of both H_3O^+ and OH^- is equal to about 10^{-7} M at 25°C.

The most important property of water is its ability to form multiple hydrogen bonds.

1.2 THE HYDROGEN BOND

The hydrogen bond in water (Fig. 1.5) was first suggested by Latimer and Rodebush in 1920. Hydrogen bonding occurs when an atom of hydrogen is attracted by rather strong forces to two atoms instead of only one, so that it may be considered to be acting as a bond between them. Typically hydrogen bonding occurs where a partially positively charged hydrogen atom lies between partially negatively charged oxygen or nitrogen atoms, but is also found elsewhere, such as between fluorine atoms in HF_2^- (i.e., $(F\cdots H\cdots F)^-$). Hydrogen bonding is characterized by its preferred dimensions, molecular orientation, approximate linearity, and changes in infrared frequency and intensity. The hydrogen bond includes significant attractive electrostatic effects and a mutual penetration of atoms within their van der Waals radii.

The hydrogen bond in water is part (about 90%) electrostatic and part (about 10%) covalent. On forming the hydrogen bond, the donor hydrogen atom stretches away from its covalently bonded oxygen atom and the acceptor lone pair stretches away from its oxygen atom and toward the donor hydrogen atom, both oxygen atoms being pulled toward each other. The hydrogen-bonded proton has lower electron density relative to the other protons. Note that, even at temperatures as low as a few kelvin, there are considerable oscillations (<ps) in the hydrogen bond length and angles.

A hydrogen bond is naturally formed from a complex combination of interdependent interactions. In order to understand it better, we may consider contributions from the following:

(a) **electrostatic attraction** due to the opposite charges of the oxygen atom on one water molecule and a hydrogen atom on another. This is a relatively strong long-range interaction inversely proportional to the molecular separation with an effective range less than 3 nm. It is based on the nuclear and electronic charges, or on dipoles plus quadrupoles, octupoles, and so on.

(b) **polarization attraction** due to net attractive effects between nuclear charges and distortable electron clouds with an effective range less than 0.8 nm. These may form relatively strong short-range interactions, which increase cooperatively depending on the local environment. They may be considered as inversely varying with the fourth power of the molecular separation. This net attractive effect may contain a small repulsive element due to the consequentially slightly increased electron cloud overlap.

(c) **covalency attraction** due to the formation of molecular orbitals encompassing nuclei from two or more water molecules. This is a highly directional moderate-range attractive interaction that increases on further cooperative hydrogen bond formation. It is very dependent on the spatial arrangement of the molecules within the local environment with an effective range less than 0.6 nm. There is still some controversy surrounding this partial covalency, with arguments being given for and against it in the recent literature. However, if the water hydrogen bond is considered within the context of the complete range of molecular hydrogen bonding, then it appears most probable that it is partially covalent, and this covalency increases with the approach of the oxygen atoms from neighboring water molecules. Covalency increases the network stability relative to purely electrostatic effects.

(d) **dispersive attractive interaction** due to the coordinated effects of neighboring electron clouds. This is a weak, short-range interaction, with an effective range less than 0.6 nm, that may be considered to vary inversely with the sixth power of the molecular separation.

(e) **electron repulsion** due to electron cloud overlap. This is a very short-range interaction, with an effective range less than 0.4 nm, that may be considered to vary inversely with the twelfth power of the molecular separation.

(f) **nuclear quantum effects** due to zero point vibrational energy in the water molecule's covalent bonds. These bias the length of the O–H covalent bond longer than its "equilibrium" position length, as the shorter HO–H····OH_2 hydrogen bonds are stronger, and so also increase the average dipole moment of water molecules in the liquid phase. They are the cause of many of the significant differences in behavior between light water (H_2O) and heavy water (D_2O) and cause variation in the interaction between molecules, depending on their relative orientations. Hydrogen bonds within heavy water are stronger because of the lesser vibrational displacement of their atoms. It is possible, but unproven, that hydrogen bonds between *para*-H_2O, possessing no ground-state spin, are stronger and last longer than hydrogen bonds between *ortho*-H_2O.

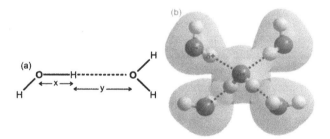

Figure 1.5 (a) In the hydrogen bond the hydrogen atom is attached by strong covalent bond to one oxygen atom, while being much more weakly attracted to another by another bond, here shown as a dashed line. This bond (length *y*) is approximately twice as long as the covalent bond (length *x*). The overall O–H····O hydrogen bond is almost straight. Generally, the strength of the bond depends on how straight it is and the presence of other interacting hydrogen bonds. The straighter the O–H····O bond, the stronger it is. If the bond is stronger, the covalent bond (*x*) lengthens, whereas lengths *y* and *x*+*y* both shorten. (b) Hydrogen bonds linking a cluster of water molecules. An important feature of the hydrogen bond is that it possesses direction; by convention this direction is that of the shorter covalent bond (the O–H hydrogen atom being donated to the O-atom acceptor atom on another H_2O molecule).

In liquid water, the hydrogen atom is covalently attached to the oxygen of a water molecule with bond energy 492 kJ/mol, but the hydrogen bond contributes an additional, much smaller attraction (maximally about 23 kJ/mol) to a neighboring oxygen atom of another water molecule (Fig. 1.5). This is almost five times the average thermal collision fluctuation at 25°C and is far greater than any included van der Waals interaction. Although such hydrogen bonds are clearly not weak compared with the thermal energy, they are constantly switching partners because of the low activation energy required for such processes (Fig. 1.6a).

The bond strength depends on its length and angle, with the strongest hydrogen bonding in ambient liquid water existing within the short linear proton-centered $H_5O_2^+$ ion at about 120 kJ/mol. However, small deviations from linearity in the bond angle (up to 20°) have a relatively minor effect. The dependency on bond length is very important, with the bond energy having been shown to exponentially decay with distance. The hydrogen bond may be considered broken if the total bond length (O–H····O) is somewhat greater than about 0.33 nm or the bond angle (O-H····O) bent to somewhat less than about 140°, although *ab initio* calculations indicate that much bonding energy may still remain and more bent but shorter bonds may be relatively strong.

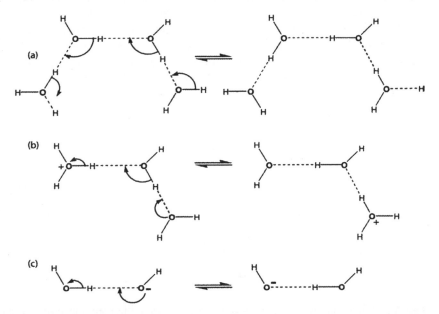

Figure 1.6 Hydrogen atoms rapidly exchange between different water molecules because of protonation/deprotonation processes (a), with the hydrogen atoms moving about 0.1 nm on each exchange. Both acids (b) and bases (c) catalyze this exchange.

The hydrogen atoms are constantly exchanging between different water molecules because of protonation/deprotonation processes (Fig. 1.6a). Both acids (Fig. 1.6b) and bases (Fig. 1.6c) catalyze this exchange, and even when at its slowest (at about pH 7), the average time that the atoms in an H_2O molecule stay together as a single molecular entity is only about a millisecond. As this brief period is, however, much longer than the timescales commonly encountered during physical investigations into water's hydrogen bonding or hydration properties, water molecules are usually described as though permanent and treated as symmetric (Fig. 1.1). However, this symmetric state does not exist for the instantaneous structure of water molecules in the liquid phase, except as an "average," due to interactions between different molecules.

The electronic structure of water (Fig. 1.2) rationalizes the formation of (approximately planar) trigonal hydrogen bonding that can be found around some restricted sites in the hydration of proteins and where the numbers of hydrogen bond donors and acceptors are necessarily unequal because of spatial restrictions. However, trigonal hydrogen bonding is also possible with two donor and one acceptor hydrogen bonds associated with individual water molecules (Fig. 1.7). This flexibility in the hydrogen bonding topology facilitates hydrogen-bonding rearrangements. Bifurcated hydrogen bonds, where both hydrogen atoms from one water molecule are hydrogen-bonding to the same other water molecule (Fig. 1.7) or one hydrogen atom simultaneously forms hydrogen bonds with two other water molecules, have just under half the strength of a normal hydrogen bond (per half the bifurcated bond) and present a low-energy route for hydrogen-bonding rearrangements. They allow the constant randomization of the hydrogen bonding within the network.

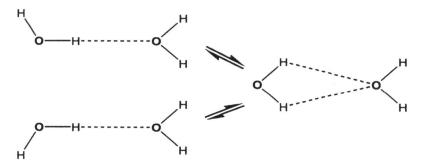

Figure 1.7 Rotation of a water molecule (*left*) through a bifurcated hydrogen bond (*right*) where two hydrogen bonds are donated from the one water molecule. This is a low activation energy process so long as no other hydrogen bonds are broken. The hydrogen-bonded structures on the left show trigonal arrangements with the hydrogen bond being accepted in a central rather than tetrahedral position.

Water's hydrogen bonding holds water molecules up to about 15% closer than they would be if water was a simple liquid with just van der Waals interactions. However, as hydrogen bonding is directional it restricts the number of neighboring water molecules to about four rather than the larger number found in simple liquids. For example, xenon atoms have 12 nearest neighbors in the liquid state. Formation of hydrogen bonds between water molecules gives rise to large, but mostly compensating, energetic changes in enthalpy (becoming more negative as the distance between the water molecules reduces and consequentially the bond strength increases) and entropy (becoming less positive as the bond strength increases and consequentially the tetrahedral organization increases). Both changes are particularly large, based on per-mass or per-volume basis, because of the small size and mass of the water molecule and the many hydrogen bonds involved. This enthalpy–entropy compensation is almost complete, however, with the result that very small enthalpic or entropic effects imposed by solutes, for example, may exert a considerable influence on aqueous systems. The strength of bonding also depends on the orientation and positions of the other neighboring bonded and nonbonded atoms and their unbonded electrons.

1.3 THE ANOMALOUS PROPERTIES OF WATER

The behavior of liquid water is quite different from what is found with other liquids. Although it is an apparently simple molecule (H_2O), it has a highly complex and anomalous character because of its intramolecular hydrogen bonding. As liquid water is so commonplace in our everyday lives, it is often regarded as a "typical" liquid. In reality, water is most atypical as a liquid. In particular, the high cohesion between molecules gives it high freezing and melting points, such that our planet and we are bathed in liquid water. The large heat capacity, high thermal conductivity, and high water content in organisms contribute to thermal regulation and prevent local temperature fluctuations, thus allowing us to more easily control our body temperature. The high latent heat of evaporation gives resistance to dehydration and considerable evaporative cooling. Water is an excellent solvent because of its polarity and small size, particularly for polar and ionic compounds and salts. It has unique hydration properties toward biological macromolecules (particularly the very important proteins and nucleic acids) that determine their three-dimensional structures, and hence their functions, in solution. This hydration forms gels that can reversibly undergo the gel–sol phase transitions that underlie many cellular mechanisms. Water ionizes and allows easy proton exchange between molecules, thereby contributing to the richness of the ionic interactions in biology.

At 4°C water expands on heating or cooling. This density maximum, together with the low density of ice, results in (i) the necessity that all of a body of freshwater (not just its surface) be close to 4°C before any freezing can occur; (ii) the freezing of rivers, lakes, and oceans from the top down, thereby permitting survival of the ecology at the bottom, insulating the water from further freezing, reflecting back sunlight into space, and eventually allowing rapid thawing; and (iii) density-driven thermal convection causing seasonal mixing in deeper temperate waters carrying life-providing oxygen into the depths. The large heat capacity of the oceans and seas allows them to act as heat reservoirs such that sea temperatures vary only a third as much as land temperatures and so moderate our climate (e.g., the Gulf stream carries tropical warmth to northwestern Europe). The high surface tension of water and its expansion on freezing encourage the erosion of rocks to give soil for our agriculture.

Notable among the anomalies of water are the opposite changes in properties between hot and cold water, with the anomalous behavior more accentuated at low temperatures where the properties of supercooled water often also diverge from those of hexagonal ice. As (supercooled) cold liquid water is heated, it shrinks and becomes difficult to compress, its refractive index increases, the speed of sound within it increases, gases become less soluble, and it becomes easier to heat and conducts heat better. In contrast, as hot liquid water is heated, it expands and becomes easier to compress, its refractive index reduces, the speed of sound within it decreases, gases become more soluble, and it becomes harder to heat and conducts heat poorly. With increasing pressure, cold water molecules move faster but hot water molecules move slower. Hot water may freeze faster than cold water, and ice melts when compressed, except at high pressures when liquid water freezes on compression. No other material is commonly found on the earth as solid, liquid, and gas.

1.4 WATER CLUSTERS AND LIQUID WATER STRUCTURE

The anomalous properties of liquid water may be explained primarily on the basis of its hydrogen bonding. When a hydrogen bond forms between two water molecules, the redistribution of electrons changes the ability for further hydrogen bonding. The water molecule donating the hydrogen atom increases electron density in its lone pair region, which encourages hydrogen bond acceptance, and the accepting water molecule reduces electron density centered on its hydrogen atoms and its remaining lone pair region, which encourages further donation but discourages further acceptance of hydrogen bonds. This electron redistribution thus results in both the cooperativity (i.e., accepting one hydrogen bond encourages the donation of another, and donating one hydrogen bond encourages the acceptance of another) and anticooperativity (i.e., accepting one hydrogen bond discourages acceptance of another, and donating one hydrogen bond discourages the donation of another) in hydrogen bond formation in the networks of hydrogen bonding within liquid water.

The cooperative nature of the hydrogen bond means that acting as an acceptor strengthens the water molecule acting as a donor. It is clear, therefore, that a water molecule with two hydrogen bonds where it acts as both donor and acceptor is somewhat stabilized relative to one where it is either the donor or acceptor of two (Fig. 1.8a).

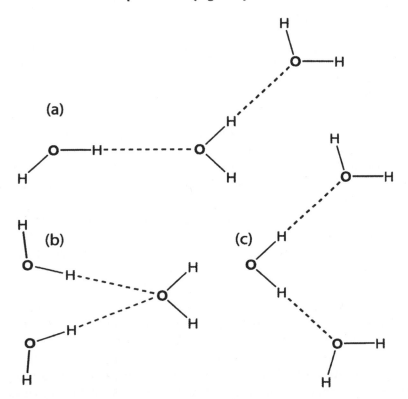

Figure 1.8 A sketch showing the cooperative and anticooperative hydrogen bonding behavior in liquid water. (a) Cooperative hydrogen bonding, where an accepted hydrogen bond by the central water molecule strengthens its donated hydrogen bond. Anticooperative behavior is shown as an accepted hydrogen bond weakens the ability to form another accepted hydrogen bond (b) and where a donated hydrogen bond weakens the ability to form another donated hydrogen bond (c).

This is the reason why the formation of the first two hydrogen bonds (donor and acceptor) gives rise to the strongest hydrogen bonds, thus encouraging water molecules to have four or two hydrogen bonds rather than zero, one, or three hydrogen bonds. Cooperative hydrogen bonding slightly increases the O–H covalent bond length while causing a much greater reduction in the H····O and O····O distances (Fig. 1.5a). There is a trade-off between the covalent and hydrogen bond strengths; the stronger the H····O bond, the weaker the O–H covalent bond and the shorter the O····O distance. The weakening of the O–H covalent bond gives rise to a good indicator of hydrogen bonding energy; the fractional increase in its length is determined by the increasing strength of the hydrogen bonding; for example, when the pressure is substantially increased (~GPa), the remaining hydrogen bonds (H····O) are forced to become shorter, causing the O–H covalent bonds to be elongated.

Although the hydrogen atoms are often shown along lines connecting the oxygen atoms (Fig. 1.5), this is now thought to be indicative of time-averaged direction only and unlikely to be found to a significant extent even in ice. Liquid water consists of a mixture of short, straight, and strong hydrogen bonds and long, weak, and bent hydrogen bonds with many intermediate between these extremes. Short hydrogen bonds in water are strongly correlated with them being straighter. Although average structures of water molecules may be described in the gaseous and some solid states, any discussion concerning the "average" structure of water molecules in the liquid state shows unrealistic properties. This is rather like trying to describe the average person as one who is neither male nor female but somewhat in between and unrepresentative of any single real person.

The instantaneous hydrogen bonded arrangement of most molecules is not as symmetrical, as shown in Fig. 1.5b. Also, the arrangement may well consist of one pair of more tetrahedrally arranged strong hydrogen bonds (one donor and one acceptor), with the remaining hydrogen bond pair (one donor and one acceptor) being either about 6 kJ/mol weaker, less tetrahedrally arranged, or bifurcated; perhaps mainly because of the anticooperativity effects. Such a division of water into higher (tetra-linked) and lower (di-linked) hydrogen bond coordinated water has been shown by modeling, and X-ray absorption spectroscopy confirms that, at room temperature, 80% of the molecules of liquid water have one (cooperatively strengthened) strong hydrogen-bonded O–H group and one non-, or only weakly, bonded O–H group at any instant (sub-femtosecond averaged and such as may occur in pentagonally hydrogen bonded clusters, Fig. 1.9), the remaining 20% of the molecules being made up of four-hydrogen-bonded tetrahedrally coordinated clusters. There is much debate as to whether such structuring represents the more time-averaged structure, which is understood by some to be basically tetrahedral (as Fig. 1.5b).

Liquid water contains by far the densest hydrogen bonding of any solvent with almost the same number of hydrogen bonds as there are covalent bonds. These hydrogen bonds can rapidly rearrange in response to changing conditions and environments (e.g., solutes). The hydrogen bonding patterns appear random in water (and hexagonal ice). For any water molecule chosen at random, there is equal probability (50%) that the four hydrogen bonds (i.e., the two hydrogen donors and the two hydrogen acceptors) are located at any of the four sites around the oxygen atoms. Water molecules surrounded by four stronger hydrogen bonds tend to clump together, forming clusters, for both statistical and energetic reasons. Hydrogen-bonded chains (i.e., O–H····O–H····O) are cooperative; the breakage of the first bond is the hardest, then the next one is weakened, and so on. Such cooperativity is a fundamental property of the networks of hydrogen-bonded water molecules in liquid water, where hydrogen bonds are up to 250% stronger than the single hydrogen bonds found in water molecule dimers (such as are found in small amounts in gaseous water), and rings of the hydrogen-bonded water molecule (e.g., cyclic pentamers rings, Fig. 1.9) occur frequently.

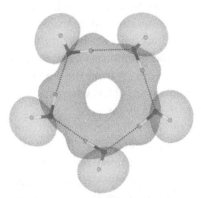

Figure 1.9 A hydrogen bonded pentameric cluster of water molecules as found in relatively high concentrations in supercooled water. *Ab initio* calculations show that some intermolecular electronic orbitals are formed, such as the one shown, thus greatly increasing the intermolecular bond strength and enabling the cooperative bond formation.

The substantial cooperative strengthening of hydrogen bonds in water is dependent on long-range interactions. Breaking one bond generally weakens those around, whereas making one bond generally strengthens those around, and this, therefore, encourages larger clusters for the same average bond density. The average hydrogen-bonded cluster size in water at 0°C has been estimated to be several hundreds. Weakly hydrogen-bonding water molecules restrict the hydrogen-bonding potential of adjacent water, so that these make fewer and weaker hydrogen bonds. Such weak bonding also persists over several layers and may cause locally changed solvation. Conversely, strong hydrogen bonding will be evident over significant distances of greater than a nanometer. Liquid water is thus a heterogeneous mixture of clusters of stronger hydrogen-bonded molecules and clusters of weaker hydrogen-bonded molecules in rapidly exchanging equilibria.

The hydrogen-bonded network is essentially complete at ambient temperatures; that is, (almost) all molecules are linked by at least one unbroken hydrogen bonded pathway to almost all, if not all, others. Hydrogen bond lifetimes are 1–20 ps, whereas broken bond lifetimes are about 0.1 ps, with the proportion of broken, "dangling" hydrogen bond capacity persisting for longer than a picosecond being insignificant. Broken hydrogen bonds are basically unstable with reactive free hydroxyl groups and will probably reform to give same hydrogen bond, particularly if the other three hydrogen bonds are in place, thus preventing easy rotation; hydrogen bond rearrangement is more dependent on the local structuring rather than the instantaneous hydrogen bond strength. If not, breakage usually leads to rotation around one of the remaining hydrogen bonds and not to translation away, as the resultant "free" hydroxyl groups and lone pairs are both quite reactive. Thus, clusters of water molecules of particular geometries may persist for significant periods. Proof of this ability may be drawn from the high degree of hydrogen bond breakage seen in the IR spectrum of ice, whereas the clustering in solid ice is taken to be lasting essentially forever.

Hydrogen bonding carries information about solutes and surfaces over significant distances in liquid water. The effect is synergistic, directive, and extensive and is reinforced by additional polarization effects and the resonant intermolecular transfer of O–H vibrational energy, mediated by dipole–dipole interactions and the hydrogen bonds. Reorientation of one molecule induces corresponding motions in the neighbors. Thus solute molecules can "sense" (e.g., affect each other's solubility) each other at distances of several nanometers, and surfaces may have effects extending to tens of nanometers. These long-range correlations of molecular orientation establish the high dielectric constant of water, and as the temperature is raised and more hydrogen bonds break, these correlations become less extensive, thereby reducing the dielectric constant.

In spite of much work, many of the properties of water are puzzling. Enlightenment comes from an understanding that water molecules form this infinite hydrogen-bonded network with localized and structured clustering.

The middling strength of the connecting hydrogen bonds seems ideally suited to life processes, being easily formed but not too difficult to break and with low activation energies for rearrangements. The important concept, which should never be overlooked, is that liquid water is not homogeneous at the nanoscopic level.

Much effort has been expended on the structure of small isolated water clusters. Typically, in the atmosphere there is about one water dimer for every thousand free water molecules. These dimers show tetrahedrality of the bonding in spite of the lack of clearly seen lone pair electrons. This tetrahedrality is primarily caused by electrostatic effects (i.e., repulsion between the positively charged nonbonded hydrogen atoms) rather than the presence of tetrahedrally placed lone pair electrons. The hydrogen bond in water dimers is sufficiently strong to result in the dimers persisting within the gas state at significant concentrations (e.g., ~0.1% H_2O at 25°C and 85% humidity) to contribute significantly to the absorption of sunlight and atmospheric reaction kinetics.

At any given time, most water molecules in liquid water form the central part of roughly tetrahedral pentameric clusters (as Fig. 1.5b). Many aspects of these basic structures are fluctuating on sub-picosecond timescales. Thus the hydrogen bonding angles, lengths, and partners are all in constant flux. Such clusters are linked to other water molecules and clusters to form an intricate network of hydrogen bonds. Such a network may be somewhat ordered at the local level but lacks long-distance order of greater than about a nanometer. The strong hydrogen bonding in these small clusters of water molecules enables their persistence for considerable time even if the constituent molecules come and go. Thus, although individual hydrogen bonds break on a picosecond timescale, they often simply reform. Even if several bonds simultaneously break and occasional members of the clusters leave the cluster, the remaining cluster retains structuring and other water molecules replace those missing in the same or equivalent positions, thereby effectively continuing the life of the cluster.

1.5 BIOLOGICAL CONSEQUENCES OF WATER CLUSTERING

Biological molecules, such as proteins and nucleic acids, and structures, such as membranes and the cytoskeleton, affect the movement and structuring of the surrounding water molecules. Of equal importance, although often ignored, is the fact that the surrounding water molecules affect the structure and activity of the biological molecules. The causal relationships are complex and inseparable. As a general theme, where the water molecules form more strongly hydrogen-bonded, tetrahedrally oriented water clusters, the local environment is more highly viscous and stiff. This reduces local structural fluctuations in the nearby biomolecules and the diffusion of small molecules. The consequential strong hydrogen-bonded network also allows for structural information to be transmitted across nanometer distances (Fig. 1.10). In contrast, where the local hydrogen bonding is weak within the water, the local viscosity is low, allowing for the easy movement of solutes and macromolecular interfaces, and any information about the biomolecular surface structuring rapidly decays away over short distances.

The structure of DNA was published by Watson and Crick way back in 1953, but they realized then that it depends on the surrounding water molecules. Now, we know that not only the local water controls the type of structure of the DNA but also the important interactions between the DNA and the proteins control its transcription and replication. Thus, the rapid processing of the genetic information within DNA is facilitated by highly discriminatory interfacial water molecules that serve as "hydration fingerprints" of given DNA sequences.

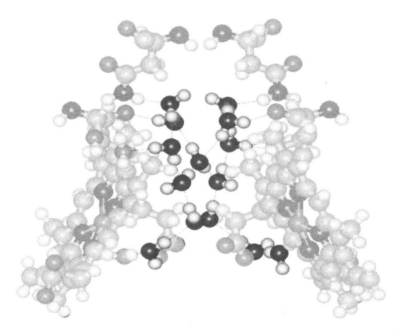

Figure 1.10 A cluster of water molecules between and connecting the heme groups and protein residues of the two subunits of *Scapharca inaequivalvis* hemoglobin. Note the symmetry in the water cluster with the formation of two pentameric rings. For clarity, only the relevant parts of the protein and heme groups are shown. The hydrogen-bonded water molecules allow the transmission of information concerning the oxygen binding of the subunits across the gap between the subunits, thus controlling the allosteric behavior of this dimeric hemoglobin.

The structural fluctuations of proteins are controlled by their hydration. Parts of their surfaces are covered by clustering of tetrahedral coordinated water molecules, thus reducing their movement and the movement of other nearby solutes, whereas other parts of the proteins' surfaces are covered by weakly interacting water molecules, thus allowing easy movement of the proteins' side chains and secondary structures and the facilitated access of other solutes to the proteins' surfaces and active sites.

The mobility of water within cells has a complex cause–effect relationship with the cell's metabolic state. Thus the intracellular water can be shifted to a more strongly hydrogen-bonded environment, likened to a "gel" state, with lower metabolic activity, low diffusion, and higher viscosity or to a less viscous "sol" state by changes in the cytoskeletal structure, in turn dependent on the energetic state of the cells. The state of the water thus controls the biological activity of the cells, and, in turn, the cellular metabolic state controls the clustering of the water. Thus the whole of a cell may be controlled almost instantaneously to behave in the same way at the same time.

1.6 CONCLUSIONS CONCERNING THE STRUCTURE OF LIQUID WATER

Liquid water is basically heterogeneous, consisting of volumes of either more strongly hydrogen-bonded and tetrahedrally oriented water molecules with mostly four nearest neighbors or containing more weakly hydrogen-bonded water molecules with variable numbers of nearest neighbors. This structuring is in constant flux and easily affected by solutes and surfaces. The anomalous properties of liquid water are explained by the changes in the relative amounts of these clusters and, particularly, by the increase in the more tetrahedral arrangements at lower temperatures.

Part II

WATER AS A SOLVENT

Chapter 2

PROTEIN DYNAMICS AND HYDRATION WATER

Marie-Claire Bellissent-Funel

Laboratoire Léon Brillouin (CEA-CNRS), CEA Saclay,
91191 Gif-sur-Yvette, France
marie-claire.bellissent-funel@cea.fr

2.1 INTRODUCTION

Water is the most abundant fluid on earth and is indispensable for life. More than 60% of the volume of living cells is occupied by water, and the loss of 10–20% of its volume causes cell death. In cells, hydrophilic and hydrophobic effects are of prime importance. These effects are related to the solvent properties of water and to the fact that some compounds are more or less soluble in water. The solvent abilities of water arise primarily from the two properties: its tendency to form hydrogen bonds (very short characteristic life time, from 10^{-13} to 10^{-12} s) and its dipolar character [1]. Water has a high dielectric constant ($\varepsilon \approx 80$), which results from its dipolar character.

Amphipathic molecules exhibit both hydrophilic and hydrophobic properties simultaneously. When one attempts to dissolve them in water, amphipathic substances form peculiar structures. Possible structures are a monolayer on the water surface, a micelle and a bilayer vesicle, with water both inside and outside. Examples of other structures that impose spatial restrictions on water molecules include polymer gels and microemulsions. In these cases, since the hydrophobic effect is the primary cause for the self-organization of these structures, obviously the configuration of water molecules near the hydrophilic–hydrophobic interfaces is of considerable relevance [2].

Water: The Forgotten Biological Molecule
Edited by Denis Le Bihan and Hidenao Fukuyama
Copyright © 2011 by Pan Stanford Publishing Pte. Ltd.
www.panstanford.com

An important field of interest of the subtle hydrophilic–hydrophobic interplay is that of associated water-soluble polymers. Their unique properties in aqueous systems make them very useful materials as rheology modifiers, suspension stabilizers, and drug carriers in pharmaceutical applications [3].

However, to answer why water is indispensable to life, we have to understand the physical interactions between water and biomolecules and, in particular, proteins. Hydration plays a major role in the structure and function of proteins. The stability of protein structures results from the subtle interplay between the hydrophilic and hydrophobic phenomena. The crystallographic structure of CO myoglobin determined by Cheng and Schoenborn [4] illustrates this subtle interplay.

2.2 WATER CONFINED IN MODEL SYSTEMS

In many technologically important situations, water is not present in its bulk form, but instead is attached to some substrates or filling small cavities. Common examples are water in porous media such as rocks or sandstones, biological materials such as the interior of cells, or surfaces of biological macromolecules and membranes. This is what we define here as the "confined" or the "interfacial water."

The structure and dynamics of water are modified by the presence of surfaces, by a change of hydrogen bonding, but also by modification of the molecular motion that depends on the distance of water molecules from the surface. Therefore, a detailed description of these properties must take into account the nature of the substrate and its affinity to form bonds with water molecules, and also the number of water layers or hydration shells. How are the properties of water modified when water is in contact with hydrophilic or hydrophobic interfaces, or both? How does the intracellular water behave? In this section, we give some examples of model systems developing either hydrophilic or hydrophobic interactions, or both, with water. In what follows, the structural and dynamic properties of confined water relative to various systems are presented.

2.2.1 Water in Hydrophilic Systems

Water in porous materials such as Vycor glass, silica gel and zeolites [5] has been actively under investigation because of its relevance in catalytic and separation processes. In particular, the structure of water near layer-like clay minerals [6 7], condensed on hydroxylated oxide surfaces [8], confined in various types of porous silica, has been studied by neutron and/or

X-ray diffraction [5]. Water molecules adsorbed on ionic surfaces have been investigated by FT-IR, quasi-elastic neutron scattering (QENS) and dielectric relaxation techniques [9]. Water in cement (hydrated tricalcium silicate) has been the subject of several studies by QENS [10].

The structure of water in Vycor, a porous silica glass [11], has been investigated as functions of level of hydration and temperature. We present here some of our results for two levels of hydration: full hydration (0.25 g water/g dry Vycor) [11] and 25% hydration (0.0625 g water/g dry Vycor) [12]. The latter corresponds roughly to a monolayer coverage.

2.2.1.1 Water in Vycor at full hydration

Results for the two levels of hydration of Vycor [11] demonstrate that the fully hydrated case is almost identical to the bulk water and the partially hydrated case is not much different. It seems that the confinement of water favors the nucleation of cubic ice that appears superimposed on the spectrum of liquid water, and its proportion can be deduced from the intensity of the (111) Bragg peak. The proportion of cubic ice increases with decreasing temperature. In fact, at –100°C, the spectrum of confined water looks similar to that of cubic ice (Fig. 2.1). This is in sharp contrast to bulk water, which always nucleates into hexagonal ice. The spectrum shown in Fig. 2.2 gives a clear evidence of water that is present below the Bragg peaks at –18°C, obtained by subtraction of the weighted spectrum of the same sample cooled down to –100°C.

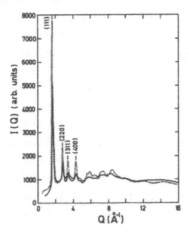

Figure 2.1 Spectrum of cubic ice (–198°C) (dotted line) compared with that of confined D_2O at –100°C from fully hydrated Vycor (continuous line). Reproduced from Ref. [11] by permission.

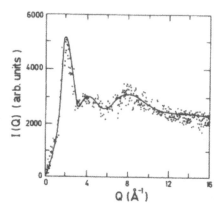

Figure 2.2 Spectrum of confined D_2O at $-18°C$ from fully hydrated Vycor after subtraction of Bragg peaks. There is 23% liquid water. Reproduced from Ref. [11] by permission.

2.2.1.2 Interfacial water in Vycor: Low-temperature phase transitions

Results relative to a 25% hydrated Vycor glass [12] corresponding to a monolayer of water molecules indicate that, at room temperature, interfacial water has a structure similar to that of bulk supercooled water at a temperature of about 0°C, which corresponds to a shift of about 30 K [5]. Therefore, the structure of interfacial water is characterized by an increase of long-range correlations that corresponds to the building of the H-bond network as it appears in low-density amorphous ice [13]. There is no evidence of ice formation when the sample is cooled from room temperature down to −196°C (liquid nitrogen temperature). Nevertheless, despite the fact that interfacial water and LDA show the same position of the first sharp diffraction peak (FSDP), their structures are not strictly identical. Such structural differences could be related to a topologically distorted hydrogen bond network, as already invoked in the case of other interfacial water systems. Moreover, by combining calorimetry, diffraction and high-resolution QENS we have put in evidence peculiar properties of interfacial water [12]. Interfacial water at the surface of Vycor, a hydrophilic inert material (chemically and dynamically), experiences different dynamic crossovers. As far as the rotational motion of water is concerned, transitions are detected at 150 and 220 K, as shown in Fig. 2.3a. At 150 K, the hydrogen bond network becomes softer. But no change in the hydrogen bond strength has been detected at 220 K. The 220 K dynamic crossover could then be associated to a structural change in hydrogen bond connectivity.

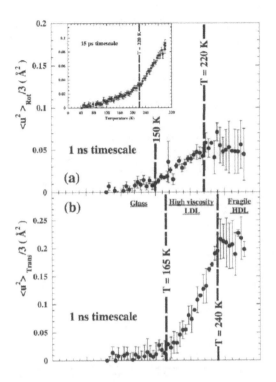

Figure 2.3 Temperature dependence of rotational and translational mean-square displacements (MSD) for interfacial water $<u^2>_{Rot}$ (a) and $<u^2>_{Trans}$ (b) have been extracted from the two Q ranges defined in figure 3 of Ref. [12]. The measurements were performed at 1 μeV resolution; i.e., the dynamics is determined up to 1 ns. Inset: temperature dependence of $<u^2>_{Rot}$ measured at 80 μeV resolution (QENS, ANL/IPNS); i.e., the dynamics is measured up to 15 ps. At this resolution, the 220 K rotational MSD of water is fully detected. Data from Ref. [12].

As far as the translational motion of water is concerned, transitions are detected at 165 and 240 K. After exhibiting a glass transition at 165 K, interfacial water experiences a first-order liquid–liquid transition at 240 K from a low-density to a high-density liquid, as shown in Fig. 2.3b. Further evidence of these transitions has been obtained by high-energy inelastic neutron scattering [14, 15].

2.2.1.3. Residence time and hydrogen bond life time of water confined in various media

Figure 2.4 shows the Arrhenius plots of the residence time τ_0 for water of hydration at the surface of a C-phycocyanin protein (CPC) [16] as compared with those of water in Vycor at different levels of hydration [17] and bulk water [18].

Figure 2.5 shows the Arrhenius plots of the hindered rotations characteristic time, τ_1, for water in Vycor at two levels of hydration [17] and bulk water [18]. We have shown that this time can be associated with the hydrogen bond life time [16–18] of water.

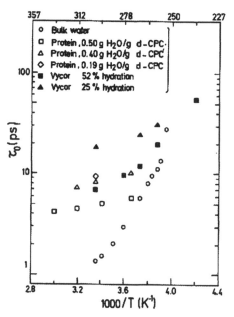

Figure 2.4 Arrhenius plot of the residence time τ_0 for different levels of hydration: at the surface of H_2O-hydrated d-CPC protein (empty symbols); contained in hydrated Vycor (solid symbols); as compared with bulk water (empty circles). Reproduced from Ref. [16] by permission of The Royal Society of Chemistry.

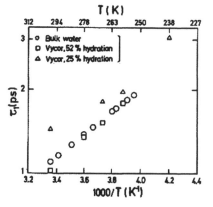

Figure 2.5 Arrhenius plot of the hydrogen bond life time τ_1 as compared with bulk water (empty circles). Reproduced from Ref. [16] by permission of The Royal Society of Chemistry.

The residence times τ_0 of confined water from 25% hydrated Vycor and from hydrated proteins are always longer than the residence time of bulk water at the same temperature. They increase rapidly as either the temperature or the level of hydration decreases. For example, for the 25% hydrated Vycor sample, $\tau_0 = 25$ ps at −15°C.

The hydrogen bond life times τ_1 for confined water are close to those of bulk water [18]. They have Arrhenius temperature dependence (Fig. 2.5), whereas the residence time τ_0 does not exhibit such a behavior (Fig. 2.4).

2.2.2 Water in Hydrophobic Systems

Among hydrophobic model systems, one experimental investigation of particular interest concerns the structure of water contained in carbon powder [5, 23]. The structure of water has been determined both by X-ray and ND, as functions of hydration, from room temperature down to 77 K. In agreement with previous work [24], this study gave support to the existence of a region near the interface where the properties of water are markedly different from those of the bulk liquid. A crude determination from the specific area indicates that, for a hydration equal to 50%, the thickness does not exceed 5 Å. This value must be compared with the computer simulations data [24], which indicate that structural modifications do not extend beyond 10 Å from the solid surface.

When partially hydrated samples are cooled down to 77 K, no crystallization peak is detected by differential thermal analysis. X-ray and neutrons show that an amorphous form is obtained and that its structure is different from those of low- and high-density amorphous ices already known [13]. This absence of crystallization in low hydrated samples and the appearance of an amorphous phase of ice seem to be a common behavior for hydrophilic and hydrophobic systems, but as reported below, the nature of amorphous phases is different.

2.2.3 Water in Mesoporous Hydrophilic/Hydrophobic Sol–Gel Silicas

The phase transformation of water and ice in two samples of mesoporous sol–gel silicas with pore diameters of ~50 Å has been recently investigated [25]. One sample was modified by coating with a layer of trimethylchlorosilane, giving a predominantly hydrophobic internal surface, whereas the unmodified sample was hydrophilic. The pore structure was characterized by nitrogen gas adsorption and NMR cryoporometry, and the melting/freezing behavior of water and ice in the pores was studied by DSC and ND for cooling and heating cycles covering the range from 300 to 200 K. Measurements were made for filling factors of 0.9, 0.45, and 0.2. The results show a systematic difference in the form of ice created in each of the samples. The nonmodified sample gives similar results to previous studies with hydrophilic silicas, exhibiting a defective form of cubic ice superimposed on a more disordered pattern that changes with temperature and that has been characterized as "plastic" ice. The modified sample has similar general features but displays important variability in the ice transformation features, particularly for the case of the low filling factor ($f = 0.2$). The results exhibit a complex temperature-dependent variation of the crystalline and disordered components that are substantially altered for the different filling factors.

2.2.4 Water in Hydrophilic/Hydrophobic Systems

The water/dimethyl sulphoxide (DMSO) mixture is a model system to study the structure and dynamics of water under simultaneous hydrophilic/hydrophobic hydration.

The local quasi-tetrahedral structure of pure water has been found to be largely preserved in the presence of DMSO (X-ray and ND). However, there is a transfer of hydrogen bonds from water to DMSO, in agreement with a greater hydrogen bond affinity for water/DMSO than water/water. Water HH pair correlation functions of pure water and at the eutectic composition have been determined: the local coordination number is maintained [26]. The dynamic behavior of water in DMSO solutions has been studied [27]. There is evidence of some slowing down of the translational diffusive motion of water. In contrast, the effect on hydrogen bonding is not significant. A similar behavior has been observed in water/trehalose [28] and water/pyridine solutions [29].

2.3 WATER IN BIOLOGICAL SYSTEMS

Why is it so important to understand the effects of water on the shapes of biological molecules and/or that of biological molecules on water? On a practical level, understanding the relationship between the structure, dynamics, and function of biological molecules in water may one day help researchers design new drugs that act by blocking or enhancing various biochemical pathways.

In cells, because of the considerable macromolecular crowding and the high values of the macromolecular surface/volume ratio, the properties of water are influenced by hydrophilic/hydrophobic interactions [30].

2.3.1 Effect of Biological Macromolecules on Water

The assessment of perturbation of liquid water structure and dynamics by hydrophilic and hydrophobic molecular surfaces is fundamental to the quantitative understanding of the stability and enzymatic activity of globular proteins and functions of membranes.

The surface exposed by macromolecules to their aqueous environment consists mainly of hydrophilic domains that attract water dipoles [31, 32]. However, superficial apolar domains also induce the perturbation of structure of water in contact with them, generally into a clathrate-like form. According to Wiggins and MacClement [33], significant amounts of water are retained in a highly "structured form" inside hydrophobic pockets. The same authors defined this water as "structured water," or "interfacial water." Because this form of water generally is difficult to extract, it is also called "bound water." The properties of this water have been described extensively in Refs. [32–35]. This distinction in terms of two classes of water, bound water and free water, has been often used.

Both the effect of water of hydration on the structure and dynamics of proteins and the effect of proteins on the dynamics of water have been studied using ND and QENS. It is now well admitted that a protein is a flexible entity and that this flexibility is required for protein to function. As reported by Rupley and Careri [36], it is known that a monolayer of water molecules ($h = 0.4$) is necessary for protein to initiate its function.

The dynamics of water molecules on the surface of a CPC [16, 37] is slowed down. A similar behavior is observed for water confined in a porous hydrophilic model system [17, 38] (see Figs. 2.4 and 2.5).

Water–biomolecules interactions have been extensively studied in dilute solutions, crystals and rehydrated powders, but none of these model systems may capture the behavior of water in the highly organized intracellular milieu. Because of the experimental difficulty of selectively probing the structure and dynamics of water in intact cells, different views about the properties of cell water have proliferated. The recent studies of dynamics of water in cells by elastic and inelastic neutron scattering [39] and by NMR spectroscopy [40] need to be mentioned.

2.3.2 Effect of Water on Biological Macromolecules: From Hydrated Powders to Solutions

In the field of biology, the effects of hydration on equilibrium protein structure and dynamics are fundamental to the relationship between structure and biological function [41–45].

Hydrophilic–hydrophobic interactions [46] control the equilibrium of biological systems. It is worth noting that, in presence of biological macromolecules, such as peptides, enzymes, proteins, and DNA, all these behaviors can be found depending on the nature more or less hydrophilic or hydrophobic of each site or residue. Bonds, in particular, play a major role in the structure of these macromolecules.

The dynamics and function of bio-macromolecules are intimately related. Thus, an understanding of physiological phenomena requires a comprehensive knowledge of internal molecular dynamics [47] as they occur under the crowded macromolecular conditions that characterize the intracellular environment. Such an environment gives rise to a high degree of dispersion of the intracellular water molecules at the surface of the different macromolecular components. A wide variety of internal motions exists in biological macromolecules. At physiological temperatures, motions are partly vibrational and partly diffusive [48].

The description of internal diffusion in proteins is complicated by the variety of existing motions. These involve groups of atoms undergoing a plethora of continuous or jumplike diffusion. Neutron spectroscopy permits the investigation of motions in the time ranges from 10^{-14} to 10^{-9} s (time of flight, backscattering, spin echo techniques). Because of the large incoherent scattering cross section of hydrogen nuclei (about 40 times larger than the cross section of other elements), this technique is sensitive to the motions of the nonexchangeable hydrogen atoms in a sample. Moreover, as the hydrogen atoms are distributed "quasi-homogeneously" in the structure of a biological macromolecule, the neutron technique provides us with a description of the average dynamics, in terms of internal motions (vibrational and relaxational dynamics).

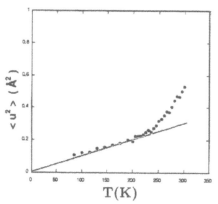

Figure 2.6 Evolution of the mean-square displacements (MSD), $<u^2>$, as a function of the temperature for a hydrated CPC ($h = 0.5$ g D2O/g protein). Reproduced from Ref. [48] by permission.

Diffusive motions are of particular interest, in the context of dynamical or "glass transition" determined between 180 and 220 K for many soluble proteins [49].

Figure 2.6 gives an example of the dynamical transition occurring at a temperature close to 220 K for the CPC. Below 220 K there are only vibrations. Above 220 K the diffusive motions are activated [48].

It should nevertheless be noted that the MSD values are not direct dynamic quantities since they just provide an estimate of the extension induced by a mode with no information on its characteristic time. Time-dependent quantities, like diffusion coefficients, residence times or vibrational correlation times, can be obtained by a full analysis of the energy dependence of the neutron scattered intensity, $S(Q, \omega)$.

However, such analysis is somewhat difficult because an inelastic neutron spectrometer, depending on its specific time resolution, provides access only to correlation times lying in a specific time window.

2.3.2.1 Hydrated powders of proteins: Effect of level of hydration

In order to limit the contribution of bulk water, powders of protein hydrated at a level of hydration h (g water/g protein) between 0 and 1.5 have been used. According to the work of Rupley and Careri [36], the enzymatic function of lysozyme is activated for a minimum level h of hydration equal to 0.25,

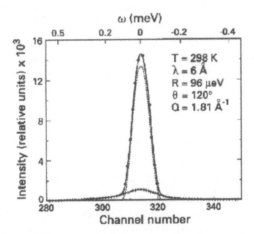

Figure 2.7 Elastic (dotted line) and quasi-elastic (crossed line) components of a neutron scattering time-of-flight spectrum at 6 Å wavelength, and 298 K of a parvalbumin $PaCa_2HHd$ sample hydrated at $h = 0.31$ g D_2O/g dry protein. R is the energy resolution. The filled circles are the experimental points; the solid line is the model fit (see equation 4 in Ref. [51]).

while the full water coverage of the protein is reached at $h = 0.3$. Dynamics of hydrated powdered samples of a parvalbumin protein [50, 51] and a photosynthetic CPC [52] has been studied as functions of level of hydration and temperature.

The parvalbumin is a small-sized (11.5 kDa) globular protein, namely, the Ca^{2+}/Mg^{2+} binding EF-hand protein, which plays a critical role in regulating the Ca^{2+} signal in excitable cells (neurons and muscle cells). The influence of hydration on the internal dynamics of parvalbumin has been investigated using incoherent quasi-elastic neutron scattering and solid state ^{13}C NMR. An example of quasi-elastic spectrum is given in Fig. 2.7.

In particular, the quasi-elastic component of the spectrum of the fully Ca-loaded parvalbumin, $PaCa_2$, hydrated with D_2O reveals dynamical aspects related to diffusive motions that might be functionally important because they contribute to the general flexibility of the protein. The fitting of the quasi-elastic component is based on the model of Volino and Dianoux [53] using a delta function and a single Lorentzian line. As shown in Fig. 2.8, the half-width at half-maximum of the Lorentzian line, Γ, is plotted as a function of Q^2, at two h values, 0.31 and 0.65. The invariance of Γ as a function of Q^2, at $h = 0.31$, is to be interpreted as the result of reorientational motions of protons (exclusively from the protein), at the observation time scale of 10 ps. At higher hydration, $h = 0.65$ (415 water molecules, approximately exceeding a continuous monolayer), Γ is apparently dependent on Q^2. This indicates that diffusive motions of protein protons occur within a confined volume under these conditions. A fraction $p = 0.72$ of immobile protons has been estimated, which indicates that about 25% of the protons in the protein are involved in short-time (in the 10 ps range) diffusive motions. These protons belong to the residues at the surface of protein such as lysines, aspartic acids and glutamic acids. This level of hydration corresponds to more than one monolayer of water molecules for which the protein is functional; the volume of diffusion is found to be close to that of a sphere of radius 1.7 Å.

The first steps of the hydration ($h < 0.5$) are characterized by a progressive increase in local mobility until diffusive motions (with a spatial extent not exceeding 1.7 Å and a diffusion coefficient of 0.68×10^{-5} cm^2/s) are observed at a higher hydration value of 0.65 [50, 51].

The natural abundance solid state ^{13}C NMR experiments have been carried out on a Brucker 400 NMR spectrometer operating at 100 MHz under conditions of cross-polarization and magic-angle spinning. Selective motions are apparent by NMR at time scales of 10 ns, or less, at the level of the charged and externally located lysil side chains, as well as the level of more internally located side chains (from Ala and Ile residues), whereas incoherent quasi-elastic neutron scattering monitors constrained diffusive motions of

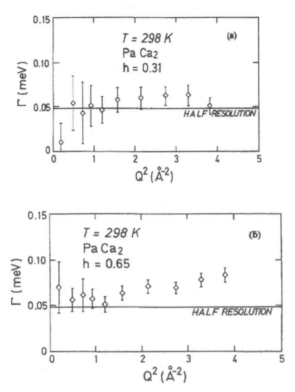

Figure 2.8 (a) Variation of the half-width at half-maximum of the Lorenzian line versus Q^2 for a parvalbumin $PaCa_2HHd$ sample hydrated at $h = 0.31$ g D_2O/g dry protein. (b) Same conditions with the exception that $h = 0.65$ g D_2O/g dry protein. All measurements are performed at 298 K. Reproduced from Ref. [51] by permission.

hydrogen atoms with a spatial extent not exceeding 1.7 Å at time scales up to 20 ps. To conclude, peripheral water–protein interactions influence the internal dynamics of globular proteins in a global manner. This was the first determination by QENS of the picosecond dynamics of a hydrated powder of protein. It has to be mentioned that if the neutron technique provides us with an average of internal dynamics over the protein protons, the ^{13}C NMR technique allows us to access lysine selective motions. Both experimental approaches indicate an increase in protein dynamics upon progressive hydration of the dry protein. When energy resolution of the neutron technique is increased, the strong decrease (nearly to zero) of the fraction of protons seen as immobile confirms that not only the polar side chains at the surface of the protein but also the internal nonpolar side chains and backbone are primarily affected during the early steps of hydration. These results confirm our previous results that hydration acts on the dynamics of the protein at both local and global levels. This is probably essential for the biological function of the protein. As shown by the present results, inelastic neutron scattering enables us to probe the dynamics of almost all the secondary structure elements, including the backbone.

An example of studies of dynamics combining the neutron results with those of computer molecular dynamics (CMD) simulations concerns the CPC [52]. Both techniques cover a similar time range (a few picoseconds to nanoseconds) and space range of a few angstroms. This strategy offers a unique opportunity to validate the potential of MD simulations and get a detailed knowledge of protein dynamics. The dynamics of water also becomes accessible in a detailed way. Mode-coupling theory (MCT), which is appropriate to describe supercooled liquids, provides us with a model for confined or interfacial water [37, 54].

By combining incoherent QENS and CMD, some physical picture of internal diffusive motions of a globular protein arises. Specific differences are seen between the backbone and side-chain dynamics. The time dependence for the backbone is less nonexponential, indicating a narrower spread of relaxation times. A smooth, "depth-dependent" dynamics has been detected that may have important consequences for protein function [52]. It may allow local reorganization of the structure for efficient ligand binding without affecting the internal stability.

By combining coherent inelastic neutron scattering and CMD, some physical picture of collective motions of protein arises. The protein backbone and side chains present the same slow collective dynamics [55, 56]. This could be explained by the fact that these two parts are linked via covalent bonding [48, 57].

2.3.2.2 Protein solutions

The number of studies on dynamics of protein solutions, in the native state [58, 59] and in chemical [60–63], heat, and pressure denatured states [64–66] and using quasi-elastic incoherent neutron scattering, has increased during the last years.

The dynamics of myoglobin and lysozyme in solution has been compared with that in hydrated powders [58]. In the case of solution samples, global motions in protein have been identified and the diffusion constants are in good agreement with the results of classical measurements. The geometry of local diffusive motions and their characteristic average relaxation times have been derived. From dry powder to coverage by one water layer, the surface side chains progressively acquire the possibility to diffuse locally. The finding that the side-chain motion in globular proteins contains a strong, liquid-like, nonvibrational confined diffusion component at physiological temperatures is consistent with the use of this model for proteins [67, 68].

On subsequent hydration, the main effect of water is to improve the rate of these diffusive motions. Motions with higher average amplitude occur in solutions about three times more than for a hydrated powder at complete coverage, with a shorter average relaxation time, about 4.5 ps compared with 9.4 ps for one water monolayer.

The local dynamics of *E. coli* aspartate transcarbamylase (ATCASE) solution, in T and R conformations, has been investigated by incoherent inelastic and QENS experiments [59], and a global view of protein dynamics as sensed via the individual motions of its hydrogen atoms has been obtained.

ATCase is a 310 kDa allosteric enzyme that catalyses the first committed step in pyrimidine biosynthesis. The binding of its substrates, carbamylphosphate and aspartate, induces significant conformational changes. This enzyme shows homotropic cooperative interactions between the catalytic sites for the binding of aspartate. This property is explained by a quaternary structure transition from a conformation with a low affinity for aspartate (T state) to a conformation with a high affinity for this substrate (R state). The same quaternary structure change is observed upon binding of the bisubstrate analogue PALA (*N*-(phosphonacetyl)-L-aspartate). This transition from the T state to the R state results in a 5% increase in the radius of gyration of the protein. In *E. coli* ATCase, the increase in surface exposed to the solvent during the T to R transition was calculated from crystallography [59]. However, this calculation does not provide information about the change in molecular dynamics that is associated with this transition. The present study allows a direct experimental quantitative determination of this change. The increased mobility in the R state might facilitate the migration of the substrates towards the catalytic site. We give evidence of some partial mechanism, detected at the picosecond time scale and resulting from the dynamic coupling between the surface water molecules and the protein residues occurring during the T to R transition. This is surprising because the transition is of the order of seconds, whereas our present study is in 15 picoseconds. Such a picosecond time coupling between water dynamics and protein dynamics has already been elucidated by molecular dynamics simulation [59]. However, to get a further understanding of the role of water molecules in the enzyme mechanism, two directions must be considered: (i) QENS experiments have to be performed at various temperatures and in relation with thermodynamic properties, and (ii) H/D substituted samples have to be designed. To get specially prepared isotope-substituted samples is not easy.

2.3.2.3 Hydration-coupled dynamics in protein

The existence of a protein dynamical transition around 220 K is notorious, and the central role of the protein hydration shell is now widely recognized as the driving force for this transition [69]. It should be noted, however, that *stricto sensu*, a transition should be associated to a thermal event as detected by calorimetry. In our knowledge, this is not the case for the so-called 220 K dynamical transition. This is why we prefer the term *dynamical crossover*. In a recent paper [70], we proposed a mechanism, at the molecular level, for the contribution of hydration water. In particular, we identified the importance of rotational motion of the hydration water as a source of configurational entropy triggering (i) the 220 K protein dynamical crossover (the so-called dynamical transition) [71], but also (ii) a much less intense and scarcely reported protein dynamical crossover, associated with a calorimetric glass transition, at 150 K.

The strong correlation shown in Fig. 2.9 strongly suggests that the transition at 150 K in rotational dynamics of the interfacial water is intimately connected to the protein transition at 150 K. In our knowledge, such a dynamical transition in hydrated proteins has only been scarcely reported [72]. This latter phenomenon is the more direct evidence of a direct change in entropy fluctuations of the protein–water system at 150 K. Nevertheless, by calorimetry alone, it is not possible to determine whether the transition is due to the protein or hydration water, or both. A firm conclusion could be reached in following the MSD temperature dependence on a fully deuterated protein [34, 37] sample hydrated by H_2O.

To summarize, we have shown a strong parallel evolution at 150 and 220 K between the MSDs related (i) to interfacial water rotational dynamics (Figs. 2.3a and 2.9a) and (ii) to proton dynamics of a hydrated protein Fig. 2.9b. This connection is made at the local scale (a few angstroms) and in the time scale of a few nanoseconds. We interpret these observations as evidences that interfacial water rotational dynamics is the real source of entropy driving protein dynamics.

Altogether, we reach this final view of the protein–water interaction and how this interaction can drive the protein function: the protein external side-chain short time motions, induced by fast water reorientational motion (leading here to $<u^2>_{Rot\ water}$, Fig. 2.9a), propagate in a hierarchical way along the protein structure from the residue side chains down to the protein core to induce the longer time scale protein backbone motion necessary for protein function. The dynamical crossovers experienced by water at 150 and 220 K are also detected in the protein dynamics, even though the time scales of the crossovers can be different (longer times for protein than for interfacial water).

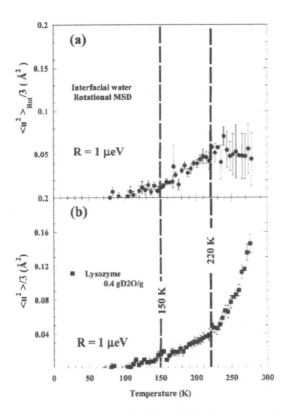

Figure 2.9 Temperature dependence of (a) the rotational mean-square displacement (MSD) of the interfacial water from Vycor (from Ref. [12]) and (b) the dynamic transition in 0.4 g D_2O/g lysozyme. At the 1 μeV resolution used presently (spectrometer IN16, ILL), motions are probed up to a few nanoseconds. A correlation is shown between the onset of rotational motion of interfacial water at 150 K and a transition in proton MSD of lysozyme. This correlation is also clearly detected at 220 K. The plateauing of the water MSD above 220 K is only apparent and due to instrumental resolution effects (see inset of Fig. 2.3). Such an MSD plateau is not seen in the case of protein because a significant fraction of the large distribution of the protein correlation times lies in the nanosecond time range. Reproduced from Ref. [70] by permission of the PCCP Owner Societies.

2.4 CONCLUSION

In this review, it appears that the structure and the dynamics of water are deeply perturbed when hydrophilic or hydrophobic interactions, or both, are developed in a medium. From the more recent findings, combining various techniques and molecular dynamics simulations, one gets the following picture of confined water. Water, in the vicinity of a hydrophilic surface, is in a state equivalent to bulk water at a lower temperature. As previously shown, the structure and dynamics of confined water depends on the degree of hydration of the sample. In particular, at room temperature, interfacial water shows a slow dynamics similar to that of bulk water at a temperature 30 K lower, so that it behaves like low-temperature water.

In order to understand the microscopic origin of the confinement and slowing down of motions of water molecules and the exact role played in this context, the theory of kinetic glass transition in dense supercooled liquids [54, 73] has been recently used. This theory leads to some description of the dynamics of confined water in terms of correlated jump diffusion [74] instead of jump diffusion [18]. This more sophisticated way shows a large distribution of residence times for water molecules in the cage formed by the neighboring molecules, which is a more realistic view than the sharp separation of water molecules into two classes according to their mobility [16].

This description looks consistent with molecular dynamics simulations of supercooled water [75] and has been confirmed by high-resolution QENS experiments of water in model systems [8–10, 12, 19–22, 27–29, 41] and from hydrated powders of protein [38, 46, 76–78]. Short-time dynamics results about hydrated myoglobin have been interpreted by using this same theory of kinetic glass transition in dense supercooled liquids [77]. Incoherent spin-echo signals of a hydrated β-lactoglobulin have been interpreted in terms of two contributions from surface water and protein, respectively [78]. The dynamics of surface water follows a KWW stretched exponential function with an exponent close to 0.5, in agreement with the previous findings from experiments and MD simulations.

Recently, this model has been extensively used by Chen and coauthors [76], who observe a fragile-to-strong dynamic crossover at 220 K, in protein hydration water for hydrated powder of lysozyme protein. The transition of hydration water at 220 K is the object of hot debates. Recently, on the basis of dielectric spectroscopy studies of hydrated lysozyme, Pawlus *et al.* [79] did not detect any transition but instead showed a smooth behavior of the whole system (protein and hydration water) over a temperature range (173–253 K). These authors, therefore, dismiss any fragile-to-strong crossover in the hydration water of protein. They suggest that the sharp rise in protein MSD clearly observed in neutron scattering experiments is indirectly an instrumental effect arising from the fact that at 220 K, the relaxation times of the system match the time window of the spectrometer.

One has to put the emphasis on the capability of QENS to selectively probe spatially resolved dynamical modes at a molecular level, not only for dynamics of hydration water but for dynamics of protein [80]. Combining

QENS with molecular dynamics (MD) simulations and other experimental techniques (solid-state ^{13}C NMR spectroscopy, dielectric spectroscopy, light scattering, etc.) is the basis of a successful methodology for such a scientific challenge. As a result, a microscopic view of the structural and dynamic events following the hydration of a globular protein can be obtained. The results of QENS combined with NMR spectroscopy [51] or MD simulations [52] suggest that peripheral water–protein interactions influence the protein dynamics in a global manner. There is a progressive induction of mobility at increasing hydration from the periphery towards the protein interior. Finally, the question arises whether the motions observed in the picosecond-to-nanosecond range through the combined experimental approach translate an adaptive trend in relation to protein function or are without any significance to function. This is a central question in structural biology in its attempts to correlate protein dynamics and function. Recently, a view of protein dynamics in relation to the function has been proposed by Frauenfelder and coworkers in which the dominant conformational motions are slaved by the hydration shell and the bulk solvent [81].

Acknowledgments

The author is grateful to the members of the Group "Biology and Disordered Systems" of Laboratoire Léon Brillouin (J.-M. Zanotti, J. Teixeira, S. Dellerue and I. Koeper) who contributed to this work.

J. Parello is acknowledged for his fruitful collaboration on dynamics of parvalbumin protein, and J. Smith and A. Petrescu for their contribution to the molecular dynamics simulations of C-phycocyanin protein.

G. Hervé is acknowledged for his fruitful collaboration on dynamics of aspartate transcarbamylase.

T. Yamaguchi, K. Yoshida, and S. Kittaka are acknowledged for their continuing collaboration in the field of confined water and protein dynamics.

References

1. Bellissent-Funel, M.-C. (2003) Status of experiments probing the dynamics of water in confinement, *Eur. Phys. J. E*, **12**, 83–92.

2. Linse, P. (1989) Molecular dynamics study of the aqueous core of a reversed ionic micelle, *J. Chem. Phys.*, **90**, 4992–5004.

3. Winnik, M. A., and Yekta, A. (1997) Associative polymers in aqueous solutions, *Curr. Opin. Colloid Interface Sci.*, **2**, 424–436.

4. Cheng, X., and Schoenborn, B. P. (1990) Hydration in protein crystals. A neutron diffraction analysis of carbonmonoxymyoglobin, *Acta Cryst.*, **B46**, 195–208.

5. Bellissent-Funel, M.-C., and Teixeira, J. (1998) Structural and dynamic properties of bulk and confined water, in *Freeze-Drying/Lyophilization of Pharmaceutical and Biological Products* (ed. Rey, L., and May, J. C.), Marcel Dekker, New York, pp. 53–77.

6. Soper, A. K. (1991) Structural Studies of water near an interface, in *Hydrogen-Bonded Liquids* (ed. Dore, J. C., and Teixeira, J.), Kluwer Academic Press, Dordrecht, C329, pp.147–158, and references therein.

7. Tardy, Y., Mercury, L., Roquin, C., and Vieillard, P. (1999) Le concept d'eau ice-like: hydratation-déshydratation des sels, hydroxydes, zéolites, argiles et matières organiques vivantes ou inertes, *C. R. Acad. Sci. Paris, Sciences de la terre et des planètes*, **329**, 377–388.

8. Kuroda, Y., Kittaka, S., Takahara, S., Yamaguchi, T., and Bellissent-Funel, M.-C. (1999) Characterization of the state of two-dimensionally condensed water on hydroxylated chromium (III) oxide surface through FT-IR, quasielastic neutron scattering and dielectric relaxation measurements, *J. Phys. Chem. B*, **103**, 11064–11073.

9. Takahara, S., Kittaka, S., Mori, T., Kuroda, Y., Yamaguchi, T., and Shibata, K. (2002) Neutron scattering study on the dynamics of water molecules adsorbed on SrF_2 and ZnO surfaces, *J. Phys. Chem. B*, **106**, 5689–5694.

10. Fratini, E., Faraone, A., Baglioni, P., Bellissent-Funel, M.-C., and Chen, S. H. (2002) Dynamic scaling of QENS spectra of glassy water in aging cement paste, *Physica A*, **304**, 1–10.

11. Bellissent-Funel, M.-C., Bosio, L., and Lal, J. (1993) Structural study of water confined in porous glass by neutron scattering, *J. Chem. Phys.*, **98**, 4246–4252.

12. Zanotti, J.-M., Bellissent-Funel, M.-C., and Chen, S.-H. (2005) Experimental evidence of liquid-liquid transition in interfacial water, *Europhys. Lett.*, **71**, 91–97.

13. Bellissent-Funel, M.-C., Teixeira, J., and Bosio, L. (1987) Structure of high-density amorphous water. II: neutron scattering study, *J. Chem. Phys.*, **87**, 2231–2235.

14. Zanotti, J.-M., Bellissent-Funel, M.-C., Chen, S. H., and Kolesnikov, A. I. (2006) Further evidence of liquid-liquid transition in interfacial water, *J. Phys. Cond. Matt.*, **18**, S2299–S2304.

15. Zanotti, J.-M., Bellissent-Funel, M.-C., and Kolesnikov, A. I. (2007) Phase transitions of interfacial water at 165 and 240K. Connections to bulk water physics and protein dynamics, *Eur. Phys. J.*, **141**, 227–233.

16. Bellissent-Funel, M.-C., Zanotti, J.-M., and Chen, S. H. (1996) Slow dynamics of water molecules on surface of a globular protein, *Faraday Discuss.*, **103**, 281–294.

17. Bellissent-Funel, M.-C., Chen, S. H., and Zanotti, J.-M. (1995) Single particle dynamics of water molecules in confined space, *Phys. Rev. E*, **51**, 4558–4569.

18. Teixeira, J., Bellissent-Funel, M.-C., Chen, S. H., and Dianoux, A. J. (1985) Experimental determination of the nature of diffusive motions of water molecules at low temperatures, *Phys. Rev. A*, **31**, 1913–1917.

19. Yoshida, K., Yamaguchi, T., Kittaka, S., Bellissent-Funel, M.-C., and Fouquet, P. (2008) Thermodynamic, structural, and dynamic properties of supercooled water confined in mesoporous MCM41 studied with calorimetric, neutron diffraction, and neutron spin echo measurements, *J. Chem. Phys.*, **129**, 054702-1-054702-11.

20. Faraone, A., Liu, L., Mou, C.-Y., Yen, C.-W., and Chen, S.-H. (2004) Fragile-to-strong liquid transition in deeply supercooled confined water, *J. Chem. Phys.*, **121**, 10843–10846.

21. Liu, L., Chen, S.-H., Faraone, A., Yen, C.-W., and Mou, C.-Y. (2005) Pressure dependence of fragile-to-strong transition and a possible second critical point in supercooled confined water, *Phys. Rev. Lett.*, **95**, 117802–1–117802–4.

22. Mallamace, F., Broccio, M., Corsaro, C., Faraone, A., Wanderlingh, U., Liu, L., Mou, C.-Y., and Chen, S. H. (2006) The fragile-to-strong dynamic crossover transition in confined water: nuclear magnetic resonance results, *J. Chem. Phys.*, **124**, 161102.

23. Bellissent-Funel, M.-C., Sridi-Dorbez, R., and Bosio, L. (1996) X-ray and neutron scattering studies of the structure of water at a hydrophobic surface, *J. Chem. Phys.*, **104**, 10023–10029.

24. Rossky, P. J. (1994) Structure and dynamics of water at interfaces, in *Hydrogen Bond Networks* (ed. Bellissent-Funel, M.-C., and Dore, J. C.), Kluwer Academic Press, Dordrecht, C435, pp. 337–338.

25. Jelassi, J., Castricum, H. L., Bellissent-Funel, M.-C., Dore, J. C., Webber, B., and Sridi-Dorbez, R. (2010) Studies of water and ice in hydrophilic and hydrophobic mesoporous silicas I: pore characterisation and phase transformations, *Phys. Chem. Chem. Phys.*, **12**, 2838 – 2849.

26. Soper, A. K., and Luzar, A. (1996) Orientation of water molecules around small polar and nonpolar groups in solution: a neutron diffraction and computer simulation study, *J. Phys. Chem.*, **100**, 1357–1367.

27. Cabral, J. T., Luzar, A., Teixeira, J., and Bellissent-Funel, M.-C. (2000) Single-particle dynamics in DMSO-water eutectic mixture by neutron scattering, *J. Chem. Phys.*, **113**, 8736–8745.

28. Köper, I., Bellissent-Funel, M.-C., and Petry, W. (2005) Dynamics from pico- to nanoseconds of trehalose in aqueous solutions as seen by quasielastic neutron scattering, *J. Chem. Phys.*, **122**, 014514–014516.

29. Almasy, L., Banki, P., Bellissent-Funel, M.-C., Bokor, M., Cser, L., Jancso, G., Tompa, K., and Zanotti, J.-M. (2002) QENS and NMR studies of 3-picoline-water solutions, *Appl. Phys. A*, **74**, S516–S518.

30. Mentré, P. (1995) *L'eau dans la cellule. Une interface hétérogène et dynamique des macromolécules*, Masson, Paris.

31. Rose, G. D., Geselowitz, A. R., Lesser, G. J., Lee, R. H., and Zehfus, M. H. (1985) Hydrophobicity of amino acids. Residues in globular proteins, *Science*, **229**, 834–838.

32. Rupley, J. A., Gratton, E., and Careri, G. (1983) Water and globular proteins, *Trends Biochem. Sci.*, **8**, 18–22.

33. Wiggins, P. M., and MacClement, B. A. E. (1987) Two states of water found in hydrophobic clefts: their possible contribution to mechanisms of action of pumps and other enzymes, *Int. Rev. Cytol.*, **108**, 249–304.

34. Bellissent-Funel, M.-C., Lal, J., Bradley, K. F., and Chen, S. H. (1993) Neutron structure factors of *in vivo* deuterated amorphous protein C-phycocyanin, *Biophys. J.*, **64**, 1542–1549.

35. Wiggins, P. M. (1988) Water structure in polymer membranes, *Prog. Polym. Sci.*, **13**, 1–35.

36. Rupley, J. A., and Careri, G. (1991) Protein hydration and function, *Adv. Prot. Chem.*, **41**, 37–172.

37. Dellerue, S., and Bellissent-Funel, M.-C. (2000) Relaxational dynamics of water molecules at protein surface, *Chem. Phys.*, **258**, 315–325.

38. Zanotti, J.-M., Bellissent-Funel, M.-C., and Chen, S. H. (1999) Relaxational dynamics of supercooled water in porous glass, *Phys. Rev. E*, **59**, 3084–3093.

39. Tehei, M., Franzetti, B., Wood, K., Gabel, F., Fabiani, E., Jasnin, M., Zamponi, M., Oesterhelt, D., Zaccai, G., Ginzburg, M., and Ginzburg, B.-Z. (2007) Neutron scattering reveals extremely slow cell water in a Dead Sea organism, *PNAS*, **104**, 766–771.

40. Persson, E., and Halle, B. (2008) Cell water dynamics on multiple time scales, *PNAS*, **105**, 6266–6271.

41. Rupley, J. A., Gratton, E., and Careri, G. (1983) Water and globular proteins, *Trends Biochem. Sci.*, **8**, 18–22.

42. Colombo, M. F., Rau, D. C., and Parsegian, V. A. (1992) Protein solvation in allosteric regulation: a water effect on hemoglobin, *Science*, **256**, 655–659.

43. Steinbach, P. J., and Brooks, B. R. (1993) Protein hydration elucidated by molecular dynamics simulation, *Proc. Natl. Acad. Sci. USA*, **90**, 9135–9139.

44. Lounnas, V., Pettitt, B. M., and Phillips, G. N., Jr. (1994) A global model of the protein–solvent interface, *Biophys. J.*, **66**, 601–614.

45. Nakasako, M. (2004) Water–protein interactions from high-resolution crystallography, *Phil. Trans. R. Soc. London B*, **359**, 1191–1206.

46. Bellissent-Funel, M.-C. (2005) Interactions hydrophiles-hydrophobes: des systèmes modèles aux systèmes vivants, *C. R. Acad. Sci. Geosci.*, **337**, 173–179.

47. Smith, J. C. (1991) Protein dynamics: comparison of simulations with inelastic neutron scattering experiments, *Q. Rev. Biophys.*, **24**, 227–291.

48. Bellissent-Funel, M.-C. (2006) Internal diffusive and collective motions in proteins studied by neutron scattering, *J. Mol. Liquids*, **129**, 44–48.

49. Doster, W., Cusack, S., and Petry, W. (1989) Dynamical transition of myoglobin revealed by inelastic neutron scattering, *Nature*, **337**(23), 754–756.

50. Zanotti, J.-M., Parello, J., and Bellissent-Funel, M.-C. (2002) Influence of hydration and cation binding on parvalbumin dynamics, *Appl. Phys. A*, **74**, S1277–S1279.

51. Zanotti, J.-M., Parello, J., and Bellissent-Funel, M.-C. (1999) Hydration-coupled dynamics in proteins studied by neutron scattering and NMR: the case of the typical EF-hand calcium-binding parvalbumin, *Biophys. J.*, **76**, 2390–2411.

52. Dellerue, S., Petrescu, A. J., Smith, J. C., and Bellissent-Funel, M.-C. (2001) Radially softening diffusive motions in a globular protein, *Biophys. J.*, **81**, 1666–1676.

53. Volino, F., and Dianoux, A. J. (1980) Neutron incoherent scattering law for diffusion in a potential of spherical symmetry, *Mol. Phys.*, **41**, 271–279.

54. Gotze, W., and Sjogren, L. (1992) Relaxation processes in supercooled liquids, *Rep. Prog. Phys.*, **55**, 241–376.

55. Bellissent-Funel, M.-C. (2006) Internal diffusive and collective motions in proteins studied by neutron scattering, *J. Mol. Liquids*, **129**, 44–48.

56. Dellerue, S., Petrescu, A. J., Smith, J. C., Longeville, S., and Bellissent-Funel, M.-C. (2000) Collective dynamics of a photosynthetic protein probed by neutron spin-echo spectroscopy and molecular dynamics simulation, *Physica B*, **276–278**, 514–515.

57. Kneller, G., Hinsen, K., Petrescu, A.-J., Dellerue, S., Bellissent-Funel, M.-C., and Kneller, G. R. (2000) Harmonicity in slow protein dynamics, *Chem. Phys.*, **261**, 25–37.

58. Perez, J., Zanotti, J.-M., and Durand, D. (1999) Evolution of the internal dynamics of two globular proteins from dry powder to solution, *Biophys. J.*, **77**, 454–469.

59. Zanotti, J.-M., Herve, G., and Bellissent-Funel, M.-C. (2006) Aspartate transcarbamylase T and R forms dynamics: an inelastic neutron scattering study, *Biochim. Biophys. Acta*, **1764**, 1527–1535.

60. Receveur, V., Calmettes, P., Smith, J. C., Desmadril, M., Coddens, G., Durand, D. (1997) Picosecond dynamical changes on denaturation of yeast phosphoglycerate kinase revealed by quasielastic neutron scattering, *Proteins*, **28**, 380–387.

61. Russo, D., Pérez, J., Zanotti, J.-M., Desmadril, M., Durand, D. (2002) Dynamic transition associated with the thermal denaturation of a small beta protein, *Biophys. J.*, **83**, 2792–2800.

62. Appavou, M.-S., Gibrat, G., Bellissent-Funel, M-C. (2009) Temperature dependence on structure and dynamics of Bovine Pancreatic Trypsin Inhibitor (BPTI): a neutron scattering study, *Biochim. Biophys. Acta*, **1764**, 414–423.

63. Gibrat, G., Blouquit, Y., Craescu, C. T., Bellissent-Funel, M.-C. (2008) Biophysical studies of thermal denaturation of calmodulin protein: dynamics, *Biophys. J.*, **95**(11), 5247–5256.

64. Appavou, M.-S., Gibrat, G., and Bellissent-Funel, M.-C. (2006) Influence of pressure on structure and dynamics of bovine pancreatic trypsin inhibitor (BPTI): small angle and quasi-elastic neutron scattering studies, *Biochim. Biophys. Acta*, **1764**, 414–423.

65. Calandrini, V., Hamon, V., Hinsen, K., Calligari, P., Bellissent-Funel, M.-C., Kneller, G. (2008) Relaxation dynamics of lysozyme under pressure: combining molecular dynamics and quasi-elastic neutron scattering, *Chem. Phys.*, **345**, 289–297.

66. Filabozzi, A., Di Bari, M., Deriu, A., Di Venere, A., Andreani, C., and Rosato, N. (2005) Pressure dependence of protein dynamics investigated using elastic and quasi-elastic neutron scattering, *J. Phys. Condens. Matter*, **17**, S3101–S3109.

67. Kneller, G. R., and Smith, J. C. (1994) Liquid like side-chain dynamics in myoglobin, *J. Mol. Biol.*, **242**, 181–185.

68. Hinsen, K., Petrescu, A. J., Dellerue, S., Kneller, G. R., and Bellissent-Funel, M.-C. (2002) Liquid-like and solid-like motions in proteins, *J. Mol. Liquids*, **98–99**, 381–398.

69. Doster, W. (2008) The dynamical transition of proteins, concepts and misconceptions, *Eur. Biophys. J.*, **37**, 591–602.

70. Zanotti, J.-M., Gibrat, G., and Bellissent-Funel, M.-C. (2008) Hydration water rotational motion as a source of configurational entropy driving protein dynamics. Crossovers at 150 and 220 K, *Phys. Chem. Chem. Phys.*, **10**, 4865–4870.

71. Green, J. L., Fan, J., and Angell, C. (1994) The protein-glass analogy: new insight from homopeptide comparisons, *J. Phys. Chem.*, **98**, 13780–13790.

72. Miyazaki, Y., Matsuo, T., and Suga, H. (2000) Low-temperature heat capacity and glassy behavior of lysozyme crystal, *J. Phys. Chem. B*, **104**, 8044–8052.

73. Cummins, H. Z., Li, G., Du, W. M., and Hernandez, W. M. (1994) Relaxational dynamics in supercooled liquids: experimental tests of the mode-coupling theory, *Physica A*, **204**, 169–201.

74. Chen, S. H., Gallo, P., and Bellissent-Funel, M.-C. (1996) Slow dynamics of water in supercooled states and near hydrophilic surfaces, in *Non Equilibrium Phenomena in Supercooled Fluids, Glasses and Materials* (ed. Giordano, M., Leporini, D., and Tosi, M. P.), World Scientific Publication, Singapore, pp. 186–194.

75. Gallo, P., Sciortino, F., Tartaglia, P., and Chen, S. H. (1996) Slow dynamics of water molecules in supercooled water, *Phys. Rev. Lett.*, **76**, 2730–2733.

76. Chen, S. H., Liu, L., Fratini, E., Baglioni, P., Faraone, A., and Mamontov, E. (2006) Observation of fragile-to-strong dynamic cross over in protein hydration water, *PNAS*, **103**, 9012–9016.

77. Settles, M., and Doster, W. (1996) Anomalous diffusion of adsorbed water: a neutron scattering study of hydrated myoglobin, *Farad. Discuss.*, **103**, 269–279.

78. Yoshida, K., Yamaguchi, Y., Bellissent-Funel, M.-C., and Longeville, S. (2007) Hydration water in dynamics of a hydrated beta-lactoglobulin, *Eur. Phys. J. Spec. Top.*, **114**, 223–226.

79. Pawlus, S., Khodadadi, S., and Sokolov, A. P. (2008) Conductivity in hydrated proteins: no signs of the fragile-to-strong crossover, *Phys. Rev. Lett.*, **100**, 108103–10806.

80. Sakai, V. G., and Arbe, A. (2009) Quasielastic neutron scattering in soft matter, *Curr. Opin. Colloid Interface Sci.*, **14**, 381–390.

81. Frauenfelder, H., Chen, G., Berendzen, J., Fenimore, P. W., Jansson, H., McMahon, B. H., Stroe, I. R., Swenson, J., and Young, R. D. (2009) A unified model of protein dynamics, *PNAS*, **106**, 5129–5134.

Chapter 3

EFFECT OF HYDRATION ON PROTEIN DYNAMICS

Mikio Kataoka[a,b] and Hiroshi Nakagawa[b]

[a] *Graduate School of Materials Science,*
  Nara Institute of Science and Technology,
  8916-5 Takayama, Ikoma, Nara 630-0192, Japan
[b] *Neutron Biophysics Group, Japan*
  Atomic Energy Agency, 2-4 Shirakata Shirane,
  Tokai-mura, Naka-gun, Ibaraki 319-1195, Japan
  kataoka@ms.naist.jp

3.1 INTRODUCTION

Most proteins function in an aqueous environment. Two types of water molecules exist in a protein–water system: bulk water and hydration water. On the surface of the protein, water molecules interact strongly with the protein; this interaction is termed "protein hydration" [1]. Protein hydration affects the structural flexibility of a protein, as well as its biological functions [2, 3]. A number of studies have suggested that protein dynamics are essential for expression of function [4–6]. Most of these results have shown that function-related dynamics are activated by hydration water [4]. Single-molecule experiments demonstrate that proteins utilize the thermal fluctuation of the protein (protein structural mobility and/or Brownian motion) for biological functions, rather than the chemical energies derived from ATP hydrolysis [7, 8]. *In vivo*, such a thermal fluctuation is substantially provided by water molecules in the cell. The internal environment of the cell is usually quite crowded, because of the high densities of the numerous biomolecules. In such a context, most water molecules exist in hydration water rather than

Water: The Forgotten Biological Molecule
Edited by Denis Le Bihan and Hidenao Fukuyama
Copyright © 2011 by Pan Stanford Publishing Pte. Ltd.
www.panstanford.com

bulk water [9]. In this section, we will review the effect of hydration water on protein dynamics on a molecular scale, as determined by an inelastic neutron scattering (INS) experiment. The time scale of the hydration-coupled protein dynamics is in the picosecond to nanosecond range [10]. INS covers this dynamics range. For a long time, INS has been a powerful tool in multiple fields within physics, e.g., solid-state physics, magnetism, and soft matter physics. The applications of INS to biology have been quite limited, but the studies about the dynamics of protein, DNA, and biomembrane have increased year by year during this decade. Some of these studies make essential contributions to molecular biology, and not only to biophysics [5, 11].

3.2 NEUTRON SCATTERING

Neutrons are hadrons and are experimentally obtained from reactor-based or pulsed accelerator–based (spallation) neutron sources [12]. Neutrons are electrically neutral and are scattered by the atomic nucleus. (This is in contrast to X-rays, which are scattered by electrons.) Neutrons have a spin, which is another important feature. Neutrons are also scattered by nuclear spin, which causes incoherent scattering. The distinctive feature of neutron scattering is that the incoherent scattering cross section of hydrogen (80 barn, 1 barn = 10^{-24} cm^2) is extraordinarily larger than any other element [12]. Coherent scattering gives information about the interatomic correlation function, whereas incoherent scattering gives information about the self-correlation function [12]. We must consider another type of scattering, in which the energy of the incident neutrons is similar to that of molecular motion. Therefore, neutrons can activate or deactivate molecular motions, which cause inelastic scattering. Coherent elastic scattering and incoherent inelastic scattering are experimentally important combinations. The former gives information about the structure, while the latter gives information about internal dynamics [12]. Almost half of the atoms in a protein are hydrogen atoms, which are uniformly distributed in the protein structure. Thus, neutrons scattered by a protein give information on the global dynamics of the protein. The D$_2$O-hydrated protein is used in the INS experiment for the selective observation of protein dynamics and not for hydration water because of the small incoherent scattering cross section of deuterium (2.0 barn).

3.2.1 Neutron Spectrum

The typical neutron scattering spectra along an energy axis are shown schematically in Fig. 3.1 [13]. The spectra are categorized into the following three scattering types: elastic, quasi-elastic, and inelastic. Elastic scattering is the spatial correlation of an atom between time 0 and time infinity. At cryogenic temperatures, the molecular motions are thermally suppressed, and consequently strong elastic scatterings are observed. At higher temperatures, since the probability of finding an atom at time 0 decreases, the elastic scattering becomes smaller.

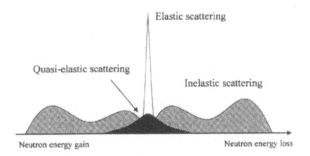

Figure 3.1 Schematic diagram of three observed types of incoherent neutron scattering—elastic, quasi-elastic, and inelastic—at three typical temperatures [13].

Figure 3.2 shows the temperature dependence of the elastic scattering intensities of hydrated and unhydrated proteins. As the temperature increases, the elastic intensity decreases, which indicates that the protein becomes more flexible.

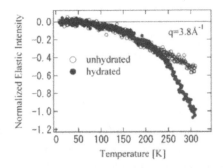

Figure 3.2 Temperature dependence of elastic neutron scattering intensity at $q = 3.8$ Å^{-1} for unhydrated and hydrated SNase.

The strong decrease of intensity at around 240 K for the hydrated protein suggests the appearance of additional motions by hydration, which will be described in detail in a later section. The scattering vector dependence of the incoherent elastic scattering intensity gives the atomic mean-square displacement (MSD) by Gaussian approximation [14, 15], as follows:

$$I_{el}(q) = S(q, \omega) \otimes R(\omega)|_{\omega = 0} = A \exp\left(- \langle u^2 \rangle\, q^2/3\right), \qquad (3.1)$$

where $I_{el}(q)$ is the elastic incoherent neutron scattering, $R(\omega)$ is the instrumental energy resolution, the symbol $\otimes$ denotes the convolution product, $\langle u^2 \rangle$ is the MSD from the center of mass (equilibrium position), A is a constant and q is the scattering vector. The decrease of the elastic scattering leads to the appearance of a quasi-elastic scattering, which comes from structural relaxation and/or diffusive motions. Inelastic scattering yields the vibration spectrum of the molecules [16, 17]. INS is thus complementary to IR or Raman spectroscopy. One advantage of INS over these spectroscopic methods is that the observed spectrum is quantitatively compared with the theoretically calculated spectrum. The estimation of transition dipole moment or changes in electronic polarizability is necessary to calculate the IR or Raman spectrum, respectively; neither calculation is easy. Moreover, all vibrational modes are not necessarily observed by optical vibrational spectroscopy. INS is free from such a selection rule, which is another advantage.

3.2.2 Calculated Neutron Spectrum by Theoretical Analysis

The vibrational density of states calculated using normal mode analysis (NMA) and the motional spectrum from molecular dynamics (MD) simulation can be easily compared with an experimentally obtained neutron spectrum [13, 18, 19]. From the NMA, the neutron scattering function $S(q, \omega)$ is calculated as follows [13]:

$$S(\vec{q}, \omega) = \sum_{l=1}^{N} \sum_{\lambda=1}^{3N-6} b_\alpha^2 e^{\hbar \omega_\lambda \beta /2} \frac{e^{-2W_\alpha(\vec{q})} \hbar \left| \vec{q} \cdot \vec{e}_{\lambda \alpha} \right|^2}{4 m_\alpha \omega_\lambda \sinh\left(\dfrac{\beta \hbar \omega_\lambda}{2}\right)} \delta(\omega - \omega_\lambda), \qquad (3.2)$$

where b_α and m_α are the neutron scattering length and the atomic mass for atom α, respectively, ω_λ is the mode angular frequency, and λ labels the mode. $\beta = 1/k_B T$, where k_B is Boltzmann's constant and T is the temperature. W_α $(\vec{q})$ is the exponent of the Debye–Waller factor, $\exp[-2W_\alpha(\vec{q})]$. Equation (3.2) can predict the neutron vibrational spectrum and assign spectral peaks for the corresponding molecular motions (see Fig. 3.4).

From MD simulation, the observed neutron scattering function of $S(q, \omega)$ is calculated as follows [13]:

$$S(q, \omega) = \frac{1}{2\pi} \int_{-\infty}^{\infty} dt \exp\{-i\omega t\} \int_{-\infty}^{\infty} dr \exp\{-i\vec{q} \cdot \vec{r}\} g(\vec{r}, t) \qquad (3.3)$$

$$g(\vec{r}, t) = N^{-1} \sum_{l=1}^{N} \left\langle \delta\{r + \vec{R}_l(0) - \vec{R}_l(t)\} \right\rangle, \qquad (3.4)$$

where $\vec{r}$ is a position vector and $R_l(t)$ is the position vector of the lth atom at time t, which is obtained by the result of the MD simulation. Thus, INS experiments and MD simulations are compatible with one another in protein dynamics research. The combination of both methods allows detailed characterization of protein dynamics.

3.3 PROTEIN DYNAMICS IN TERMS OF MEAN-SQUARE DISPLACEMENT: DYNAMICAL TRANSITION

Protein structure fluctuates thermally. Therefore, temperature is a typical variable in studies of protein dynamics. Another essential factor is solvent effects. Under physiological conditions, most proteins exist in an aqueous environment, and water molecules strongly affect the protein fluctuations. To examine the hydration effect on protein dynamics, the hydration content is experimentally controlled in powder samples of a protein. Figure 3.3 shows the temperature dependence of the MSD of a protein, staphylococcal nuclease (SNase), in the hydrated and unhydrated state. SNase is a globular protein and a model for protein dynamics [20–22]. At a lower temperature, the MSD increases linearly. This indicates harmonic motions in a basin of the structural energy landscape (see inset in Fig. 3.3), and the slope of the line corresponds to the force constant of harmonic oscillator [15]. The force constants for the hydrated and unhydrated protein are 3.13 and 2.56 N/m, respectively. The force constant increases by hydration, indicating that hydration stiffens the harmonic properties of protein. As the temperature increases, a deviation from linearity is observed at ~150 K, indicating the appearance of anharmonic motions [2, 20]. The responsible motion is assigned to the methyl rotation and is independent of hydration [2, 20].

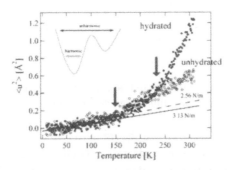

Figure 3.3 Temperature dependence of MSDs of unhydrated and hydrated SNase. The MSDs were obtained from the q dependence of elastic scattering. Inset shows the asymmetric double-well potential model [23].

In contrast, hydration-dependent anharmonic motions appear at ~240 K, which is the so-called dynamical transition [23]. It has been reported that the appearance of the dynamical transition is correlated with protein biological activity [3, 5]. Therefore, function-related structural fluctuations are induced by both thermal energy and protein hydration. In the following sections, these protein dynamics will be characterized in the spectral region.

3.4 VIBRATIONAL SPECTRUM

3.4.1 Hydration-Uncoupled Modes

Figure 3.4 shows the neutron spectra of SNase in the hydrated and unhydrated state at 20 K. As shown in the inset of Fig. 3.3, at cryogenic temperatures protein motions are trapped within a basin of local minima on the energy landscape. Thus, the measured spectra describe the distribution of the vibrational modes in the protein structure. The peaks of the spectrum were assigned by the normal mode calculation [16]. The sharp peak at 235 cm^{-1} is identified as methyl torsions. The vibrations observed in the 350–500 cm^{-1} range are the skeletal angle and dihedral displacements. The vibrations observed between 700 and 1000 cm^{-1} are mostly assigned as CH$_3$ rotation, CH$_2$ rotation, and CH bending. The vibrations observed around 1300 cm^{-1} are CH$_2$ twist, CH$_3$ symmetric bend, and so on.

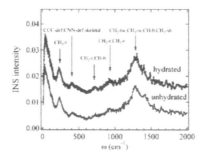

Figure 3.4 INS spectra of unhydrated and hydrated SNase. The spectrum for the hydrated sample is shifted 0.01 units along the ordinate for the sake of clarity.

In Fig. 3.4, the INS spectra are shown with two hydration levels, 0.12 and 0.44 g D$_2$O/g protein for the unhydrated and hydrated specimen, respectively. The high-energy vibrational modes are almost identical between the two specimens, indicating that the vibrations of the covalent bonds are not markedly affected by hydration. These modes are rather localized and not damped significantly by hydration friction.

3.4.2 Hydration-Coupled Modes

As shown in Fig. 3.4, the spectral peak position below 100 cm^{-1} in the hydrated state seems to shift to a higher energy compared with that of the unhydrated state. Figure 3.5 shows the comparison of the spectra in the low-energy region. The strong elastic peak at 0 cm^{-1} is observed for both samples, and a broad peak is observed around a few tens of cm^{-1}. This is the so-called boson peak [20]. This is also commonly observed for nonbiological substances such as synthetic polymer and glassy materials, but the origin of the spectral peak is still debated [24]. In the case of proteins, it has been suggested that the boson peak is sensitive to protein secondary or tertiary structure and that low-frequency modes correspond to the internal deformation of secondary structure, such as accordion-like modes of α-helices [25]. Experimentally, the boson peak position of a globular protein depends on the molecular weight, indicating that the low-frequency modes responsible for the peak are extended over the whole protein [26].

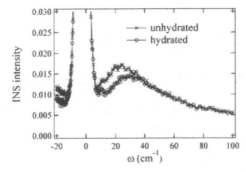

Figure 3.5 INS spectrum of unhydrated and hydrated SNase in the low-energy region. A shift of the boson peak toward higher energy is observed upon hydration.

Figure 3.5 shows that the scattering intensity of the hydrated sample lower than 40 cm^{-1} is depressed and that the boson peak position shifts from 25 to 30 cm^{-1} upon hydration. A hydration-induced shift of the boson peak was observed for other proteins, suggesting a general dynamical property of proteins [27, 28]. At cryogenic temperatures, protein motions are harmonic. The frequency up-shift of the boson peak upon hydration suggests an increase of the average force constant of harmonic motions because of an increase in the number and the strength of protein–water hydrogen bonds. A computational study demonstrated that hydration makes the energy landscape rugged and suggested that the boson peak originates from falling into the local minima on this rugged surface [29]. It is also expected that protein fluctuation is restricted in the small space of a local minimum. Figure 3.3 shows that the

MSDs of hydrated protein are slightly smaller than those of unhydrated protein, and the force constant of the protein increases upon hydration. Thus, hydration suppresses the harmonic motions of the protein in both time and space. Glycerol and trehalose can also cause a protein to become rigid at a low temperature [30]. This suggests that the hydrogen bond interaction of glassy solvent with the embedded protein affects the flexibility of the protein structure. The interaction with hydration water through hydrogen bonds on the protein surface is an essential factor in determining protein dynamics.

3.5 STRUCTURAL FLEXIBILITY AND HYDRATION

Hydration suppresses protein harmonic motions at cryogenic temperatures. On the other hand, as shown in Fig. 3.3, hydration induces a significant increase in the protein fluctuation in MSD at ~240 K. It was reported that the protein function appears above this temperature and that protein anharmonic fluctuations activated at the dynamical transition are necessary for biological function [4]. A hydrated protein undergoes such functionally relevant structural fluctuations under physiological conditions, such as those occurring in the cell at ambient temperatures.

3.5.1 Solvent-Activated Dynamics and Dynamical Transition

The appearance of the anharmonic dynamics suggests the itinerancy of protein conformation among the minima on the protein energy landscape (see inset in Fig. 3.3). This structural relaxation and/or diffusion dynamics should be reflected in the appearance of quasi-elastic scattering. At 100 K, protein dynamics are harmonic; therefore, the excess intensity over the spectrum at 100 K represents the contribution from quasi-elastic scattering [20, 31].

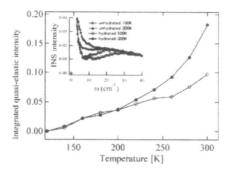

Figure 3.6 Temperature dependence of integral intensity of quasi-elastic scattering for unhydrated and hydrated SNase. Integration was taken between 8 and 32 cm⁻¹. Inset shows INS spectra of the unhydrated and hydrated SNase at 100 and 300 K.

Figure 3.6 shows the contribution of the quasi-elastic scattering at various temperatures. The inset in Fig. 3.7 shows the INS in the low-energy region with the hydrated and unhydrated samples at 100 K and 300 K.

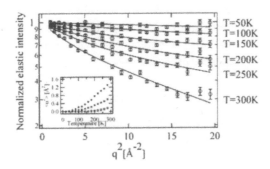

Figure 3.7 Elastic neutron scattering intensity in a wide Q range from SNase at various temperatures. The solid curves show fitting functions with a bimodal distribution. Inset shows the temperature-dependent MSDs of the large component (–), small component (∘), and averaged value (×).

The integrated quasi-elastic scattering intensities are estimated as follows:

$$S_{qel}(q,\omega)\Big|_T = S(q,\omega)\Big|_T - S(q,\omega)\Big|_{T_0} \frac{e^{-q^2 U_{vib}(T)}\left(e^{\omega/k_B T_0} - 1\right)}{e^{-q^2 U_{vib}(T_0)}\left(e^{\omega/k_B T} - 1\right)}, \quad (3.5)$$

The quasi-elastic intensities for both states increased on raising the temperature, suggesting the activation of some anharmonic motions. The quasi-elastic intensities up to approximately 200 K were not markedly affected by hydration. The anharmonicity is responsible for the transition at ~150 K commonly observed for hydrated and unhydrated proteins. Above 200 K, the increase in the quasi-elastic intensity of the unhydrated sample was roughly proportional to the temperature. On the other hand, the hydrated sample gained additional strong quasi-elastic scattering. This increase in the quasi-elastic scattering indicates the activation of hydration-dependent anharmonic motions, which corresponds to the onset of the hydration-induced dynamical transition.

The hydration water is confined on the protein surface and interacts directly with the outer part of the protein rather than the inner core. In order to separate the dynamics of the protein surface and the inner core, the dynamical heterogeneity was determined from the presence of a non-Gaussian character in the elastic scattering data [32]. The divergence from the approximation is observed in the wide q region of the elastic scattering. Figure 3.7 is the elastic scattering profile in the wide q range of the hydrated SNase. The Gaussian approximation is reasonable up to ~5 Å^{-1}, whereas the non-Gaussianity becomes remarkable in the wider q region. We demonstrated that the bimodal distribution model is simple but reasonable and practically useful to analyze the protein dynamics among various dynamical heterogeneity models [32, 33]. The distribution function for MSD should be explicitly included to explain the dynamical heterogeneity for the calculation of elastic scattering as follows:

$$S_{el}(Q) = \int_0^\infty g(\alpha) \exp(-\alpha q^2) d\alpha, \qquad (3.6)$$

where $g(\alpha)$ is the distribution function and α is $\langle u^2 \rangle /3$.

For the bimodal distribution, $g_B(\alpha)$ is given by

$$g_B(\alpha) = (1 - p) \, \delta(\alpha - \alpha_s) + p\delta(\alpha - \alpha_{a1}). \qquad (3.7)$$

The final form of $S_{el}(Q)$ for a bimodal distribution is

$$S_{el}(q) = (1 - p) \exp(- \langle u_s^2 \rangle q^2/3) + p \exp (- \langle u_l^2 \rangle q^2/3), \qquad (3.8)$$

where $\langle u_s^2 \rangle$ and $\langle u_l^2 \rangle$ are the small and large amplitude components with the corresponding population fractions $1 - p$ nd p, respectively. Two components are interpreted as the average of fluctuations of the inner core and the outer shell of protein structure, respectively. The inset of Fig. 3.7 shows the temperature dependence of MSDs of the separated components with the average MSD for the hydrated SNase. For the unhydrated SNase, the separated MSDs were almost the same as for the hydrated protein. On the other hand, the hydration affects the heterogeneity in terms of the change in the fraction factor, p. The values were $p = 0.27$ and 0.38 for the unhydrated and the hydrated SNase, respectively. The change in p indicates that the mobile fraction of the protein surface increases upon hydration. A "radial relaxation model" is proposed for the internal dynamics of the hydrated protein so far [34]. In this model, the atomic fluctuation increases from the center of the protein to its surface. Bimodal distribution analysis suggests that the relaxation of the outer shell is a function of hydration water. The dynamical transition originates from the activation of anharmonic motions driven by translational hydration water diffusion [35]. These results indicate that interaction between protein surface and hydration water is essential for the dynamical transition.

3.5.2 Effect of Solvent on Harmonic and Anharmonic Motions

The hydration environment of a protein significantly affects its dynamics. Hydration affects both harmonic and anharmonic protein dynamics. In order to examine the hydration effects in detail, the inelastic spectrum of a partially hydrated protein was observed and compared with those of hydrated and unhydrated proteins. Figure 3.9 shows the INS intensity of the hydrated, partially hydrated and unhydrated samples at 100 and 300 K. At 100 K, the boson peak was observed in the low-frequency region around a few tens of cm^{-1}. The boson peak shift was also observed for the partially hydrated sample, but the shift was slightly smaller than that of the hydrated sample. It has been reported that the boson peak position is dependent on the hydration level of the protein [3, 36], suggesting that the effect of hydration on harmonic motion is additive. It should be noted that partial hydration is sufficient to affect protein harmonic motions. On the other hand, at 300 K the spectrum of the partially hydrated sample is virtually the same as that of the unhydrated sample, whereas the spectrum of the hydrated sample exhibits large quasi-elastic scattering. The appearance of a dynamical transition requires a hydration level of more than 0.29 but less than 0.44 g D$_2$O/g protein hydration (Fig. 3.8) [20].

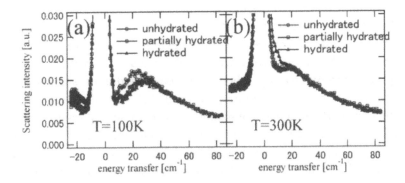

Figure 3.8 Inelastic neutron spectra of unhydrated (0.12 g D$_2$O/g protein), partially hydrated (0.29 g D$_2$O/g protein), and hydrated SNase (0.44 g D$_2$O/g protein) at (a) 100 K and (b) 300 K.

This result indicates that hydration effects at cryogenic temperatures may be additive, but that a dynamical transition has a threshold hydration level. Thus, hydration water affects both harmonic and anharmonic dynamics of proteins, but the origins of these effects are completely different.

3.6 CONCLUSION

Protein hydration strongly affects the protein structural fluctuation, suggesting that protein fluctuation is coupled with hydration structure. Figure 3.9 shows the interconversion of protein and hydration structure between two states.

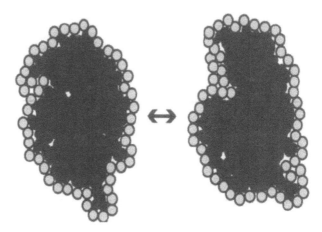

Figure 3.9 Schematic diagram of the interconversion of protein and hydration structure.

The hydration water molecules are ubiquitous on the protein surface, not only on flat surfaces but also in clefts. Protein conformational changes are accompanied with rearrangements of the hydration structure. At cryogenic temperatures, hydration waters are strongly bound to the hydration sites on the protein through hydrogen bonds; therefore, hydration stiffens protein structure due to the hydrogen bond interaction. Hydration relaxes the structure through activation of anharmonic motion at physiological temperatures. It was shown that fluctuations in the hydration shell control fast fluctuations in the protein [37]. The hydrogen bond–breaking dynamics contribute mainly to the internal molecular flexibility. These results point out that in order to understand protein dynamics, we must reveal hydration water dynamics. Another interesting result is that the boson peak up-shift is observed even in partially hydrated samples, but a dynamical transition is not observed. The difference in hydration effects at cryogenic and physiological temperatures is probably due to the dynamical properties of the protein hydration water, which are dependent on both hydration level and temperature. The hydration water dynamics and their dynamical coupling with the protein are likely to be essential for protein dynamics and biological function.

References

1. Nakasako, M. (1999) Large scale networks of hydration water molecules around bovine beta-trypsin revealed by cryogenic X-ray crystal structure analysis, *J. Mol. Biol.*, **289**, 547–564.

2. Roh, J. H., *et al.* (2005) Onset of anharmonicity in protein dynamics, *Phys. Rev. Lett.*, **95**, 038101.

3. Kurkal, V., *et al.* (2005) Enzyme activity and flexibility at very low hydration, *Biophys. J.*, **89**, 1282–1287.

4. Rasmussen, B. F., *et al.* (1992) Crystalline ribonuclease-A loses function below the dynamic transition at 220K, *Nature*, **357**, 423–424.

5. Ferrand, M., *et al.* (1993) Thermal motions and function and bacteriorhodopsin in purple membranes: effects of temperature and hydration studied by neutron-scattering, *Proc. Natl. Acad. Sci. USA*, **90**, 9668–9672.

6. Frauenfelder, H. (1989) New looks at protein motions, *Nature*, **338**, 623–624.

7. Kitamura, K., *et al.* (1999) A single myosin head moves along an actin filament with regular steps of 5.3 nanometres, *Nature*, **397**, 129–134.

8. Okada, Y., *et al.* (1999) A processive single-headed moter: Kinesin superfamily protein KIF1A, *Science*, **283**, 1152–1157.

9. Fulton. A. B. (1982) How crowded is the cytoplasm, *Cell*, **30**, 345–347.

10. Kitao, A., *et al.* (1999) Investigating protein dynamics in collective coordinate space, *Curr. Opin. Struct. Biol.*, **9**, 164–169.

11. Tehei, M., *et al.* (2007) Neutron scattering reveals extremely slow cell water in a Dead Sea organism, *Proc. Natl. Acad. Sci. USA*, **104**, 766–771.

12. Bee, M. (1988) *Quasielastic Neutron Scattering: Principles and Applications in Solid State Chemistry, Biology and Materials Science* (ed. Hilger, A.), Taylor & Francis, Bristol and Philadelphia.

13. Smith, J. C. (1991) Protein dynamics: comparison of simulations with inelastic neutron-scattering experiments, *Q. Rev. Biophys.*, **24**, 227–291.

14. Lehnert, U., *et al.* (1998) Thermal motions in bacteriorhodopsin at different hydration levels studied by neutron scattering: correlation with kinetics and light-induced conformational changes, *Biophys. J.*, **75**, 1945–1952.

15. Zaccai, G. (2000) How soft is a protein? A protein dynamics force constant measured by neutron scattering, *Science*, **288**, 1604–1607.

16. Goupil-Lamy, A. V., *et al.* (1997) High-resolution vibrational spectroscopy of a globular protein: inelastic neutron scattering and normal mode analysis of staphylococcal nuclease, *J. Am. Chem. Soc.*, **119**, 9268–9273.

17. Kataoka, M., *et al.* (2003) Neutron inelastic scattering as a high-resolution vibrational spectroscopy: new tool for the study of protein dynamics, *Spectroscopy*, **17**, 529–535.

18. Tarek, M., *et al.* (2000) Amplitudes and frequencies of protein dynamics: analysis of discrepancies between neutron scattering and molecular dynamics simulations, *J. Am. Chem. Soc.*, **122**, 10450–10451.

19. Joti, Y., *et al.* (2008), Hydration effects on low-frequency protein dynamics observed in simulated neutron scattering spectra, *Biophys. J.*, **94**, 4435–4443.

20. Nakagawa, H., *et al.* (2008) Hydration affects both harmonic and anharmonic nature of protein dynamics, *Biophys. J.*, **95**, 2916–2923.

21. Joti, Y., *et al.* (2008) Hydration-dependent protein dynamics revealed by molecular dynamics simulation of crystalline Staphylococcal nuclease, *J. Phys. Chem. B*, **112**, 3522–3528.

22. Meinhold, L., *et al.* (2005) Correlated dynamics determining X-ray diffuse scattering from a crystalline protein revealed by molecular dynamics simulation, *Phys. Rev. Lett.*, **95**, 218103.

23. Doster, W., *et al.* (1989) Dynamical transition of myoglobin revealed by inelastic neutron-scattering, *Nature*, **337**, 754–756.

24. Kanaya, T., *et al.* (1988) Low-energy excitations in polyethylene: comparison between amorphous and crystalline phases, *Chem. Phys. Lett.*, **150**, 334–338.

25. Chou, K. C. (1984) Biological functions of low-frequency vibrations (phonons). 3. Helical structures and microenvironment, *Biophys. J.*, **45**, 881–890.

26. Kataoka, M., *et al.* (1999) Low energy dynamics of globular proteins studied by inelastic neutron scattering, *J. Phys. Chem. Solids*, **60**, 1285–1289.

27. Fitter, J. (1999) The temperature dependence of internal molecular motions in hydrated and dry alpha-amylase: the role of hydration water in the dynamical transition of proteins, *Biophys. J.*, **76**, 1034–1042.

28. Steinbach, P. J., *et al.* (1991) The effects of environment and hydration on protein dynamics: a simulation study of myoglobin, *Chem. Phys.*, **158**, 383–394.

29. Joti, Y., *et al.* (2005) Protein boson peak originated from hydration-related multiple minima energy landscape, *J. Am. Chem. Soc.*, **127**, 8705–8709.

30. Caliskan, G., *et al.* (2003) Protein dynamics in viscous solvents, *J. Chem. Phys.*, **118**, 4230–4236.

31. Diehl, M., *et al.* (1997) Water-coupled low-frequency modes of myoglobin and lysozyme observed by inelastic neutron scattering, *Biophys. J.*, **73**, 2726–2732.

32. Nakagawa, H., *et al.* (2004) Protein dynamical heterogeneity derived from neutron incoherent elastic scattering, *J. Phys. Soc. Jpn.*, **73**, 491–495.

33. Tokuhisa, A., *et al.* (2007) Non-Gaussian behavior of elastic incoherent neutron scattering profiles of proteins studied by molecular dynamics simulation, *Phys. Rev. E*, **75**, 041912.

34. Dellerue, S., *et al.* (2001) Radially softening diffusive motions in a globular protein, *Biophys. J.*, **81**, 1666–1676.

35. Tournier, A. L., *et al.* (2003) Transitional hydration water dynamics drives the protein glass transition. *Biophys. J.*, **85**, 1871–1875.

36. Kurkal, V., *et al.* (2005) Low frequency enzyme dynamics as a function of temperature and hydration: a neutron scattering study, *Chem. Phys.*, **317**, 267–273.

37. Fenimore, P. W., *et al.* (2004) Bulk-solvent and hydration-shell fluctuations, similar to alpha- and beta-fluctuations in glasses, control protein motions and functions, *Proc. Natl. Acad. Sci. USA*, **101**, 14408–14413.

Chapter 4

STATISTICAL-MECHANICS THEORY OF MOLECULAR RECOGNITION: WATER AND OTHER MOLECULES RECOGNIZED BY PROTEIN

Norio Yoshida,[a,b] **Yasuomi Kiyota,**[b] **Thanyada Rungrotmongkol,**[c] **Saree Phongphanphanee,**[a] **Takashi Imai,**[d] **and Fumio Hirata**[a,b]

[a] *Department of Theoretical and Computational Molecular Science,*
   Institute for Molecular Science, Okazaki 444-8585, Japan

[b] *Department of Functional Molecular Science,*
   The Graduate University for Advanced Studies, Okazaki 444-8585, Japan

[c] *Department of Chemistry,*
   Faculty of Science, Chulalongkorn University, Bangkok 10330, Thailand

[d] *Computational Science Research Program,*
   RIKEN, Wako, Saitama 351-0198, Japan
   hirata@ims.ac.jp

4.1 INTRODUCTION

The phenomenon of life is a temporal and spatial network of chemical reactions, which are controlled by genetic information inherited from generation to generation. The genetic information itself in turn is generated and transmitted through a series of chemical processes [1]. In each of those reactions, some characteristic process takes place, which distinguishes biochemical reactions from ordinary chemical reactions in solutions. These processes are commonly referred to as "molecular recognition" (MR). For example, in order for the enzymatic reaction to occur, substrate molecules

Water: The Forgotten Biological Molecule
Edited by Denis Le Bihan and Hidenao Fukuyama
Copyright © 2011 by Pan Stanford Publishing Pte. Ltd.
www.panstanford.com

should be accommodated first by the protein in its reaction pocket to form the so-called enzyme–substrate (ES) complex [2]. MR is extremely selective and specific at atomic level. The selectivity and specificity are the key for living systems to maintain their life. Imagine what would happen if a calcium-binding protein binds, say, potassium ion erroneously. In that respect, MR is an elementary process of the phenomena of life.

The MR process can be defined as a molecular process in which one or few guest molecules are bound in high probability at a particular site, a cleft or a cavity, of a host molecule in a particular orientation [3, 4]. In this regard, MR is a molecular process determined by specific interactions between atoms in host and guest molecules. On the other hand, MR is a thermodynamic process as well, with which the chemical potential or the free energy of guest molecules in the recognition site and in the bulk solution are concerned. As an example, let us think about binding of an ion at some reaction pocket of a host protein. Usually, the reaction pocket is likely to be filled with one or few water molecules when there is no substrate. In order for a substrate molecule to come into the reaction pocket, one or some of the water molecules should be disposed from the pocket, whereas the substrate molecule itself should be partially or entirely dehydrated. The free energy changes associated with the processes are commonly called "dehydration penalty." When a guest molecule comes into a cleft or a cavity of a host molecule, it has to overcome a high entropy barrier, because the space or the degree of freedom allowed to the guest molecule is too small compared with those in the bulk solution. The conformation of the host molecule should be fluctuated in order to accommodate the guest molecule dynamically. The hinge bending motion of protein to accommodate a ligand is an example of such induced fitting. The conformational fluctuation of biomolecules is also dominated by the free energy.

The reason why MR is so challenging for any theoretical mean lies in the fact that it is a molecular process governed by thermodynamic laws. The "docking simulation" often employed in the drug design uses essentially a trial-and-error scheme to find a "best-fit complex" of host and guest molecules on the basis of geometrical and/or energetic criteria [5, 6]. However, the best-fit complex in geometrical sense will never be the most stable one in terms of thermodynamics, because it cannot account for the solvent; so neither the dehydration penalty nor the entropy barrier mentioned above is taken into account. The so-called implicit solvent model (ISM), the generalized Born (GB) [7], and the Poisson–Boltzmann (PB) equations [8], which have been used most popularly for evaluating the solvation thermodynamics of biomolecules, are much less accurate and are not insightful at all for the problem under concern, because by definition these do not have a molecular view for solvent. It is impossible to define a dielectric constant of a solvent inside a host cavity. Therefore, the dielectric constant cannot account for the dehydration penalty, especially that occurring from the host cavity. At best, those quantities can be calculated by fitting the empirical parameters such as the boundary conditions and the dielectric constants with experimental data, but then it loses credibility as a theory to predict the phenomena.

The molecular simulation, on the other hand, can provide the most detailed molecular view for the process. However, the "let-it-do" type simulation does not work for the problem at all, because MR is usually a slow and rare event. A common strategy adopted by the simulation community to overcome the difficulty is a non-Boltzmann type sampling that defines a "reaction coordinate" or an "order parameter" onto which all other degrees of freedoms are projected. The best example is the "umbrella" sampling to realize the potential of mean force, or the free energy along a conduction path of an ion in an ion channel [9]. The method is quite powerful for sampling the configuration space around an order parameter if the parameter is unique. Unfortunately, the problems in the biochemical processes are not that simple to be described by a unique order parameter. So, it is often the case that the results of the simulation depend on the choice of order parameter, and on "scheduling" of the sampling. The other methodology employed to accelerate the sampling is to apply the artificial external force on the system, for example, the external pressure applied to water molecules in aquaporin (AQP) [10]. Such simulations should verify that the configuration of water satisfies the Boltzmann distribution, otherwise the simulation may end up with just a "science fiction."

Recently, a new theoretical approach to MR has been launched on the basis of the three-dimensional reference interaction site model (3D-RISM) method, a statistical mechanics theory of liquids [11–13]. The 3D-RISM equation was derived from the molecular Ornstein–Zernike (MOZ) equation, the most fundamental equation to describe the density pair "correlation" of liquids, for a solute–solvent system in the infinite dilution by taking a statistical average over the orientation of solvent molecules [14, 15]. By solving the combined 3D-RISM with RISM equations, the latter providing the solvent structure in terms of the site–site density pair correlation functions, one can get the "solvation structure" or the solvent distributions around a solute. The solvation structure so produced retains the atomic information, because it starts from a Hamiltonian in which the information of atom–atom interactions among molecules is embedded just as in a molecular simulation. The method produces naturally all the solvation thermodynamics as well, including energy, entropy, free energy, and their derivatives such as the partial molar volume and compressibility. Unlike the molecular simulation, there is no necessity for concern about size of the system and "sampling" of the configuration space, because the method treats essentially the infinite number of molecules and integrates over the entire configuration space of a system.

The power of the 3D-RISM theory has been demonstrated fully in the solvation structure and thermodynamics of protein. The partial molar volumes of proteins in aqueous solutions calculated by Imai *et al.* [16] have exhibited quantitative agreement with corresponding experimental results. This turns out to be the first quantitative results obtained for the thermodynamics of protein entirely from the statistical mechanics theory. It was a great accomplishment by itself in the sense that it gave great confidence on the 3D-RISM to explore the stability of protein in solutions. However, it was only a prelude to a discovery that will have even bigger impact on the science. When we were analyzing the 3D distribution of water around the hen egg-white lysozyme, we found conspicuous peaks inside small cavities in the protein, which no doubt reveal the water molecules trapped inside the macromolecule [17]. In fact, the number of water molecules and the positions inside the cavity coincide with those found by the X-ray crystallography. This implies that the 3D-RISM is capable of "detecting" the molecules "recognized" by protein, or the host molecule. This is nothing but the realization of the MR.

In the present article, we review our recent studies on MR by protein on the basis of the RISM and 3D-RISM theories, which have been carried out in our group during last 5 years [18].

4.2 OUTLINE OF THE RISM AND 3D-RISM THEORIES

Let us begin the section with the following questions:
- What is the structure of liquid?
- How can the structure of liquid be characterized?

These questions are nontrivial, because unlike molecules and crystal, the liquid state does not form a structure of definite shape. One can readily define the structure of a molecule by giving the bond lengths, bond angles, and dihedral angles even for the most complex molecule like protein. The crystalline structure of solid can also be defined unambiguously by giving the lattice constants. However, molecules in liquids are in continuous diffusive motion, and thereby the definite geometry among the molecules cannot be defined. In such a case, we can only use the statistical or probabilistic language.

The probabilistic language to characterize the structure of liquids is the distribution functions that are nothing but the moments of the density field $v(\mathbf{r}) = \sum_i \delta(\mathbf{r} - \mathbf{r}_i)$ with respect to the Boltzmann weight. If there is no field applied to the system, the first moment or the average density is just constant everywhere in the system, namely, $\rho(\mathbf{r}) \equiv \langle v(\mathbf{r}) \rangle = \rho = N/V$, where V and N are the volume of the container and the number of molecules in the system, respectively, and $\langle v(\mathbf{r}) \rangle$ indicates the thermal average. So, the average density does not convey any information with respect to the liquid. However, if you look at the second moment $\rho(\mathbf{r}, \mathbf{r}') = \langle v(\mathbf{r}) v(\mathbf{r}') \rangle$, this quantity carries the structural information of liquids. The quantity is referred to as the density pair distribution function, which has essentially the same physical meaning with the radial distribution function (RDF) obtained from X-ray diffraction measurement. The density pair distribution function $\rho(\mathbf{r}, \mathbf{r}')$ is proportional to the probability density of finding two molecules at the two positions $\mathbf{r}$ and $\mathbf{r}'$ at the same time, and it becomes just a product of the average density when the distance of the two position becomes so large that there is no correlation between the density of the two positions.

$$\lim_{|\mathbf{r}-\mathbf{r}'|\to\infty} \rho(\mathbf{r},\mathbf{r}') \to \rho(\mathbf{r})\rho(\mathbf{r}') \ (= \rho^2 \text{ in uniform liquids}) \tag{4.1}$$

The quantity $g(\mathbf{r}, \mathbf{r}') = \rho(\mathbf{r}, \mathbf{r}')/\rho^2$ represents a correlation of the density at the two positions $\mathbf{r}$ and $\mathbf{r}'$. So, it is referred to as the pair correlation functions (PCF), or the RDFs when the liquid density is uniform and the translational invariance is implied. We further define a function called the "total correlation function" by $h(\mathbf{r}, \mathbf{r}') = g(\mathbf{r}, \mathbf{r}') - 1$, which represents the correlation of the density "fluctuations" at the two positions $\mathbf{r}$ and $\mathbf{r}'$,

$$h(\mathbf{r}, \mathbf{r}') = \langle \delta v(\mathbf{r}) \delta v(\mathbf{r}) \rangle / \rho^2, \tag{4.2}$$

where $\delta v(\mathbf{r}) \ (= v(\mathbf{r}) - \rho)$ denotes the density fluctuation. The main task of the liquid state theory is to find an equation that governs the function $g(\mathbf{r}, \mathbf{r}')$ or $h(\mathbf{r}, \mathbf{r}')$ on the basis of statistical mechanics, and to solve it.

As is briefly described in the Introduction, an "exact" equation referred to as the Ornstein–Zernike equation, which relates $h(\mathbf{r}, \mathbf{r}')$ with another correlation function called the direct correlation function $c(\mathbf{r}, \mathbf{r}')$, can be "derived" from the grand canonical partition function by means of the functional derivatives. Our theory to describe MR starts from the Ornstein–Zernike equation generalized to a solution of polyatomic molecules, or the MOZ equation [14],

$$\mathbf{h}(1,2) = \mathbf{c}(1,2) + \int \mathbf{c}(1,3)\rho\mathbf{h}(3,2)d(3), \tag{4.3}$$

where $\mathbf{h}(1, 2)$ and $\mathbf{c}(1, 2)$ are the total and direct correlation functions, respectively, and the numbers in the parenthesis represent the coordinates of molecules in the liquid system, including both the position $\mathbf{R}$ and the orientation Ω. The boldface letters of the correlation functions indicate that they are matrices consisting of the elements labeled by the species in the solution. In the simple case of a binary mixture, the equation can be written down labeling the solute by "u" and solvent by "v" as follows. (It is straightforward to generalize the equations to the multicomponent mixtures.)

$$h_{vv}(1,2) = c_{vv}(1,2) + \int c_{vv}(1,3)\rho_v h_{vv}(3,2)d(3)$$
$$+ \int c_{vu}(1,3)\rho_u h_{uv}(3,2)d(3), \qquad (4.4)$$

$$h_{uv}(1,2) = c_{uv}(1,2) + \int c_{uv}(1,3)\rho_v h_{vv}(3,2)d(3)$$
$$+ \int c_{uu}(1,3)\rho_u h_{uv}(3,2)d(3), \qquad (4.5)$$

$$h_{uu}(1,2) = c_{uu}(1,2) + \int c_{uv}(1,3)\rho_v h_{vu}(3,2)d(3)$$
$$+ \int c_{uu}(1,3)\rho_u h_{uu}(3,2)d(3). \qquad (4.6)$$

By taking the limit of infinite dilution ($\rho_u \to 0$), one gets

$$h_{vv}(1,2) = c_{vv}(1,2) + \int c_{vv}(1,3)\rho_v h_{vv}(3,2)d(3), \qquad (4.7)$$

$$h_{uv}(1,2) = c_{uv}(1,2) + \int c_{uv}(1,3)\rho_v h_{vv}(3,2)d(3). \qquad (4.8)$$

The equations depend essentially on six coordinates in the Cartesian space, and it includes a sixfold integral. This integral is the one which prevents the theory from applications to polyatomic molecules. It is the interaction site model and the RISM approximation proposed by Chandler and Andersen [19] that enables one to solve the equations. The idea behind the model is to project the functions onto the one-dimensional space along the distance between the interaction sites, usually placed on the center of atoms, by taking the statistical average over the angular coordinates of molecules with fixing the separation between a pair of interaction site.

$$f_{\alpha\gamma}(r) = \frac{1}{\Omega^2} \int \delta\left(\mathbf{R}_1 + \mathbf{l}_1^\alpha\right)\delta\left(\mathbf{R}_2 + \mathbf{l}_2^\gamma - \mathbf{r}\right) f(1,2)d(1)d(2), \qquad (4.9)$$

where $\mathbf{l}_i^\alpha$ is the vector displacement of site α in molecule i from the molecular center $\mathbf{R}_i$. It follows that $\mathbf{R}_i + \mathbf{l}_i^\alpha = \mathbf{r}_i^\alpha$ denotes the position of site α in molecule i. The angular average of the second terms in Eqs. (4.7) and (4.8) is formidable, but the approximation

$$c(1,2) \approx \sum_{\alpha\gamma} c_{\alpha\gamma}\left(\left|\mathbf{r}_1^{\alpha} - \mathbf{r}_2^{\gamma}\right|\right) \tag{4.10}$$

allows one to perform the angular average to lead the RISM equation

$$\rho h\rho = \omega * c * \omega + \omega * c * \rho h\rho, \tag{4.11}$$

where the asterisk denotes the convolution integrals, that is,

$$f * g = \int f(\mathbf{r}_1, \mathbf{r}_3) g(\mathbf{r}_3, \mathbf{r}_2) d\mathbf{r}_3. \tag{4.12}$$

Hereafter, solvent density is denoted by ρ instead of ρ_v. The new function ω that appeared in Eq. (4.11) during its derivation is called the "intramolecular" correlation function, which is defined for a pair of atoms α and γ in a molecule by

$$\omega_{\alpha\gamma}(r) = \rho\delta_{\alpha\gamma}\delta(r) + (1 - \delta_{\alpha\gamma})\delta(r - l_{\alpha\gamma}), \tag{4.13}$$

in which $\delta_{\alpha\gamma}$ and $\delta(r)$ are the Kronecker and Dirac delta functions, respectively. By means of the Dirac delta function, the term $\delta(r - l_{\alpha\gamma})$ imposes a distance constraint $l_{\alpha\gamma}$ between the pair of atoms. So, giving the distance constraints to all pairs of atoms in a molecule defines the molecular structure or geometry in terms of trigonometry. This is the way in which the molecular structure is incorporated into the RISM theory.

The 3D-RISM equation for the solute–solvent system at infinite dilution can be derived from Eq. (4.8) by taking the statistical average for the angular coordinate of "solvent," not for that of solute [12, 13, 20]. The equation reads

$$h_{\gamma}(\mathbf{r}) = \sum_{\gamma'} \int c_{\gamma'}(\mathbf{r}')\left(\omega_{\gamma'\gamma}^{vv}(|\mathbf{r}'-\mathbf{r}|) + \rho h_{\gamma'\gamma}^{vv}(|\mathbf{r}'-\mathbf{r}|)\right)d\mathbf{r}', \tag{4.14}$$

where $h_{\gamma}(\mathbf{r})$ and $c_{\gamma}(\mathbf{r}')$ are the total and direct correlation functions of sites γ and γ', respectively, of solvent molecules at two positions $\mathbf{r}$ and $\mathbf{r}'$ in the Cartesian coordinate, the origin of which is placed at an arbitrary position, generally inside the protein. The functions $\omega_{\gamma'\gamma}^{vv}(\mathbf{r})$ and $h_{\gamma'\gamma}^{vv}(\mathbf{r})$ are the correlation functions for solvent molecules, which appear in Eq. (4.11). It is these equations that can be applied to MR. If one views the solute molecule as a "source of external force" exerted on solvent molecules, then the $\rho g_{\gamma}(\mathbf{r})(= \rho h_{\gamma}(\mathbf{r}) + \rho)$ is identified as the density distribution of solvent molecules in the "external force." This identification, called "Percus trick," is the key concept to realize MR by means of statistical mechanics.

The equations described above contain two unknown functions, $h(\mathbf{r})$ and $c(\mathbf{r})$. Therefore, they are not closed without another equation, which relates the two functions. Several approximations have been proposed for the closure relations: HNC, PY, MSA, and so on [14]. The HNC closure can be obtained from the diagrammatic expansion of the pair correlation functions with respect to the density and discarding a set of diagrams called the "bridge diagrams," which have multifold integrals. It should be noted that the terms kept in the HNC closure relation still include those up to the infinite orders of the density. Alternatively, the relation has been derived from the linear response of a free energy functional to the density fluctuation created by a molecule fixed in the space within the Percus trick. The HNC closure relation reads

$$h_\gamma(\mathbf{r}) = \exp(-u_\gamma(\mathbf{r})/k_B T + h_\gamma(\mathbf{r}) - c_\gamma(\mathbf{r})) + 1, \qquad (4.15)$$

where k_B and T are Boltzmann's constant and temperature, respectively, and $u(\mathbf{r})$ the interaction potential between pair of atoms in the system. Equation (4.15) is the relation that incorporates the physical and chemical characteristics of the system into the theory through $u(\mathbf{r})$. The PY approximation can be obtained from the HNC relation just by linearizing the factor $\exp[h(\mathbf{r}) - c(\mathbf{r})]$. The HNC closure has been quite successful for describing the structure and thermodynamics of liquids and solutions including water. However, the approximation is notorious in the low-density regime. Sometimes the drawback becomes fatal when one tries to apply the theory for associating liquid mixtures or solutions, especially of dilute concentration, because a solution of "dilute" concentration is equivalent to low-density liquid for the minor component. In order to get rid of the problem, Kovalenko and Hirata proposed the following approximation, or the Kovalenko–Hirata (KH) closure [21]:

$$g_\gamma(\mathbf{r}) = \begin{cases} \exp(d_\gamma(\mathbf{r})) & \text{for } d_\gamma(\mathbf{r}) \le 0 \\ 1 + d_\gamma(\mathbf{r}) & \text{for } d_\gamma(\mathbf{r}) > 0 \end{cases} \qquad (4.16)$$

where $d_\gamma(\mathbf{r}) = -u_\gamma(\mathbf{r})/k_B T + h_\gamma(\mathbf{r}) - c_\gamma(\mathbf{r})$. The approximation turns out to be quite successful even for the mixture of complex liquids.

The procedure to solve the equations consists of two steps. We first solve the RISM equation (4.11) for $h_{\gamma'\gamma}^{vv}(\mathbf{r})$ of solvent or a mixture of solvents in case of solutions. Then, we solve the 3D-RISM equation, Eq. (4.14), for $h_\gamma(\mathbf{r})$ of a protein-solvent (solution) system, inserting $h_{\gamma'\gamma}^{vv}(\mathbf{r})$ for the solvent into Eq. (4.14), which has been calculated in the first step. Considering the definition $g(\mathbf{r}) = h(\mathbf{r}) + 1$, $g(\mathbf{r})$ thus obtained is the three-dimensional distribution of solvent molecules around a protein in terms of the interaction site representation of solvent or a mixture of solvents in case of solutions. The so-called solvation free energy can be obtained from the distribution function through the following equations [21, 22] corresponding, respectively, to the two closure relations described above, Eqs. (4.15) and (4.16).

$$\Delta \mu_{HNC} = \rho^v k_B T \sum_\gamma \int d\mathbf{r} \left[\frac{1}{2} h_\gamma^{uv}(\mathbf{r})^2 - c_\gamma^{uv}(\mathbf{r}) - \frac{1}{2} h_\gamma^{uv}(\mathbf{r}) c_\gamma^{uv}(\mathbf{r}) \right] \qquad (4.17)$$

$$\Delta\mu_{\rm KH} = \rho^{\rm v}k_{\rm B}T\sum_{\gamma}\int d\mathbf{r}\left[\frac{1}{2}h_{\gamma}^{\rm uv}(\mathbf{r})^2\Theta\left(-h_{\gamma}^{\rm uv}(\mathbf{r})\right)-c_{\gamma}^{\rm uv}(\mathbf{r})-\frac{1}{2}h_{\gamma}^{\rm uv}(\mathbf{r})c_{\gamma}^{\rm uv}(\mathbf{r})\right], \quad (4.18)$$

where Θ denotes the Heaviside step function. The other thermodynamic quantities concerning solvation can be readily obtained from the standard thermodynamic derivative of the free energy except for the partial molar volume.

The partial molar volume, which is a very important quantity for probing the response of the free energy (or stability) of protein to pressure, including the so-called pressure denaturation, is not a "canonical" thermodynamic quantity for the (V, T) ensemble, since the volume is an independent thermodynamic variable of the ensemble. The partial molar volume of protein at infinite dilution can be calculated from the Kirkwood–Buff equation [23] generalized to the site–site representation of liquid and solutions [24, 25],

$$\bar{V} = k_{\rm B}T\chi_{\rm T}\left[1-\rho\sum_{\gamma}\int c_{\gamma}(\mathbf{r})d\mathbf{r}\right], \quad (4.19)$$

where $\chi_{\rm T}$ is the isothermal compressibility of pure solvent or solution, which is obtained from the site–site correlation functions of solutions.

In what follows, we show applications of the theory described above in order to demonstrate the robustness of the theory.

Figure 4.1a–c exhibits the theoretical and experimental results for the distribution of water in a cavity "inside" the hen egg-white lysozyme, which is surrounded by the residues from Y53 to I58 and from A82 to S91. For simplicity, only the surrounding residues are displayed, except for A82 and L83, which are located in the front side. Figure 4.1a shows the isosurfaces of $g(\mathbf{r}) > 8$ for water oxygen (green) and hydrogen (pink) in the cavity. Four distinct peaks of water oxygen and seven distinct peaks of water hydrogen are found in the cavity, implying that four water molecules are accommodated. From the isosurface plot, we have reconstructed the most probable model of the hydration structure, as shown in Fig. 4.1b. It is interesting to compare the hydration structure obtained by the 3D-RISM theory with crystallographic water sites of X-ray structure [17]. The crystallographic water molecules in the cavity are depicted in Fig. 4.1c, showing four water sites in the cavity, much as the 3D-RISM theory has detected. Moreover, the positions of water molecules obtained by the theory and experiment are quite similar to each other. Thus, the 3D-RISM theory can predict the water-binding sites with great success.

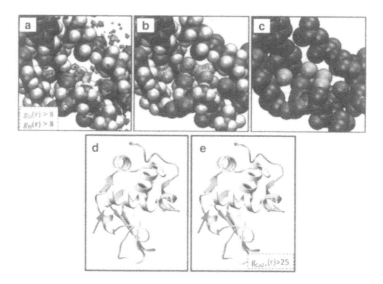

Figure 4.1 Comparison of distribution of water and Ca^{2+} ion evaluated by 3D-RISM and X-ray. The isosurfaces of water oxygen (green) and hydrogen (pink) for the three-dimensional distributions larger than 8 (a), the most probable model of the hydration structure reconstructed from the isosurface plots (b), and the crystallographic water sites (c). The position of Ca^{2+} ion measured by X-ray (d) and the isosurfaces of Ca^{2+} ion for the three-dimensional distributions larger than 25 (e) [17, 26, 27]. For color reference, see p. 361.

Figure 4.1d,e shows the position of Ca^{2+} binding site of mutated human lysozyme evaluated by X-ray crystallography and 3D-RISM, respectively [26, 27]. In Fig. 4.1e, the distribution function, $g(\mathbf{r})$, of Ca^{2+} ion greater than 25 is shown. The highest peak of Ca^{2+} around the holo-Q86D/A92D mutant is located at almost the same position as the experimental results. The results indicate that the Ca^{2+} binding site empirically observed is the one that has the highest probability of finding the ion in the theoretical description. This result also indicates that the 3D-RISM theory can predict the ion-binding sites.

In the following sections, we will demonstrate how the 3D-RISM theory is capable of describing MR.

4.3 PROTON EXCLUSION OF AQUAPORINS

The proton exclusion from AQPs is one of the most important questions to be solved in the fields of biochemistry, medicine, and pharmacology. Although the channels are extremely permeable by water, approximately a billion molecules pass through the channel per second, protons are strictly excluded from the permeation [28, 29]. In many previous works on molecular dynamics simulation of proton exclusion, AQP1 and GlpF have been used to study this problem [30–40]. There are essentially two mechanisms conceivable for the proton transfer in the channel. One is the proton jump mechanism similar to that in bulk ice or water, in which a proton transfers from one minimum to the other in the double-well potential of a hydrogen bond between two water molecules by tunneling through the barrier, which is called as the Grotthuss mechanism [41, 42]. The process requires two water molecules to be in the right mutual orientation to form the double well potential. If water molecules are prevented from the reorientational dynamics by any reason, a proton may not transport through the channel. The other mechanism of the proton transport is due to the translational motion of water molecules, that is, a proton may move in the channel by "riding" on a water molecule or making a hydronium ion. The mechanism is similar to usual ion transport in the channel. Therefore, any mechanism that prevents an ion from the translational motion through the channel can be the cause of the proton insulation: steric hindrance, electrostatic barrier, and so on. The unspecific desolvation effects proposed by Warshel is nothing but the electrostatic barrier enhanced by decreased water population or screening [36, 37]. The mechanism should be readily examined if one can calculate the distribution of the hydronium ion in the channel. The information of the hydronium ion distribution in the channel may also be useful for examining the possibility of the proton jump mechanism, because a proton should be exist most likely in the form of the hydronium ion except for the moment of barrier crossing.

Unfortunately, it is extremely difficult for the molecular simulation to examine the distribution of molecular ions in a channel, because it should sample the free energy surface of the ion including their orientations throughout the channel.

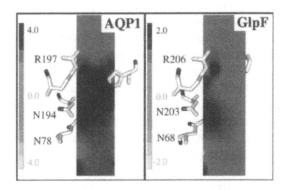

Figure 4.2 The contour map of the electrostatic potential inside the channels (esu unit) [47].

The 3D-RISM theory is a powerful tool to tackle such problems [11–13, 43]. As seen in the previous section, the 3D-RISM theory has been applied successfully to the distribution of water and ions inside protein cavity to reproduce experimental results [17, 26, 27, 44–47]. In the present section, we apply the theory to AQP1 and GlpF for elucidating the proton exclusion from those channels [46, 47].

In Fig. 4.2, the contour map of the electrostatic potential due to the channel atoms, the 3D-distribution of the water and of hydronium ions, and the one-dimensional profile of the distribution of the solution components are depicted along the channel axis. As can be readily seen from the figure, water in both the channels is continuously distributed throughout the channel. However, the distribution of hydronium ions is intermitted by gaps both for AQP1 and GlpF, although there is some difference in the distribution between the two channels: in AQP1, the hydronium ion is excluded from large area extending from R197 (or Ar/R) to NPA, while the gap in GlpF is limited in a small area around R206 (Ar/R). Note that "gap" does not mean "nothing is there," since water molecules are distributed continuously throughout the channel. It is also understood from the figure that the distribution of the hydronium ion is essentially determined by the electrostatic potential inside the channel: the hydronium ion is excluded primarily from the channel by the positive electrostatic atmosphere. The difference in the electrostatic potential between AQP1 and GlpF is originated apparently from the additional positive field produced by the residue H182 in AQP1. From those results, we can draw an important conclusion with respect to the mechanism of proton exclusion in AQP1. Needless to mention about the proton jump mechanism, the proton as a positive charge cannot pass through the large electrostatic barrier inside the channel. On the other hand, the gap of the distribution is small in GlpF, which leaves slight possibility for proton to transfer through the proton jump mechanism. Remember that water is distributed continuously even in the area where the hydronium ion is excluded. If the water molecules and hydronium ions around that area have some freedom to rotate and arrange themselves to make the double-well potential for the proton, then the proton may jump through the potential barrier via tunneling. The distribution of oxygen and hydrogen of water around the area does not indicate the particular coordination that prevents the molecule from the reorientation. Can a proton then permeate all the way through the channel via the proton jump mechanism? In order to answer the question, we have examined the water distribution around the NPA region of GlpF, where the mechanism is suspected to be broken due to the formation of the so-called bipolar orientation.

Figure 4.3 shows the distributions of oxygen and hydrogen atoms of water at the NPA region in AQP1 and GlpF. The oxygen atom of a water molecule is coordinated by the two hydrogen atoms of the residues, N203 and N68 of GlpF, N194, and N78 of AQP1. The peaks of oxygen are about the same

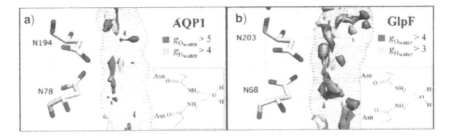

Figure 4.3 The three-dimensional distribution functions of water in the NPA region of AQP1 and G1pF: pale red and light blue surface represent the distribution function of oxygen and hydrogen of water, respectively, and the dot surface denotes the pore area. For color reference, see p. 361

distance from both the residues making hydrogen bonds. (See the illustrative picture in the insets of Fig. 4.3.) Such orientation of water molecules entirely conflicts with the configuration of the hydrogen bond network of water, thereby it excludes the possibility of the proton jump mechanism around that area.

The physics of the proton exclusion in AQP1 and GlpF stated above are illustrated schematically in Fig. 4.4. The gap is very large in the case of AQP1, extending from R197 to the NPA region. From the results, we can readily conclude in the case of AQP1 that protons are excluded from permeation primarily due to the electrostatic repulsion inside the channel. On the other hand, in the case of GlpF, the results leave slight possibility for proton to permeate through the gap around R206 by the proton jump mechanism. However, the mechanism does not work entirely throughout the channel due to the formation of the bipolar orientation at the NPA region. So, a proton has small but finite conductivity in GlpF through the combined mechanism of the proton jump and the diffusion of hydronium ions in accord with experiment [48].

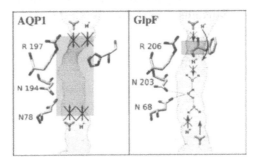

Figure 4.4 Schematic illustration of the mechanism of proton exclusion in AQP1 and GlpF channels. The transparent pale red color represents the gap region in the distribution of hydronium ions, and the molecules colored yellow represent the hydronium ions [46].

4.4 CARBON MONOXIDE ESCAPE PATHWAY IN MYOGLOBIN

Myoglobin (Mb) is a globular protein which has important biological functions such as oxygen storage. Due to its biochemical function, many researchers have made intensive efforts to identify the escape pathway of the ligand, both experimentally and theoretically. Thirty years ago, Perutz *et al.* proposed that the entry and exit pathway of carbon monoxide (CO) into the active site of Mbs involves rotation of the distal histidine to form a short and direct channel between the heme pocket and solvent [49]. It has been well-recognized that there are several intermediate states separated by activation barriers along the escape pathway, which are referred to as "Xe sites." Four major Xe sites are known, which are called Xe1, Xe2, Xe3, and Xe4. The experimental results indicate that CO spends some time at the Xe sites before escaping to solvent at room temperature [50]. In the mixed Xe and CO solutions, the difference in affinity between Xe and CO to each Xe trapping site makes the CO escape pathway different depending on the Xe concentration. It is believed that the dissociated CO escapes to the solvent through the Xe1 trapping site predominantly under the Xe-free condition. On the other hand, CO escapes through the Xe4 site in a Xe-rich solution. Terazima and co-workers have measured the partial molar volume (PMV) change of the system along the pathway of CO escape process under Xe solution on the basis of the transient grating (TG) method [51, 52]. They hypothesized that the intermediate of the pathway is through the Xe4 site, since the experiments were carried out under the Xe-rich condition.

In this section, we apply the 3D-RISM theory to investigate the CO escape pathway of Mb. As mentioned above, the ligand dissociation process of Mb occurs from heme to solvent through some specific cavities. We discuss the CO escape pathway from Mb in terms of 3D distribution functions of the ligand and the PMV change along the pathway. The 3D-RISM calculation to evaluate the 3D distribution functions is performed with the geometry optimization of Mb structure [53].

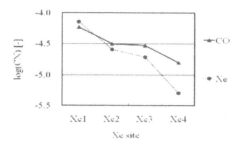

Figure 4.5 Coordination number of CO and Xe molecules in each Xe site, which are calculated from the RDF at each Xe cavity. Vertical axis is made log scale [53].

Figure 4.5 shows the coordination numbers (CNs) of Xe and CO in each Xe site, which are obtained from the distribution function. Since the structure of Mb was obtained in a Xe–water mixture by using 3D-RISM optimization procedure, these results are regarded as the ligand affinity in a Xe-rich condition. As seen in the figure, CNs of both ligands show similar behavior, namely, the Xe1 site has largest affinity, while the Xe4 site has smallest. Note that Xe affinity of the Xe1 site is larger than CO. Hence the difference of ligand affinity between Xe and CO becomes largest at Xe1 and Xe4 sites. This result indicates that CO prefers Xe4 site to Xe1 in a Xe-rich condition. In that, although the affinity of CO ligand is largest for the Xe1 site, CO trapping in the Xe1 site is obstructed by the high Xe affinity. These are consistent with experimental observations [54].

In order to evaluate the PMV change through the CO escape pathway, the PMV of the "intermediate structure" of Mb should be calculated. Here, "intermediate structure" means that one CO molecule is trapped in specific Xe site of Mb. Therefore, the CO molecule is regarded as a part of solute species. The model is referred to as "explicit ligand model." In the present study, the position of the trapped CO molecule is determined by the peaks of the 3D distribution functions.

Table 4.1. PMVs of each model and their changes

Models	PMV [cm³/mol]		
MbCO	9029.0		
Mb:CO(Xe1)	9030.4		
Mb:CO(Xe4)	9032.8		
Mb+CO	9019.6		
	Xe1	Xe4	(exptl. [52])
ΔV_1 =	1.4	3.8	(3 ± 1)
ΔV_2 =	−10.8	−13.2	(−12.6 ± 1.0)
ΔV_{total} =	−9.4	−9.4	(−10.7 ± 0.5)

Note: ΔV_1 represents partial molar volume change from MbCO to Mb:CO (Xe4), and ΔV_2 represents change from Mb:CO (Xe4) to Mb + CO. ΔV_{total} represents change from MbCO to Mb + CO.

We calculated PMV of the CO associated (MbCO), intermediate (Mb:CO), and CO dissociated (Mb + CO) states [53]. The PMV and its changes are shown in Table 4.1. The PMV of Mb:CO(Xe1) is smaller than those of Mb:CO(Xe4). Because the cavity size of the Xe1 site is large enough, it can accommodate both CO and a water molecule at the same time. On the other hand, in the case of the Xe4 site, there is no space to include a water molecule when the cavity is occupied by a CO molecule. However, a void space is created in the cavity, because the size of the Xe4 site is larger than a CO molecule. This void space contributes to the increase in PMV of Mb:CO(Xe4). The change of PMV through Xe4 site shows excellent agreement with the experimental data by Terazima *et al.* [52], which supports Terazima's conjecture regarding the escape pathway of CO from Mb.

4.5 PROTON TRANSPORT THROUGH THE INFLUENZA A M2 CHANNEL

The M2 proton channel, which is found on viral membrane of the virus, has received a great deal of attention as a target of drug development due to its important role of proton transport and viral replication [55, 56]. It is known that an amantadine drug family is effective against influenza A by blocking M2 channel, which disturbs viral proton conduction and consequently causes inhibition of viral replication [57]. However, the underlying mechanism of proton blockade activity by amantadine is   not clarified yet. In addition, many new strains of influenza virus are resistant to amantadine. Therefore, understanding the mechanism of proton transportation through the virus membrane is one of the central issues for drug design.

Several electrophysiological results have shown that the M2 channel is highly selective for proton, and that its gating is controlled by pH [58–62]. A number of researchers has explored the relationship between functions and molecular structures of the M2 channel, and have indicated important roles of His37 in gating mechanism. Pinto and co-workers have demonstrated that a mutant of M2 channel with the His residue substituted by Gly, Ser, or Thr loses the proton selectivity, and that the selectivity is restored upon addition of imidazole. This has suggested that the imidazole group plays an important role in proton selectivity [63]. Concerning the pH-controlled gating mechanism, Hu *et al.* have made important suggestions on the basis of their experiments that the pKa associated with four histidines at the gating region is different from each other, and that the open forms are dominated by triply and quadruply protonated histidine [64]. These experimental results suggest that His 37 acts as a pH sensor switch to open and close the gate and as a selective filter to allow the permeation of proton, but not other cations. The experiments also indicate that Trp 41 has an important role in the gating. Some mutants of the protein, in which the Trp41 is replaced by a residue of smaller size, Ala, Cys, or Phe, have higher proton conductivity compared with its wild type [65]. The results from UV Raman spectra show the interaction between protonated imidazole of His37 and the idole of Trp41, or cation–pi interaction [66, 67]. These investigations suggest that the indole of Trp41 has a role to occlude the channel pore. This residue behaves as a "door" to turn open or close the pore, which is controlled by protonated His due to the cation–pi interaction.

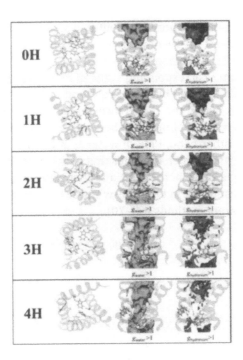

Figure 4.6 Structure of Trp41 gating (orange) of different protonated histidine tetrad, and 3D-DFs of water in the channel (cyan) and hydronium ion (red), with $g > 1$. For color reference, see p. 362.

To investigate the mechanism of proton transfer in the M2 channel, we consider the distributions of water and hydronium ion inside the channel as a function of pH, which regulates the protonated state (PS) of histidine tetrad from the non-PS (0H) to the quadruple PS (4H) in the decreasing order of pH. For each state, we randomly pick up the coordinates of M2 protein from a trajectory of the MD simulation that has been carried out previously [68]. The three-dimensional distribution (3D-DF) of water and hydronium ion with $g > 1$ in the five different protonated states of M2 channel is depicted in Fig. 4.6. The figure indicates that the accessibility of water (cyan in Fig. 4.6) to the channel pore increases with the protonated state of the channel is in the order 0H < 1H < 2H < 3H < 4H. This result can be explained readily in terms of the pore diameter, which is widened due to the electrostatic repulsion among the protonated histidines His37, and the bulky indole ring of Trp41, which can turn to block or open the gate. The results suggest that there are two distinct states in the channel conformation, or "open" and "closed" forms. The 0H, 1H, and 2H forms are considered as closed forms, since water distributions are not observed at the selective filter regions of His37 (yellow stick in Fig. 4.6) and at the gating region with the Trp41 residues (orange sphere and stick in Fig. 4.6). On the other hand, the 3H and 4H forms with a continuous water distribution along the pore are identified as the open form (Fig. 4.6). In addition, the narrowest 3D-DF of water in the channel, also the narrowest of pore, is seen at the Trp41 region.

The PMFs of water corresponding to the distribution function are shown in Fig. 4.7, in which the high barriers are found only in the 0H, 1H, and 2H states. It is obvious that in the closed form water cannot overcome the high barrier made up of the steric hindrance between the channel atoms and water molecules. On the other hand, PMF of water in 3H and 4H is negative along the entire channel pore, which indicates that water molecules in the channel are more stable than those in the bulk, and that water is permeated through the channel. The results are in harmony with the previous theoretical studies carried out by different methods [68–70].

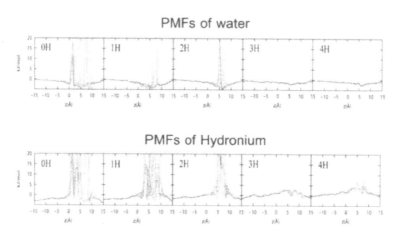

Figure 4.7 Potential of mean force of water and hydronium ion in each protonation state of histidine. A line in each state represents the PMF in a conformation.

In the three states of closed gate, or 0H, 1H, 2H, hydronium ions exhibit a behavior similar to water, but with lower distribution, and higher barrier in PMFs compared with those of water (Fig. 4.7). It indicates that a hydronium ion, or a proton, cannot distribute in the channel and is prevented from transporting across the channel. The results are consistent with those for the closed conformations of the M2 channel at high pH values reported in many experimental and theoretical studies [58–62, 70].

As was already seen, water continuously distributes throughout the 3H and 4H channels (Fig. 4.6) indicating the dramatic structural change of the channel from closed to open conformations. In contrast, the distributions of hydronium ion in the 3H and 4H states do not seem to be much different from those in 0H to 2H. However, it is apparent from Fig. 4.7 that PMF of the hydronium ion in 3H and 4H exhibits entirely different behavior than that of

0H to 2H. Protons in the 0H to 2H forms have extremely high barriers due to the same cause in the case of water, and no chance to exist in the gating region of the channel, while the barrier heights in the 3H and 4H forms are just 2–3 and 5–7 kJ/mol, respectively, which are comparative to the thermal energy. There is another interesting observation in the figure. The barrier height for protons is higher in 4H than in 3H, which is against the heuristic argument on the basis of the pore size around the gating region.

All these observations are suggestive of two competing factors working on the proton distribution as the protonated state of the channel is increased from 0H to 4H. One of these is the channel opening due to the increased repulsion among the protonated His residues, which tends to enhance the distribution of proton as well as water in the channel. The other factor is the electrostatic repulsion between protons and the protonated His residues, which will reduce the proton distribution with increasing number of the protonated His. The two effects balance at the 3H state to make the proton distribution in the channel maximum. These results are consistent with those obtained by Voth *et al.* [70].

4.6 CONCLUDING REMARKS

In this article, we have presented a new method to describe molecular recognition (MR) in biomolecules on the basis of the statistical mechanics of molecular liquids, or the RISM/3D-RISM theory. In some phenomena for which thermodynamic and structural data are available, the theoretical results have exhibited at least qualitative agreement with the experimental results. The typical example is the positions of water molecules in a cavity of the hen egg-white lysozyme for which the theoretical and experimental results exhibited quantitative agreement. In other cases where there are no experimental data to be compared with, the theory has demonstrated its predictive capability. The best example is the proton (hydronium ion) distribution in AQPs. In this case, there are no experimental data available, probably, due to the resolution of the current neutral diffraction measurement, which is not high enough. We believe that the prediction will be proven sooner or later by a neutron diffraction technique.

Although the RISM/3D-RISM theory has proven its capability of "prediction," other few summits are to be conquered before it establishes itself as the "theory of molecular recognition." The problem is concerned with the conformational fluctuation of protein. For example, the present theory still requires experimental data for structure of protein as an "input." In other words, we have not yet succeeded to "build" tertiary structure of protein from the amino acid sequence. If we are able to build the tertiary structure in different solution conditions, containing say electrolytes or other ligands, on the free energy surface produced by the RISM/3D-RISM method, we will be able to attain two goals at the same time, which are the two most-highlighted problems in the biophysics: the "protein folding" and MR. The statement, "different solution conditions," has an even deeper implication. Experimental results are clearly indicating that some of the folding processes are driven or enforced by "salt bridges" and/or "water bridges." This implies that those methodologies which do not account water molecules and electrolytes explicitly are fatal in this business. The RISM/3D-RISM theory certainly has such an ability to realize those ions and water molecules "bridging" amino-acid residues inside protein, as been demonstrated in this chapter. So, if one could sample the protein conformation on the potential of mean force or the free energy surface produced by the RISM/3D-RISM method, it should attain the two goals at the same time. We have already developed such methodologies to explore the large fluctuation of protein by combining the RISM/3D-RISM theory with the molecular dynamics and Monte Carlo method.

The experimental analysis of protein function involves time-dependent properties such as the rate of an enzymatic reaction and the conduction rate of ions in an ion channels. These properties are related to relatively small fluctuations of protein around the native conformations. In enzymatic reactions, an enzyme may have to "open" its "door" in order to accommodate substrate molecules into the reaction pocket. The ion channels have some device called "gating" mechanism to control the flow of ions into the channel pore. The mechanisms are regulated often by the conformational fluctuation of protein. Analyses of those processes require evaluation of "dynamic" or time-dependent properties of both protein as well as solvent, which are sometimes closely correlated. In such a case, the "dynamics" on the free energy surface described above is insufficient. We have to describe the dynamics of protein and solvent on an equal footing. To our best knowledge, the generalized Langevin equation is the only theory to meet such a requirement. The study to combine the RISM/3D-RISM theory with the generalized Langevin equation in order to realize the correlated dynamics of protein and solvent is in progress in our group.

Any of the methods that we have been developing requires solving the 3D-RISM equations for many conformations of a protein. Currently, it is taking few hours to solve the 3D-RISM equation for "one" conformation of protein with about 200 residues, using the best available workstation in our group, which has two 4-core 3.0 GHz Xeon CPU with 64 GB RAM. So, it is not feasible at present to solve such problems stated above, even though we could have succeeded to build the methodologies. However, there is good news in this respect. A national project to build a next-generation computer is under way in Japan. The new machine to be built and a 3D-RISM program well-tuned to the machine will hopefully solve the most important problems in the life science.

Acknowledgments

The work is supported by the Grant-in Aid for Scientific Research on Innovative Areas "Molecular Science of Fluctuations toward Biological Functions" from the MEXT in Japan. We are also grateful to the support by the grant from the Next Generation Supercomputing Project, Nanoscience Program of the ministry.

References

1. Watson, J. D., Hopkins, N. H., Roberts, J. W., Steitz, J. A., and Weiner, A. M. (eds.) (1987) *Molecular Biology of the Gene*, 4th ed., Benjamin Cummings.
2. Michaelis, L., and Menten, M. L. (1913) Die Kinetik der Invertinwirkung, *Biochem. Z.*, **49**, 333–369.
3. Lehn, J. M. (1995) *Supramolecular Chemistry*, Wiley-VCH, Weinheim, Germany.
4. Gellman, S. G. (1997) Introduction: molecular recognition, *Chem. Rev.*, **97**, 1231–1232.
5. Sotriffer, C., and Klebe, G. (2002) Identification and mapping of small-molecule binding sites in proteins: computational tools for structure-based drug design, *Il Farmaco*, **57**, 243–251.
6. Gohlke, H., and Klebe,G. (2002) Approaches to the description and prediction of the binding affinity of small-molecule ligands to macromolecular receptors, *Angew. Chem. Int. Ed.*, **41**, 2644–2676.
7. Still, W. C., Tempczyk, A., Hawley, R. C., and Hendrickson, T. (1990) Semianalytical treatment of solvation for molecular mechanics and dynamics, *J. Am. Chem. Soc.*, **112**, 6127–6129.
8. Gilson, M. K., and Honig, B. (1988) Energetics of charge: charge interactions in proteins, *Proteins Struct. Funct. Genet.*, **3**, 32–52.
9. Kato, M., and Warshel, A. (2005) Through the channel and around the channel: validating and comparing microscopic approaches for the evaluation of free energy profiles for ion penetration through ion channels, *J. Phys. Chem. B*, **109**, 19516–19522.
10. Zhu, F., Tajkhorshid, E., and Schulten, K. (2004) Theory and simulation of water permeation in aquaporin-1, *Biophys. J.*, **86**, 50–57.
11. Hirata, F. (ed.) (2003) *Molecular Theory of Solvation*, Springer-Kluwer, Dordrecht, Netherlands.
12. Kovalenko, A., and Hirata, F. (1998) Three-dimensional density profiles of water in contact with a solute of arbitrary shape: a RISM approach, *Chem. Phys. Lett.*, **290**, 237–244.

13. Beglov, D., and Roux, B. (1997) An integral equation to describe the solvation of polar molecules in liquid water, *J. Phys. Chem. B*, **101**, 7821–7826.

14. Hansen, J.-P., and McDonald, I. R. (2006) *Theory of Simple Liquids*, 3rd ed., Academic Press, London.

15. Blum, L., and Torruella, A. J. (1972) Invariant expansion for two-body correlations: thermodynamic functions, scattering, and the Ornstein–Zernike equation, *J. Chem. Phys.*, **56**, 303–310.

16. Imai, T., Kovalenko, A., and Hirata, F. (2004) Solvation thermodynamics of protein studied by the 3D-RISM theory, *Chem. Phys. Lett.*, **395**, 1–6.

17. Imai, T., Hiraoka, R., Kovalenko, A., and Hirata, F. (2005) Water molecules in a protein cavity detected by a statistical-mechanical theory, *J. Am. Chem. Soc.*, **127**, 15334–15335.

18. Yoshida, N., Imai, T., Phonphanphanee, S., Kovalenko, A., and Hirata, F. (2009) Molecular recognition in biomolecules studied by statistical-mechanical integral-equation theory of liquids, *J. Phys. Chem. B*, **113**, 873–886.

19. Chandler, D., and Andersen, H. C. (1972) Optimized cluster expansions for classical fluids. II. Theory of molecular liquids, *J. Chem. Phys.*, **57**, 1930–1937.

20. Cortis, C. M., Rossky, P. J., and Friesner, R. A. (1997) A three-dimensional reduction of the Ornstein–Zernicke equation for molecular liquids, *J. Chem. Phys.*, **107**, 6400–6414.

21. Kovalenko, A., and Hirata, F. (1999) Self-consistent description of a metal–water interface by the Kohn–Sham density functional theory and the three-dimensional reference interaction site model, *J. Chem. Phys.*, **110**, 10095–10112.

22. Singer, S. J., and Chandler, D. (1985) Free energy functions in the extended RISM approximation, *Mol. Phys.*, **55**, 621–625.

23. Kirkwood, J. G., and Buff, F. D. (1951) The statistical mechanical theory of solutions, *J. Chem. Phys.*, **19**, 774–777.

24. Imai, T., Kinoshita, M., and Hirata, F. (2000) Theoretical study for partial molar volume of amino acids in aqueous solution: implication of ideal fluctuation volume, *J. Chem. Phys.*, **112**, 9469–9478.

25. Harano, Y., Imai, T., Kovalenko, A., and Hirata, F. (2001) Theoretical study for partial molar volume of amino acids and polypeptides by the three-dimensional reference interaction site model, *J. Chem. Phys.*, **114**, 9506–9511.

26. Yoshida, N., Phongphanphanee, S., Maruyama, Y., Imai, T., and Hirata, F. (2006) Selective ion-binding by protein probed with the 3D-RISM theory, *J. Am. Chem. Soc.*, **128**, 12042–12043.

27. Yoshida, N., Phongphanphanee, S., and Hirata, F. (2007) Selective ion binding by protein probed with the statistical mechanical integral equation theory, *J. Phys. Chem. B.*, **111**, 4588–4595.

28. Preston, J. M., Carroll, T. P., Guggino, W. B., and Agre, P. (1992) Appearance of water channels in Xenopus oocytes expressing red cell CHIP28 protein, *Science*, **256**, 385–387.

29. Borgnia, M., Neilsen, S., Engel, A., and Agre, P. (1999) Cellular and molecular biology of the aquaporin water channels, *Annu. Rev. Biochem.*, **68**, 425–458.

30. de Groot, B. L., and Grubmüller, H. (2001) Water permeation across biological membranes: mechanism and dynamics of aquaporin-1 and GlpF, *Science*, **294**, 2353–2357.

31. de Groot, B. L., and Grubmüller, H. (2005) The dynamics and energetics of water permeation and proton exclusion in aquaporins, *Curr. Opin. Struct. Biol.*, **15**, 176–183.

32. Tajkhorshid, E., Nollert, P., Jensen, M. Ø., Miercke, L. J. W., O'Connell, J., Stroud, R. M., and Schulten, K. (2002) Control of the selectivity of the aquaporin water channel family by global orientational tuning, *Science*, **296**, 525–530.

33. Jensen, M. Ø., Tajkhorshid, E., and Schulten, K. (2003) Electrostatic tuning of permeation and selectivity in aquaporin water channels, *Biophys. J.*, **85**, 2884–2899.

34. Ilan, B., Tajkhorshid, E., Schulten, K., and Voth, G. A. (2004) The mechanism of proton exclusion in aquaporin channels, *Proteins*, **55**, 223–228.

35. Chen, H., Ilan, B., Wu, Y., Zhu, F., Schulten, K., and Voth, G. A. (2007) Charge delocalization in proton channels, I: the aquaporin channels and proton blockage, *Biophys. J.*, **92**, 46–60.

36. Burykin, A., and Warshel, A. (2003) What really prevents proton transport through aquaporin? Charge self-energy versus proton wire proposals, *Biophys. J.*, **85**, 3969–3706.

37. Burykin, A., and Warshel, A. (2004) On the origin of the electrostatic barrier for proton transport in aquaporin, *FEBS Lett.*, **570**, 41–46.

38. Chakrabarti, N., Roux, B., and Pomès, R. (2004) Structural determinants of proton blockage in aquaporins, *J. Mol. Biol.*, **343**, 493–510.

39. Chakrabarti, N., Tajkhorshid, E., Roux, B., and Pomès, R. (2004) Molecular basis of proton blockage in aquaporins, *Structure*, **12**, 65–74.

40. Kato, M., Pisliakov, A. V., and Warshel, A. (2006) The barrier for proton transport in aquaporins as a challenge for electrostatic models: the role of protein relaxation in mutational calculations, *Proteins*, **64**, 829–844.

41. von Grotthuss, C. J. D. (1806) Sur la décomposition de l'eau et des corps qu'elle tient en dissolution à l'aide de l'électricité galvanique, *Ann. Chim.*, **LVIII**, 54–73.

42. Agmon, N. (1995) The Grotthuss mechanism, *Chem. Phys. Lett.*, **244**, 456–462.

43. Kovalenko, A., and Hirata, F. (1999) Potential of mean force between two molecular ions in a polar molecular solvent: a study by the three-dimensional reference interaction site model, *J. Phys. Chem. B*, **103**, 7942–7957.

44. Imai, T., Hiraoka, R., Kovalenko, A., and Hirata, F. (2007) Locating missing water molecules in protein cavities by the three-dimensional reference interaction site model theory of molecular solvation, *Protein*, **66**, 804–813.

45. Phongphanphanee, S., Yoshida, N., and Hirata, F. (2007) The statistical-mechanics study for the distribution of water molecules in aquaporin, *Chem. Phys. Lett.*, **449**, 196–201.

46. Phongphanphanee, S., Yoshida, N., and Hirata, F. (2008) On the proton exclusion of aquaporins: a statistical mechanics study, *J. Am. Chem. Soc.*, **130**, 1540–1541.

47. Phongphanphanee, S., Yoshida, N., and Hirata, F. (2009) The potential of mean force of water and ions in aquaporin channels investigated by the 3D-RISM method, *J. Mol. Liquid*, **147**, 107–111.

48. Saparov, S. M., Tsunoda, S. P., and Pohl, P. (2005) Proton exclusion by an aquaglyceroprotein: a voltage clamp study, *Biol. Cell*, **97**, 545–550.

49. Perutz, M. F., and Matthews, F. S. (1966) An X-ray study of azide methaemoglobin, *J. Mol. Biol.*, 21, 199–202.

50. Schotte, F., Soman, J., Olson, J. S., Wulff, M., and Anfinrud, P. A. (2004) Picosecond time-resolved X-ray crystallography: probing protein function in real time, *J. Struc. Biol.*, **147**, 235–246.

51. Sakakura, M., Yamaguchi, S., Hirota, N., and Terazima, M. (2001) Dynamics of structure and energy of horse carboxymyoglobin after photodissociation of carbon monoxide, *J. Am. Chem. Soc.*, **123**, 4286–4294.

52. Nishihara, Y., Sakakura, M., Kimura, Y., and Terazima, M. (2004) The escape process of carbon monoxide from myoglobin to solution at physiological temperature, *J. Am. Chem. Soc.*, **126**, 11877–11888.

53. Kiyota, Y., Hiraoka, R., Yoshida, N., Maruyama, Y., Imai, T., and Hirata, F. (2009) Theoretical study of CO escaping pathway in myoglobin with the 3D-RISM theory, *J. Am. Chem. Soc.*, **131**, 3852–3853.

54. Tilton, R. F., Singh, U. C., Kuntz, I. D., and Kollman, P. A. (1998) Cavities in proteins: structure of a metmyoglobin–xenon complex solved to 1.9 Å, *J. Mol. Biol.*, **199**, 195–211.

55. Lamb, R. A., Holsinger, L. J., and Pinto, L. H. (1994) in Receptor-Mediated Virus Entry into Cells (ed. Wimmer, E.), Cold Spring Harbor Laboratory Press, Cold Spring Harbor, NY, pp. 303–321.

56. Helenius, A. (1992) Unpacking the incoming influenza virus, *Cell*, **69**, 577–578.

57. Pinto, L. H., Holsinger, L. J., and Lamb, R. A. (1992) Influenza virus M2 protein has ion channel activity, *Cell*, **69**, 517–528.

58. Lear, J. D. (2003) Proton conduction through the M2 protein of the influenza A virus; a quantitative, mechanistic analysis of experimental data, *FEBS Lett.*, **552**, 17–22.

59. Mould, J. A., Li, H., Dunlak, C. S., Lear, J. D., Pekosz, A., Lamb, R. A., and Pinto, L. H. (2000) Mechanism for proton conduction of the M2 ion channel of influenza A virus, *J. Biol. Chem.*, **275**, 8592–8599.

60. Mould, J. a., Drury, J. E., Frings, S. M., Kaupp, U. B., Pekosz, A., Lamb, R. A., and Pinto, L. H. (2000) Permeation and activation of the M2 ion channel of influenza A virus, *J. Biol. Chem.*, **275**, 31038–31050.

61. Chizhmakov, I. V., Geraghty, F. M., Ogden, D. C., Hayhurst, A., Antoniou, M., and Hay, A. J. (1996) Selective proton permeability and pH regulation of the influenza virus M2 channel expressed in mouse erythroleukaemia cells, *J. Physiol.*, **494**, 329–336.

62. Vijayvergiya, V., Wilson, R., Chorak, A., Gao, P. F., Cross, T. A., and Busath, D. D. (2004) Proton conductance of influenza virus M2 protein in planar lipid bilayers, *Biophys. J.*, **87**, 1697–1704.

63. Venkataraman, P., Lamb, R. A., and Pinto, L. H. (2005) Chemical rescue of histidine selectivity filter mutants of the M2 ion channel of influenza A virus, *J. Biol. Chem.*, **280**, 21463–21472.

64. Hu, J. Fu, R., Nishimura, K., Zhang, L., Zhou, H., Busath, D. D., Vijayvergiya, V., and Cross, T. A. (2006) Histidines, heart of the hydrogen ion channel from influenza A virus: toward an understanding of conductance and proton selectivity, *PNAS*, **103**, 6856–6870.

65. Tang, Y., Zaitseva, F., Lamb, R. A., and Pinto, L. H. (2002) The gate of the influenza virus M2 proton channel is formed by a single tryptophan residue, *J. Biol. Chem.*, **277**, 39880–39886.

66. Okada, A., Miura, T., and Takeuchi, H. (2001) Protonation of histidine and histidine–tryptophan interaction in the activation of the M2 ion channel from influenza A virus, *Biochemistry*, **40**, 6053–6060.

67. Takeuchi, H., Okada, A., and Miura, T. (2003) Roles of the histidine and tryptophan side chains in the M2 proton channel from influenza A virus, *FEBS Lett.*, **552**, 35–38.

68. Intharathep, P., Laohpongspaisan, C., Rungrotmongkol, T., Loisruangsin, A., Malaisree, M., Decha, P., Aruksakunwong, O., Chuenpennit, K., Kaiyawet, N., Sompornpisut, P., Pianwanit, S., and Hannongbua, S. (2008) How amantadine and rimantadine inhibit proton transport in the M2 protein channel, *J. Mol. Graph. Model.*, **27**, 342–348.

69. Kass, I., and Arkin, T. (2005) How pH opens a H+ channel: the gating mechanism of influenza A M2, *Structure*, **13**, 1789–1798.

70. Chen, H., Wu, Y., and Voth, G. A. (2007) Proton transport behavior through the influenza A M2 channel: insights from molecular simulation, *Biophys. J.*, **93**, 3470–3479.

Chapter 5

HYDROPHOBIC MOLECULES IN WATER

B. Cabane

Physique et Mécanique des Milieux Hétérogènes,
Centre National de la Recherche Scientifique, UMR 7636,
École Supérieure de Physique et de Chimie Industrielles,
10 Rue Vauquelin, 75231 Paris Cedex 05, France
bcabane@pmmh.espci.fr

5.1 INTRODUCTION

Liquid water owes its cohesion and most of its extraordinary properties to a network of H bonds that connect the molecules in the liquid state [1]. Molecules such as sugars, ethanol, and urea that can take part in this H-bond network are called hydrophilic. They can be dissolved by water in very large amounts, limited only by the eventual crystallization of the solute. Molecules such as hydrocarbons that do not take part in this network are called hydrophobic. These molecules *should not be* in water, not because they hate water, as the word "hydrophobic" would suggest, but because water "dislikes them." In this sense the word "hydrophobic" is a misnomer.

In this chapter I will argue that the solubility of such molecules in water is not as low as simple models would let us believe, and give a simple view of how such molecules can get into water at low concentrations. Then I shall present two examples of large hydrophobic molecules that have very low solubility in water and yet have important biological functions.

5.1.1 Small Hydrophobic Molecules

This section presents a simple view of hydrophobic molecules and how they can be dissolved in water. These molecules may be small apolar molecules such as those of the common gasses oxygen (O_2), nitrogen (N_2), carbon dioxide (CO_2), and methane (CH_4) or those of hydrocarbons that do not take part in the H-bond network of water, or they may be large molecules that carry a few apolar groups with same property.

A naive view of the dissolution of such molecules into water would let us believe that this process would be nearly impossible. Indeed, if you try to apply regular solution theory, in order to dissolve a small apolar molecule, you would have to make a cavity by extracting at least one water molecule, and this would break at least four H bonds (Fig. 5.1). The energy cost would be at least 30 kT, causing the solubility to be extremely small, lower than 10^{-13} mole/mole. Moreover, this solubility would rise extremely fast with temperature.

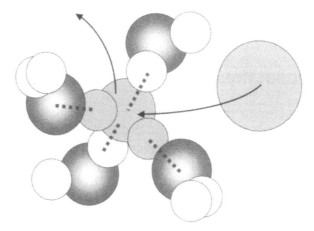

Figure 5.1 To create a cavity inside liquid water, at least one water molecule must be extracted, at the cost of breaking four H bonds. For color reference, see p. 362.

Common observations show that this model is wrong: we know that fish breathe oxygen that is dissolved in water, that divers have to cope with nitrogen that is dissolved in their blood, and that CH_4 is exchanged between oceans and the atmosphere. So the solubility is not as low as predicted; in fact it is in the range of 10^{-5} to 10^{-3} mole/mole. Not only are the solubilities much higher than predicted by the regular solution model, but they also have the opposite temperature variation. Indeed, the solubilities of salts rise with temperature, whereas those of apolar gases decrease at higher temperatures (Fig. 5.2). A simple experiment consists in filling a spectrometer cell with cold water and heating it up to 50°C; gas bubbles nucleate on the walls of the cell, indicating that the solubility of gasses was higher at a lower temperature and has been reduced by heating to 50°C.

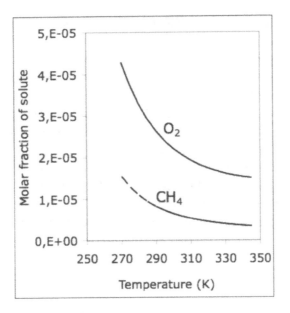

Figure 5.2 The solubility of oxygen and CH4 in water [2].

A better model consists in looking for cavities that form spontaneously in the structure of the liquid and inserting the solute into these cavities [3]. In this case, the energetic cost of the operation is made of two terms. The first one is the cost of the reorganization of the H bonds around the cavity; the second one is the variation of energy that results from the insertion of the solute in the cavity:

$$\Delta U = \Delta U_{H\ bonds} + \Delta U_{insertion} \tag{5.1}$$

The creation of the cavity has an entropic cost as well, which consists of two terms. The first one is the entropic cost (or gain) that results from the reorganization of the H bonds around the cavity; the second one is the entropic cost due to the fact that the centers of all water molecules are excluded from the cavity:

$$\Delta S = \Delta S_{H\ bonds} + \Delta S_{cavity} \tag{5.2}$$

For spontaneous cavities, the energetic and entropic terms associated with the reorganization of H bonds compensate each other exactly in the free energy of the cavity, which reduces to:

$$\Delta G = \Delta G_{insertion} + \Delta U_{cavity} \tag{5.3}$$

Usually (for apolar solutes), $\Delta G_{insertion}$, originating from van der Waals forces, is small compared with ΔG_{cavity}, which is related to the equation of the state of liquid water. Then the penalty for dissolution into water is essentially an entropic penalty, due to the fact that water molecules are excluded from these spontaneous cavities.

It is interesting to figure out what these spontaneous cavities look like. For small cavities we have a good model, which is the structure of crystalline gas hydrates in which water molecules form an H-bond cage around the gas molecules in such a way that not a single H bond is lost: all H bonds are directed either parallel to the cavity or away from it (Fig. 5.3). CH_4 hydrates are important in oil production because they tend to form crystals in pipelines that carry mixtures of oil, water, and gas from off-shore fields. These crystals can aggregate and block the flow of the three-phase mixture. It is also known that large deposits of CH_4 hydrates exist at large depths under the seas, and they may be either a source of energy or an environmental problem in the event that they would be released into the atmosphere.

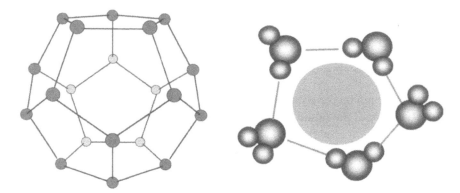

Figure 5.3 The structure of a cavity in a chlorine hydrate crystal. Water molecules are at the corners of a pentagonal dodecahedron. H bonds are formed along edges of the dodecahedron and also between the dodecahedron and adjacent dodecahedra and between the dodecahedron and interstitial water molecules [4]. For color reference, see p. 363.

For large cavities, it is no longer possible for water molecules to orient all their H bonds, either parallel to the cavity or away from it, so the energy balance is no longer as favorable as in the case of small cavities. Moreover, the entropy cost becomes larger, as it grows proportionally with the surface area of the cavity. Figure 5.4 shows some numbers for hydrocarbons. The free-energy balance, G, of transferring such molecules into water, grows regularly with the number of carbons in the molecule by about 1.5 kT with each additional CH_2 group. For large hydrocarbon molecules with about 12 carbon atoms, the resulting solubility in water is quite small, on the order of 10^{-8} mole/mole.

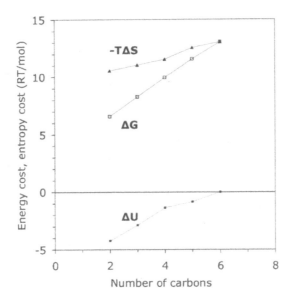

Figure 5.4 Energy cost, entropy cost, and free-energy balance of transferring 1 mole of hydrocarbon into water, as functions of the number of carbon atoms in the hydrocarbon molecule [5].

Even for molecules with polar groups that do take part in the H-bond network of water, such as fatty alcohols and fatty acids, the solubility is still low, on the order of 10^{-5} to 10^{-3} mole/mole. Moreover, if these molecules are rigid, they may crystallize, and then the free energy of dissolution will carry an extra penalty, due to the cost of melting the crystal structure. Thus, crystals of large hydrophobic molecules may be nearly insoluble. This raises a practical question: can these molecules have any biological function, or may we consider that their solubility is too low for them to do anything significant?

The next two sections of this chapter present examples of large hydrophobic molecules that have important biological functions despite their rather low solubility: drug molecules and plant tannins.

5.1.2 Large Hydrophobic Molecules: Drug Molecules

There is a general classification scheme of drug molecules according to their solubility in water and their solubility in the membranes of cells (Fig. 5.5). Indeed, these characteristics determine the bioavailability of the drug after it has been administered, for example, though the oral route [6].

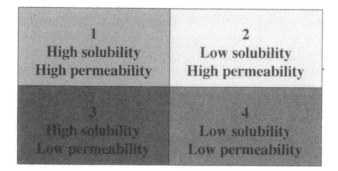

Figure 5.5 General classification scheme for drugs administered through the oral route [6]. For color reference, see p. 363.

Interestingly, more than half of the drugs that are now developed are in class 4; that is to say, they have low solubility in water and low solubility in lipids. Christopher Lipinski has discussed the reasons for this low solubility in a famous paper [7]. He analyzed all the current molecules that had been tested as drug candidates by two drug companies, and found a systemic trend toward a higher molar mass over the years (Fig. 5.6). The reason for this trend is the requirement for higher specificity or maybe just a simple fact that smaller molecules have been tested already. A higher molar mass means lower solubility, especially if the molecule contains many hydrophobic groups, such as six- or five-membered rings.

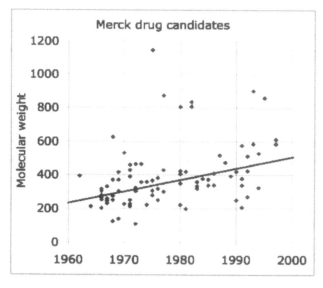

Figure 5.6 Molecular weights of Merck-advanced drug candidates over the years.

In turn, if a drug is administered orally, low solubility means that a very small fraction of the drug may be able to dissolve, get across the epithelial cells of the gut, and reach its biological target. A typical dosage is 100 mg of the drug, which has to dissolve in a few tenths of a liter of liquid. This would require solubility on the order of 10^{-3} g/g. If the actual solubility is only 10^{-5} g/g, only 1% of the drug will be dissolved. Moreover, if the drug is in the form of large crystals, these crystals will not have time to dissolve during their transit through the gut. Therefore, the bioavailability of the drug may be extremely low.

So how can we improve the bioavailability of drugs that are going to be administered orally? One way is to produce the drug in the form of nanoparticles that are dispersed in water. Then, even if the equilibrium solubility is low, the rate of transfer to the epithelial cells of the gut may be high because dissolved drug molecules are pumped efficiently by the epithelial cells, and are quickly resupplied by the nanoparticles, thanks to their high surface area. Thus, if the diffusion distances are short, transfer may be efficient even if the solubility is low (Fig. 5.7).

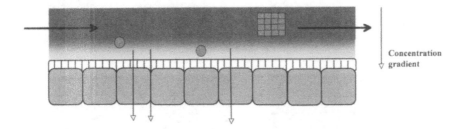

Concentration gradient

Figure 5.7 Transport of sparingly soluble solids through the gut. The layer of epithelial cells is represented in grey, supporting the mucus layer (in vertical lines). Large crystals (solubility 10^{-6} g/g to 10^{-5} g/g) go through the digestive tract with almost no dissolution. Amorphous particles have much higher solubility. Nanoparticles have a large specific area (10–100 m^2/g) and can supply solute molecules at a high rate if these molecules are taken up efficiently by the epithelial cells. For color reference, see p. 363.

The other way is to produce drugs as amorphous particles. Indeed, an amorphous solid is essentially a supercooled liquid, and its dissolution does not require the melting of a crystal lattice. Typically, the solubility of an amorphous organic solid is an order of magnitude above that of the crystalline form. Organic solids can be prepared in the amorphous state through processes in which the molecules are unable to organize into a crystal lattice. One way to make amorphous particles is called the solvent-shifting method: the drug is dissolved in a polar solvent such as ethanol, and then this solution is dumped into a large volume of water. The polar solvent molecules mix with water, and the drug molecules find themselves stranded in an aqueous solution. Figure 5.8 presents some images of what happens next. In this experiment, the solvent with dissolved molecules has been injected through a needle into the middle of an aqueous solution. The circular ring showing up in the first picture is located at the interface between water and the polar solvent. The white dots are crystals of the drug that have nucleated after the drug solution was invaded by water. In subsequent images, interdiffusion of water and the polar solvent progresses over large distances; consequently the ring becomes thicker while the drug crystals keep growing.

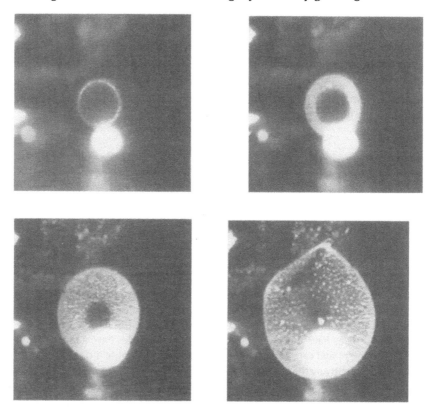

Figure 5.8 Video images of the precipitation of an organic solute through solvent shifting [8]. A solution of cholesteryl acetate in acetone was injected through a glass capillary into a large drop of water. The organic solution is seen as a dark drop at the center of the image, surrounded by a ring of bright particles. These particles are crystals of cholesteryl acetate that have nucleated at the acetone–water interface. Outside this ring is the aqueous solution, which also contains amorphous nanoparticles of choleteryl acetate, too small to be seen through optical microscopy.

Upon closer inspection of these images, one may notice that the crystals are larger at the interior of the ring, where the concentration of the polar solvent is higher and the super saturation is lower: this is what classical nucleation theory predicts. Conversely, on the water ridge side (toward the exterior of the ring), the particles become smaller, until you can't see them anymore. However, observation of this water-rich solution through light scattering and X-ray scattering shows that it does contain nanoparticles of the drug and that these particles are amorphous. This diffusion-and-stranding method is one of the simplest processes that can be used to produce a dispersion of amorphous particles.

The real problem is not, however, to make amorphous nanoparticles of hydrophobic molecules but to keep them in that amorphous state. Indeed, the difference in solubility between amorphous and crystalline particles is a fantastic drive for recrystallization. This is clearly seen in Fig. 5.9, where crystals can be seen growing at the expense of the nanoparticle dispersion in such recrystallization processes. A classical question in such recrystallization processes is whether the amorphous solid reorders to make a crystal structure or whether it dissolved to supply the growth to crystals that have nucleated somewhere else. Here the answer is quite clear: crystals have nucleated on an impurity or a surface defect, they mop up all the dissolved molecules from the solution around them, and they cause the nanoparticles to dissolve and vanish from the dispersion.

Figure 5.9 Recrystallization of a dispersion of amorphous particles in water. The original dispersion is seen as an array of pointlike particles in the upper half of the image. Crystals have nucleated in the lower half of the image. They grow by capturing solute molecules that are released into the solution by the nanoparticles. The recrystallization process is driven by the difference in equilibrium solubility between the crystals and the amorphous particles.

At this point it is appropriate to reconsider the questions that were raised at the beginning of this chapter: (i) Even though these large drug molecules are hydrophobic, there is a small but significant fraction of them that dissolves in an aqueous phase; (ii) these dissolved molecules control the fate of the drug particles in aqueous solutions; and (iii) when the drug is taken orally, they control its bioavailability.

5.1.3 More Large Hydrophobic Molecules: Plant Tannins

Another important class of hydrophobic molecules is the tannins that are produced by plants and find their way into our food and drinks. Plant tannins are water-soluble polyphenols, produced by plants, which can interact with proteins and precipitate them [9]. The simplest tannin, epigallocatechin gallate (EGCG), is extracted from the leaves of green tea; its chemical structure is presented in Fig. 5.10. Because this molecule contains large hydrophobic groups, and because it is rigid, its solubility in water is low, despite the H-bond donors. Yet it has significant physiological effects. On the one hand, it does bind to proteins, which can cause some severe antinutritional effects. Indeed, animals that are fed a diet containing more than 5% of tannins may lose weight and die. On the other hand, in low concentrations, it may have beneficial effects because the *enol* to *keto* equilibrium of the hydroxyls enables them to react with free radicals and mop them up from the aqueous solution.

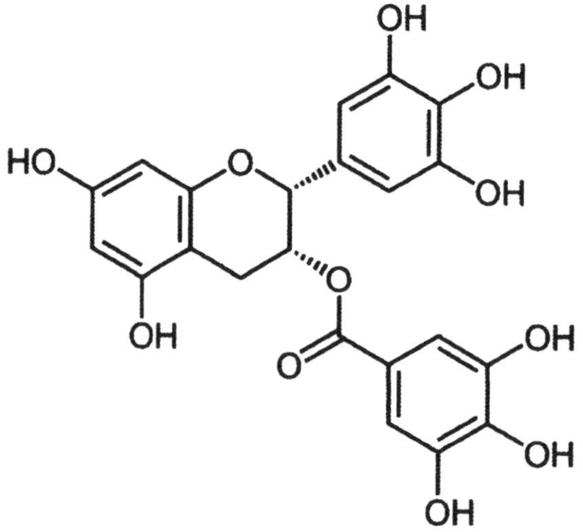

Figure 5.10 The chemical structure of EGCG, a tannin (polyphenol) that is present in tea leaves.

Tannins are found throughout the plant kingdom: in the surface wax of leaves, in fruits, e.g., most berries but also grapes and apples, and in the bark of some trees. Plants produce tannins for protection against all sorts of parasites and predators, including us. In the fight against parasites such as fungi or bacteria, plants try to keep the parasites out. This is the function of cell walls that are made of pectin and cellulose. To invade a plant, a parasite may release enzymes such as cellulases and pectinases, which break down cellulose and pectin polymers. To fight this invasion, the plant releases tannins, polyphenols that bind to the enzymes and precipitate them. The production of tannins is also helpful to plants in order to deter predators (herbivores or humans) who would eat the plants and then use digestive enzymes to break down the food proteins. In this case, the tannins will bind to the digestive enzymes so that the predator becomes unable to digest the food.

This brings the next question: how do we cope with tannins in our diet, and why don't we die from drinking red wine? This is again a question related to the bioavailability of sparingly soluble, hydrophobic molecules. Obviously we need to catch the tannins before they reach our digestive enzymes. Our tool for this is the presence, in our saliva, of proteins that have this function. They are basic proline-rich proteins, abbreviated as *basic PRP*. They are present in the saliva of all herbivores and absent in carnivores that do not need them. They are thought to have two functions. One is to regulate the concentration of free tannins that reach our digestive system. The other is to contribute to the feeling of astringency, which you experience when you drink wine that is too young or tea that is too strong—it is a tactile feeling in the mouth similar to stickiness, loss of lubrication, or the feeling that your mouth has dried up, which serves as a warning signal that this drink may contain too much tannins.

The proteins that have these functions are somewhat unusual. Figure 5.11 presents the amino acid sequences of two of these proteins: the first one, called IB-5, has an incredibly high content of the amino acid called proline—about 41%. This amino acid has a five-membered ring, available for hydrophobic interactions, and a carbonyl group, which is an H-bond acceptor. Also the high proline content prevents the formation of alpha helices so that the protein is unstructured according to the criteria of structural biology. In the presence of tannins it binds the tannin molecules and precipitates. The other protein, called II-1, has a similar content of proline, but it is glycosylated. In the presence of tannins, it binds the tannins but remains in solution.

1 SPPGKPQGPPQQE	1 FLISGKPVGRRPQGGNQPQRPPPPP
14 GNKPQGPPPP	26 GKPQGPPPQGGNQSQGPPPPP
24 GKPQGPPPA	47 GKPEGRPPQGRNQSQGPPPHP
33 GGNPQQPQAPPA	68 GKPERPPPQGGNQSQGTPPPP
45 GKPQGPPPPPQ	89 GKPERPPPQGGNQSHRPPPPP
56 GGRPPRPAQGQQPPQ	110 GKPERPPPQGGNQSRGPPPHR
	131 GKPEGPPPQEGNKSRSAR

"IB-5"
Prolin rich (41 %)
Not glycosylated

"II-1"
Prolin rich (35 %)
Glycosylated

Figure 5.11 Amino acid sequences of the human salivary proteins IB5 and II-1. Both have repeated proline sequences that prevent the formation of alpha helices and that are involved in hydrophobic bonding to other solutes, such as plant tannins. IB5 is not glycosylated; in the presence of tannins, such as EGCG, it precipitates. II-1 is glycosylated on one or a few of the N amino acids; in the presence of tannins, such as EGCG, it remains in solution. For color reference, see p. 364.

To understand how these proteins are adapted to their functions, we have determined their average structures using small-angle X-ray scattering [10]. We used the X-ray beam produced by a synchrotron, sent it on the sample, and collected the scattered rays according to the scattering angle or the scattering vector q, which is a combination of the angle and wavelength.

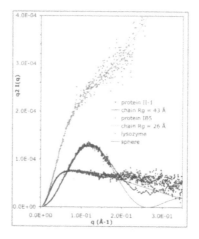

Figure 5.12 Small-angle X-ray-scattering spectra of proteins in solution, in the Kratky-Porod representation. Horizontal scale: Scattering vector q, in Å^{-1}. Vertical scale: Scattered intensity multiplied by the square of the scattering vector, $q^2I(q)$. Symbols and lines: Data for lysozyme and theoretical scattering curve for dense spheres (black dots and black line); data for salivary protein II-1 and theoretical curve for a random chain (blue dots and pink curve); data for salivary protein IB-5 and scattering curve for an elongated chain (pink dots and blue curve). For color reference, see p. 364.

Figure 5.12 shows small-angle X-ray-scattering spectra of proteins in solution in the Kratky-Porod representation, that is, $q^2I(q)$ versus q, together with theoretical scattering curves. The black spectrum is that of the globular protein lysozyme, and it is fitted quite well by the scattering curves for dense spheres. The blue spectrum is that of the salivary protein II-1, which is fitted by the theoretical curve for a random chain. Finally the pink spectrum is that of the salivary protein IB-5, and it corresponds to an object that is even more elongated. These spectra demonstrate that the structures of salivary proteins are not at all like those of ordinary globular proteins—they do not form alpha helices, they do not fold, and instead they remain "unstructured" in the aqueous solution (Fig. 5.13) [11]. It is interesting to consider whether this random chain configuration is an evolutionary adaptation to the capture of other solutes such as plant tannins.

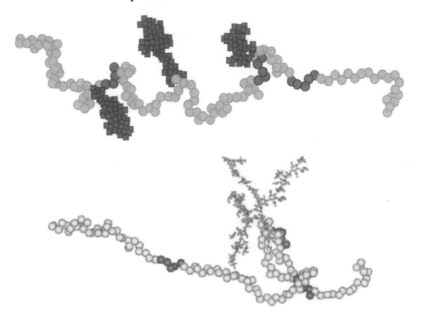

Figure 5.13 Conformations of glycosylated salivary proteins II-1, constructed using a mathematical algorithm that looks for structures that provide the best fit to the small-angle X-ray scattering spectrum [11]. The blue balls are amino acids in the protein chain, and the orange side groups are the branched sugar chains. The top panel presents a protein with three sugar side groups, and the bottom panel a protein with a single sugar side group. For color reference, see p. 365.

Next we need to understand how these proteins perform their biological function, which is to catch the tannin molecules before they reach our digestive enzymes. We do not yet have direct evidence of the mechanism by which tannins bind to proteins, but we can suppose that the hydrophobic rings of the polyphenols interact with the five-membered proline ring in the proteins. Moreover, the hydroxyl groups of the polyphenols may bind to the H-bond acceptors in the proteins. The essential information we obtain from scattering is that all these groups on the proteins are available for binding because the proteins are not folded.

Finally we must examine the consequences of tannin-to-protein binding, that is, whether tannins bind to individual proteins or whether they bridge neighboring proteins together. Figure 5.14 shows scattering curves of solutions of the glycosylated protein II-1 in the presence of increasing concentration of tannins. Remarkably, low concentrations of tannins have no effect on the scattering by the protein. However, beyond a threshold of about 10 tannin molecules per protein, there is a sudden rise of intensity at the low-scattering vector q, and the excess scattering is typical of scattering by dense spheres. Thus we may conclude that tannins cause the formation of dense protein aggregates. Therefore the function of the protein must be to regulate the concentration of free tannin molecules: as soon as this concentration rises above the threshold, the excess tannins are mopped up by the protein.

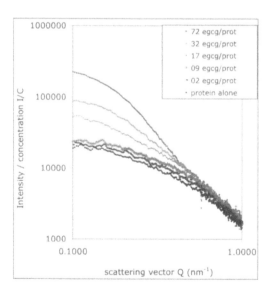

Figure 5.14 Spectra from solutions of the glycosylated salivary protein II-1 in the presence of increasing amounts of the tannin EGCG. Horizontal scale: Scattering vector q, in nm^{-1}. Vertical scale: Scattered intensity $I(q)$. At low molar ratios of tannin per protein, the spectra are identical with the spectrum of the protein alone in solution, indicating that the tannins do not cause protein aggregation. At high ratios (threshold 10 tannins/protein) there is a large excess of scattered intensity at a small q, which reflects the aggregation of the proteins, caused by the tannins.

The main finding is that the tannin-protein aggregates are dense globular objects with a radius equal to the length of a protein molecule. This has a simple explanation: since the protein is glycosylated, it must keep its sugar groups at the surface of the aggregate, and therefore the aggregates cannot grow larger than the length of a protein molecule. This is the structure expected for micelles formed by amphiphilic molecules. Moreover the fact that there is a threshold for the formation of the aggregates is also in agreement with the idea that the protein II-1 behaves as an amphiphile and forms micelles at a high tannin concentration.

We also suspect that glycosylated and nonglycosylated proteins may have different functions. Glycosylated proteins mop up the tannins but remain in solution. We think that their function is to regulate tannin concentrations. Nonglycosylated proteins also bind tannins, but they precipitate out of solution. We suppose that this could be part of the mechanism leading to loss of lubrication in the mouth and the feeling of astringency. These mechanisms illustrate how nonstructured proteins can regulate the solution concentration of specific solutes (in this case the tannins), and provide a warning signal when a threshold in solute concentration is exceeded.

5.2 CONCLUSION

Large molecules that carry hydrophobic groups, such as plant tannins, are not necessarily excluded from water. In many instances, there is a small but significant fraction of them that can dissolve in aqueous media. These dissolved molecules bind proteins and therefore have very significant biological effects.

References

1   Cabane, B., and Vuilleumier, R. (2005) The physics of liquid water, *C. R. Geosci.*, **337**, 159.

2   Lide, D. R. (ed.) (2004) *CRC Handbook of Chemistry and Physics*, 85th ed., CRC Press, Boca Raton, FL.

3   Guillot, B., and Guissani, Y. (1993) *J. Chem. Phys.*, **99**, 8075.

4   Pauling, L. (1960) *The Nature of the Chemical Bond*, Cornell University Press, Ithaca, NY.

5   Tanford, C. (1980) *The Hydrophobic Effect: Formation of Micelles and Biological Membranes*, Wiley, New York.

6 Amidon, G. L., Lennernäs, H., Shah, V. P., and Crison, J. R. (1995) A theoretical basis for a biopharmaceutic drug classification: the correlation of in vitro drug product dissolution and in vivo bioavailability, *Pharm. Res.*, **12**, 413.

7 Lipinski, C. A. (2000) Drug-like properties and the causes of poor solubility and poor permeability, *J. Pharmacol. Toxicol. Methods*, **44**, 235.

8 Lannibois, H. (1995) *Des molécules hydrophobes dans l'eau*, thesis, Université Paris 6.

9 Haslam, E. (1998) *Practical Polyphenolics: From Structure to Molecular Recognition and Physiological Action*, Cambridge University Press, New York.

10 Zanchi, D., Vernhet, A., Poncet-Legrand, C., Cartalade, D., Tribet, C., Schweins, R., and Cabane, B. (2007) Colloidal dispersions of tannins in water-ethanol solutions, *Langmuir*, **23**, 9949.

11 Durand, D., private communication.

Part III

WATER AT INTERFACES

Chapter 6

HOW WATER MEETS A HYDROPHOBIC INTERFACE

Yingxi (Elaine) Zhu

Department of Chemical and Biomolecular Engineering,
182 Fitzpatrick Hall , University of Notre Dame,
Notre Dame, IN 46556, USA
yzhu3@nd.edu

6.1 INTRODUCTION

The role of water in physical situations from biology to geology is almost universally thought to be important, but the details are disputed [1–20]. For example, as concerns proteins, the side chains of roughly half the amino acids are polar, while the other half are hydrophobic; the nonmixing of the two is a major mechanism steering the folding of proteins and other self-assembly processes. As a second example, it is an everyday occurrence to observe beading-up of raindrops on raincoats or leaves of plants. Moreover, it is observed theoretically and experimentally that when the gap between two hydrophobic surfaces becomes critically small, water is ejected spontaneously [2 ,3, 18–20], whereas water films confined between symmetric hydrophilic surfaces are stable [1]. Despite its importance, water exhibits much anomalous behaviors compared with other fluids. Particularly, it presents some even more puzzling behaviors when it is under spatial confinement at hydrophobic surfaces. This chapter is adapted from discussions in several primary accounts published previously [3, 4].

In its liquid form, water consists of an ever-changing three-dimensional network of hydrogen bonds. Hydrophobic surfaces cannot form hydrogen bonds, and the hydrogen-bond network must be disrupted. So what happens when water is compelled to be close to a hydrophobic surface? Energetically,

Water: The Forgotten Biological Molecule
Edited by Denis Le Bihan and Hidenao Fukuyama
Copyright © 2011 by Pan Stanford Publishing Pte. Ltd.
www.panstanford.com

it is expected that the system forms as many hydrogen bonds as possible, resulting in a preferential ordering of the water. Entropically, it is expected the system orients randomly and thus samples the maximum number of states. Which of these two competing interactions dominates? What effect does the competition have on the dynamic and equilibrium properties of the system? The answers to these questions are still hotly debated. To help resolve this debate, static and dynamic interaction of water confined to a hydrophobic surface was studied by Zhang and coworkers using surface forces apparatus (SFA) [3, 4].

6.2 STATIC INTERACTION OF WATER BETWEEN TWO HYDROPHOBIC SURFACES

A puzzling aspect of hydrophobic attraction is that its intensity and range appear to be qualitatively different as concerns extended surfaces of large area and small molecules of modest size [2, 8, 15, 21]. One difference is fundamental: the hydrogen-bond network of water is believed for theoretical reasons to be less disrupted near a single alkane molecule than near an extended surface [2, 8, 15, 21]. A second difference is phenomenological: direct measurement shows attractive forces between extended surfaces, starting at separations too large to be reasonably explained by disruption of the hydrogen-bond network. This conclusion comes from 20 years of research using the SFA and, more recently, atomic force microscopy (AFM). The onset of attraction, ~10 nm in the first experiments [18, 22–24], soon increased by nearly an order of magnitude [25–27] and has been reported, in the most recent work, to begin at separations as large as 500 nm [19]. This has engendered much speculation because it is unreasonably large compared to the size of the water molecule (~0.25 nm). The range of interaction lessens if the system (water and the hydrophobic surfaces) are carefully degassed [5, 28–32]. Water in usual laboratory experiments is not degassed, however, and so it is relevant to understand the origin of long-range attraction in that environment. A recent review summarizes the experimental and theoretical situation [27].

In the course of experiments intended to probe the predicted slip of water over hydrophobic surfaces [9, 33], it is observed that the long-range hydrophobic force was weakened to the point of vanishing was observed when the solid surfaces experienced low-level vibrations around a mean static separation. The effect and its dependence of vibration velocity is quantified below (Fig. 6.1) with a discussion of the possible implication.

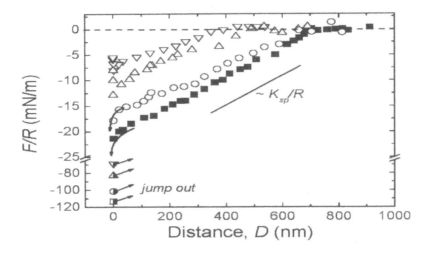

Figure 6.1 Force–distance profiles of deionized water between hydrophobic surfaces (OTE monolayers on mica). Forces were measured during approach from static deflection of the force-measuring spring while simultaneously applying small-amplitude harmonic oscillations in the normal direction with peak velocity $v_{peak} = d \times 2\pi f$, where d denotes displacement amplitude and f denotes frequency. This was 0 (solid squares), 7.6 nm/s (d = 1.6 nm, f = 0.76 Hz; circles), 26 nm/s (d = 3.2 nm, f = 1.3 Hz; up triangles), and 52 nm/s (d = 3.2 nm, f = 2.6 Hz; down triangles). The pull-off adhesion forces ("jump out"), measured at rest and with oscillation, are indicated by respective semifilled symbols. The approach data follow the straight line with slope K_{sp}/R, indicating that they represented a spring instability (jump in) such that the gradient of attractive force exceeds the spring constant (K_{sp}), 930 N/m. Reproduced with permission from Ref. [3].

The attraction recorded during the approach of octadecyltriethoxysiloxane (OTE)-coated surfaces, which were self-assembled at molecularly smooth mica surfaces, with a droplet of deionized water in between is plotted in Fig. 6.1 as a function of surface–surface separation (D). D = 0 here refers to monolayer–monolayer contact in air. In water, the surfaces jumped into adhesive contact at 0±2 Å. This "jump in" was very slow to develop, however. The pull-off force to separate the surfaces from contact at rest (113 mN/m in Fig. 6.1) implies, from the JKR theory [23], a surface energy of about 12 mJ/m² (and up to about 30% less than this when oscillations were applied). The onset of attraction at 650 nm for hydrophobic surfaces at rest is somewhat larger than in any past study. However, it is worth emphasizing that the level of pull-off force was consistent with the prior findings of other groups using other systems [1, 8–20].

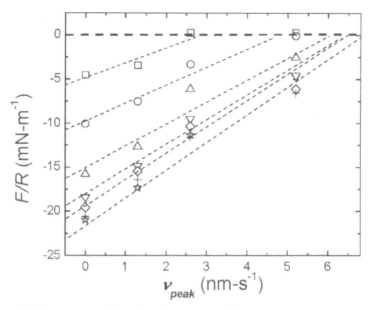

Figure 6.2 The attractive force (F/R) at seven different surface separations (D) is plotted again peak velocity. The film thickness was D = 720 nm (squares), 540 nm (circles), 228 nm (up triangles), 116 nm (down triangles), 63 nm (diamonds), 17 nm (crosses), and 5 nm (stars). Reproduced with permission from Ref. [3].

These observations clearly imply some kind of rate-dependent process. As shown in Fig. 6.2, the force F diminished with increasing velocity and its magnitude at a given D appeared in every instance to extrapolate smoothly to 0. The possible role of hydrodynamic forces was considered but discarded as a possible explanation. Similar results were also obtained when the surfaces were vibrated parallel to one another rather than in the normal direction. Some precedence is found in a recent AFM study that reported weakened hydrophobic adhesion force with increasing approach rate [31].

These observations remove some of the discrepancy between the range of hydrophobic forces between extended surfaces of macroscopic size [1, 8–20] and the range that is expected theoretically [2, 8, 15, 21]. A tentative explanation is based on the frequent suggestion that the long-range hydrophobic attraction between extended surfaces stems from the action of microscopic or submicron-sized bubbles that arise either from spontaneous cavitation or the presence of adventitious air droplets that form bridges between the opposed surfaces [5, 27, 29–32]. The experiments reported here show that this effect required time to develop. Hydrophobic attraction at long range was softened to the point of vanishing when the solid surfaces were not stationary.

6.2 DYNAMIC INTERACTION OF WATER AT A JANUS HYDROPHOBIC INTERFACE

As it has been observed theoretically and experimentally (see above), when the gap between two hydrophobic surfaces becomes critically small, water is spontaneously ejected [34–40], whereas water films confined between symmetric hydrophilic surfaces are stable apart from the (short range) effects of van der Waals attraction [41]. It is then interesting to consider the antisymmetric situation, with a hydrophilic surface on one side (Fig. 6.3a) to contain the water and a hydrophobic surface (Fig. 6.3b) on the other to force it away. This Janus situation is shown in the cartoon in Fig. 6.3.

The main result of this experimental study is that when water is confined between these two competing tendencies, the result is neither simple wetting nor dewetting. Instead, giant fluctuations (of the dynamical shear responses) is observed around a well-defined mean. This noise and fluctuation is peculiar to water and not observed with nonpolar fluids [42]. Aqueous films in the confined symmetrically hydrophilic situation also give stable dynamical responses [43]. The implied spatial scale of fluctuations is enormous compared with the size of a water molecule and lends support to the theoretical prediction that an ultrathin gas gap forms spontaneously when an extended hydrophobic surface is immersed in water [34, 44, 45].

The atomically smooth clay surfaces used in this study, muscovite mica (hydrophilic) and muscovite mica blanketed with a methyl-terminated organic self-assembled monolayer (hydrophobic), allowed the surface separation to be measured by multiple beam interferometry. Pairs of hydrophilic–hydrophobic surfaces were brought to the spacings described below, using an SFA [41] modified for dynamic oscillatory shear [46, 47]. A droplet of water was placed between the two surfaces oriented in crossed-cylinder geometry. Piezoelectric bimorphs were used to produce and detect controlled shear motions. In experiments using degassed water, the water was either first boiled and then cooled in a sealed container or subjected to vacuum for 5–10 hours in an oven at room temperature. The temperature of measurements was 25°C.

To determine firmly that findings did not depend on details of surface preparation, three methods were used to render one surface hydrophobic. In order of increasing complexity, these were (a) atomically smooth mica coated with a self-assembled monolayer of condensed OTE, using methods described previously [47]; (b) mica coated using the Langmuir–Blodgett (LB) technique with a monolayer of condensed OTE; and (c) a thin film of silver sputtered onto atomically smooth mica and then coated with a self-assembled thiol monolayer. In method (a), the monolayer quality was improved by distilling the OTE before self-assembly. In method (b), OTE was spread onto aqueous HCl (pH = 2.5), 0.5 hour was allowed for hydrolysis, the film was slowly compressed to the surface pressure π = 20 mN/m (3–4 hours), and the close-packed film was transferred onto mica by the LB technique at a creep-up speed of 2 mm/min. Finally the transferred films were vacuum-baked at 120°C for 2 hours. In method (c), 650 Å of silver was sputtered at 1 Å/s onto mica that was held at room temperature, and then octadecanethiol was deposited from a 0.5 mM ethanol solution. In this case, AFM (nanoscope II) showed the root mean square (rms) roughness to be 0.5 nm. All three methods led to the same conclusions. The contact angle of water with the hydrophobic surface was θ = 110±2° (OTE surfaces) and θ = 120±2° (octadecanethiol). In shear experiments, forces were applied to the hydrophobic surface.

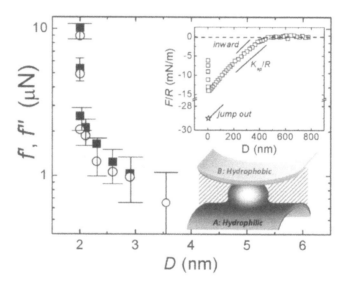

Figure 6.3 Deionized water confined between a hydrophilic surface on one side and a hydrophobic surface on the other (see cartoon, which is not to scale). The main figure shows the time-averaged viscous (circles) and elastic (squares) shear forces measured at 1.3 Hz and 0.3 nm deflection, plotted semilogarithmically against surface separation for deionized water confined between OTE deposited onto mica using the Langmuir–Blodgett technique (shear impulses were applied to this hydrophobic surface). The inset shows the static force–distance relations. Force, normalized by the mean radius of curvature ($R \approx 2$ cm) of the crossed cylinders, is plotted against thickness of the water film (D = 0 refers to contact in air). The pull-off adhesion at $D \approx 5.4$ nm is indicated by a star. Following this "jump" into contact, films of stable thickness resulted, whose thickness could be varied in the range D = 1–4 nm with applied compressive force. Reproduced with permission from Ref. [4].

The starting point was to measure the force–distance profile. The inset of Fig. 6.3 shows force normalized by the mean radius of curvature of the crossed cylinders ($R \approx 2$ cm) plotted against surface separation (D). Each datum refers to an equilibration of 5–10 minutes. Attraction was observed starting at very large separations, $D \approx 0.5$ μm, and the slope of the ensuing force–distance attraction was nearly proportional to the spring constant of the force-measuring spring. This indicated, by well-known arguments [3], that the measurements represented points of instability at which the force gradient exceeded the spring constant of the force-measuring device, although these attractive forces were very slow to develop [48]. The opposed surfaces ultimately sprang into contact from $D \approx 5$ nm, and upon pulling the surfaces apart, an attractive minimum was observed at $D = 5.4$ nm. The surfaces could be squeezed to lesser thickness. Knowing that the linear dimension of a water molecule is ~0.25 nm [41], the thickness of the resulting aqueous films amounted to on the order of 5–20 water molecules, although it is not clear that molecules were distributed evenly across this space.

In the shear measurements, the sinusoidal shear deformations were gentle—the significance of the resulting linear response being that the act of measurement did not perturb the equilibrium structure (linear responses were verified from the absence of harmonics). Using techniques that are well known in rheology, from the phase lag and amplitude attenuation, these were decomposed into one component in phase (the elastic force, f') and one component out of phase (the viscous force, f'') [49]. Fig. 6.3 (main portion) illustrates responses at a single frequency and variable thickness. The shear forces stiffened by more than an order of magnitude as the films were squeezed. When molecularly thin aqueous films are confined between clay surfaces that are symmetrically hydrophilic, deviations from the response of bulk water appear only at lesser separations [43]; evidently the physical origin is different here. Moreover, at each separation the elastic and viscous forces were nearly identical. Again this contrasts with recent studies of molecularly thin water films between surfaces that are symmetrically hydrophilic [43]. The equality of elastic and viscous forces proved to be general, not an accident of the shear frequency chosen.

Physically, the shear responses reflect the efficiency of momentum transfer between the moving (hydrophobic) surface and the adjoining water films. Fig. 6.4 illustrates the unusual result that the in-phase and out-of-phase shear forces scaled in magnitude with *the same* power law, the square root of excitation frequency. This behavior, which is intermediate between "solid" and "liquid," is often associated in other systems with dynamical heterogeneity [50, 51]. By known arguments it indicates a broad distribution of relaxation times rather than any single dominant one [52].

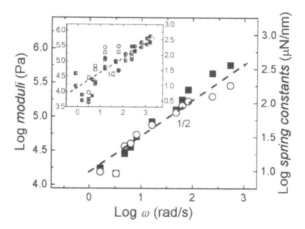

Figure 6.4 The frequency dependence of the momentum transfer between the moving surface (hydrophobic) and the aqueous film with an adjoining hydrophilic surface is plotted on log-log scales. On the right-hand ordinate scale are the viscous, g'' (circles), and elastic, g' (squares), spring constants. The equivalent loss moduli, G''_{eff} (circles), and elastic moduli, G'_{eff} (squares), are on the left-hand ordinate. The main panel, representing LB-deposited OTE, shows $\omega^{1/2}$ scaling after long-time averaging, 0.5–1 hour per datum. The inset shows comparisons with using other methods to produce a hydrophobic surface. In those experiments the thickness was generally D = 1.5–1.6 nm but occasionally as large as 2.5 nm when dealing with octadecanethiol monolayers. Symbols in the inset show data averaged over only 5–10 min with hydrophobic surfaces prepared with (a) a self-assembled monolayer of OTE (half-filled symbols), (b) LB deposition of OTE (crossed symbols), or (c) deposition of octadecanethiol on Ag (open symbols). As in the main figure, circles denote viscous forces; squares denote elastic forces. Scatter in this data reflects shorter averaging times than in the main part of this figure (cf. Fig. 6.3). Reproduced with permission from Ref. [4].

In Fig. 6.4, shear forces have been normalized in two alternative ways. The right-hand ordinate scale shows viscous and elastic forces with units of spring constant, force per nanometer of sinusoidal shear motion. Additional normalization for film thickness and estimated contact area gave the effective loss modulus, $G''_{eff}(\omega)$, out of phase with the drive, and elastic modulus, $G'_{eff}(\omega)$ (left ordinate), in phase with the drive, where ω denotes radian frequency. The slope of ½ is required mathematically by the Kramers–Kronig relations if $G'(\omega)=G''(\omega)$ [25]; its observation lends credibility to the measurements. Fig. 6.4 (main panel) illustrates this scaling for an experiment in which data were averaged over a long time. The inset shows that the same was observed using other methods to prepare a hydrophobic surface. In all of these instances, $\omega^{1/2}$ scaling was observed regardless of the method to render the surface hydrophobic but required extensive time averaging.

It is worth emphasizing that the magnitudes of the shear moduli in Fig. 6.4 are "soft"—something like those of agar or jelly. They were considerably softer than for molecularly thin aqueous films confined between surfaces that are symmetrically hydrophilic [43]. In repeated measurements at the same frequency, giant fluctuations (±30–40%) around a definite mean were observed, as illustrated in Fig. 6.5, although water confined between symmetric hydrophilic–hydrophilic surfaces (bottom panel) did not display this.

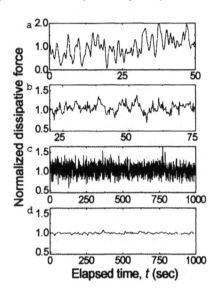

Figure 6.5 The prominence of fluctuations for water confined in a Janus interface is illustrated. In panels a–c, the viscous forces, normalized to the mean (at 1.3 Hz), are plotted against time elapsed. (a) The surfaces were first wet with ethanol to remove adsorbed gas and then flushed with degassed, deionized water; (b) the ethanol rinse was omitted; (c) this shows a long time trace for data taken under the same conditions as for (b); (d) water confined between symmetric hydrophilic–hydrophilic surfaces did not display noisy responses. Confined ethanol films likewise failed to display noisy responses (not shown). Reproduced with permission from Ref. [4].

It is extraordinary that fluctuations did not average out over the large contact area (~10 μm on a side) that far exceeded any molecular size. The structural implication is that the confined water film comprised some kind of fluctuating complex—seeking momentarily to dewet the hydrophobic side by a thermal fluctuation in one direction but unable to because of the nearby hydrophilic side; seeking next to dewet the hydrophobic side by a thwarted fluctuation in another direction; and so on. Indeed, as Weissman emphasizes [52], nearby hydrophobic and hydrophilic surfaces may produce a quintessential instance of competing terms in the free energy, to satisfy which there may be many metastable states that are equally bad (or almost equally bad) compromises [53, 54]. This suggests a physical picture of flickering capillary-type wave [4]. These proposed long-wavelength capillary fluctuations would differ profoundly from those at the free liquid–gas interface because they would be constricted by the nearby solid surface.

The possible role of dissolved gas is clear in the context of our proposed physical explanation. Indeed, submicrometer-sized bubbles resulting from dissolved air have been proposed to explain the anomalously long range of the hydrophobic attraction observed between extended surfaces [35–40] and have been observed by AFM [37–39]. Control experiments using degassed water were conducted to first test this idea. The power spectrum, included in Fig. 6.6, was nearly the same. Although it cannot be excluded that a small amount of residual dissolved gas was responsible, this method of degassing is reported to remove long-range hydrophobic attraction [36], whereas the comparison in Fig. 6.6 shows the consequence in the present situation to be minor. Next, experiments were performed to test the idea that significant amounts of trapped gas might remain on the surface even in experiments with degassed water. The hydrophobic (OTE self-assembled monolayer) surface was first wetted with ethanol, and the absence of a noisy shear response was verified. The ethanol was then flushed with copious amounts of degassed, deionized water without exposing the surfaces to air. Noisy responses to shear had reappeared when measurements were resumed after 30 minutes. It is concluded that the effects reported in this paper did not stem from the presence of dissolved gas.

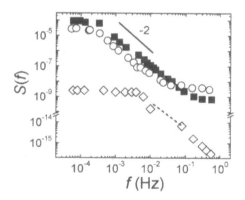

Figure. 6.6. The power spectrum for deionized water and the hydrophobized surface comprising OTE deposited by the LB technique (squares); degassed deionized water and the hydrophobized surface comprising an octadecanethiol monolayer (circles); and water containing 25 mM NaCl confined between symmetric hydrophilic-hydrophilic surfaces (diamonds). To calculate the power spectra the time elapsed was at least 10^5 seconds. The power-law exponent is close to –2. Reproduced with permission from Ref. [4]. *Abbreviation*: NaCl, sodium chloride.

To illustrate the fluctuations observed, Figs. 6.5a–d illustrate the viscous shear response of the Janus interface (see figure legend for details). Fluctuations of the elastic forces, not shown, followed the same pattern. As shown in Fig. 6.6, the power spectrum of the viscous response is the decomposition of the traces into their Fourier components whose squared amplitudes are plotted, on log–log scales, against Fourier frequency.

In the power spectrum, one observes, provided the Fourier frequency is sufficiently low, a high level of "white," frequency-independent noise. But the amplitude began to decrease beyond a threshold Fourier frequency ($f \approx$ 0.001 Hz), about 10^3 times less than the drive frequency. Other experiments (not shown) showed the same threshold Fourier frequency when the drive frequency was raised from 1.3 to 80 Hz, and therefore it appears to be a characteristic feature of the system. It defines a characteristic time for rearrangement of some kind of structure, $\sim 10^3/2\pi$ seconds, which tentatively identifies with the lifetime of bubble or vapor. The subsequent decay of the power spectrum as roughly $1/f^2$ suggests that these fluctuations reflect discrete entities, as smooth variations would decay more rapidly [52]. Noise again appeared to become "white" but with amplitude 10^4 times smaller at Fourier frequency $f > 0.1$ Hz, but the physical origin of this is not evident at this time. It did not reflect an instrumental response, as the power spectrum for water confined between symmetrically hydrophilic surfaces was much less. Whether the observed fluctuations were actually created by the drive or are also present at equilibrium has not been determined unequivocally, but the linearity of the shear moduli suggests the latter, and theoretical ideas support this [34, 45]. The noisy response appears to represent some collection of long-lived metastable configurations whose free energies are similar.

From recent theoretical analysis of the hydrophobic interaction, the expectation of dewetting emerges—it is predicted that an ultrathin gas gap, with thickness on the order of 1 nm, forms spontaneously when an extended hydrophobic surface is immersed in water [34, 44, 45]. The resulting capillary wave spectrum, as depicted in Fig. 6.7, does not appear to have yet been considered theoretically, but for the related case of the free water–vapor interface, measurements confirm that capillary waves with a broad spectrum of wavelengths up to micrometers in size contribute to its width [55]. On physical grounds, the thin gas gap suggested by our measurements should also be expected to possess soft modes with fluctuations whose wavelength ranges from small to large. From this perspective, it is then expected that the experimental geometry of a Janus-type water film, selected for experimental convenience, was incidental to the main physical effect.

This has evident connections to understanding the long-standing question of the structure of aqueous films near a hydrophobic surface and may have a bearing on understanding the structure of water films near the patchy hydrophilic–hydrophobic surfaces that are so ubiquitous in nature.

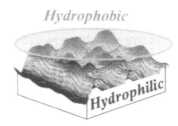

Figure 6.7 Schematic illustration of the capillary waves of water meeting the hydrophobic surface with a flickering vapor phase in between. For color reference, see p. 365.

Acknowledgments

The author is grateful for the financial support from the U.S. Department of Energy, Division of Materials Science and Engineering (DE-FG02-07ER46390), and the National Science Foundation (CBET-0730813, CBET-0651408, and CMMI-100429).

References

1. Israelachvili, J. N. (1991) *Intermolecular and Surface Forces*, 2nd ed., Academic, Press, New York.

2. Lum, K., Chandler, D., and Weeks, J. D. (1999) Hydrophobicity at small and large length scales, *J. Phys. Chem. B.*, **103**, 4570–4577.

3. Zhang, X., Zhu, Y., and Granick, S. (2001) Softened hydrophobic attraction between macroscopic surfaces in relative motion, *J. Am. Chem. Soc.*, **123**, 6736–6737.

4. Zhang, X., Zhu, Y., and Granick, S. (2002) Hydrophobicity at a Janus Interface, *Science*, **295**, 663–666.

5. Ishida, N., Inoue, Y., Miyahara, N., and Higashitani, K. (2000) Nano bubbles on a hydrophobic surface in water observed by tapping-mode atomic force microscopy, *Langmuir*, **16**, 6377–6380.

6. Kauzmann, W. (1959) Some factors in the interpretation of protein denaturation, *Adv. Prot. Chem.*, **14**, 1.

7. Tanford, C. (1973) *The Hydrophobic Effect: Formation of Micelles and Biological Membranes*, Wiley-Interscience, New York.

8. Stillinger, F. H. (1973) Structure in aqueous solutions of nonpolar solutes from the standpoints of scaled-particle theory, *J. Solution Chem.*, **2**, 141.

9. Ruckinstein, E., and Rajora, P. (1983) On the no-slip boundary-condition of hydrodynamics, *J. Colloid Interface Sci.*, **96**, 488–491.

10. Pratt, L. R., and Chandler, D. (1977) Theory of hydrophobic effect, *J. Chem. Phys.*, **67**, 3683–3704.

11. Ben-Naim, A. (1980) *Hydrophobic Interaction*, Kluwer Academic/Plenum, New York.

12. Wallqvist, A., and Berne, B. J. (1995) Computer-simulation of hydrophobic hydration forces stacked plates at short-range, *J. Phys. Chem.*, **99**, 2893–2899.

13. Hummer, G., Garde, S., Garcia, A. E., Pohorille, A., and Pratt, L. R. (1996) An information theory model of hydrophobic interactions, *Proc. Nat. Acad. Sci. USA*, **93**, 8951–8955.

14. Cheng, Y. K., and Rossky, P. J. (1999) The effect of vicinal polar and charged groups on hydrophobic hydration, *Biopolymers*, **50**, 742–750.

15. Huang, D. M., and Chandler, D. (2000) Temperature and length scale dependence of hydrophobic effects and their possible implications for protein folding, *Proc. Nat. Acad. Sci. USA*, **97**, 8324–8327.

16. Hummer, G., Garde, S., Garcia, A. E., and Pratt, L. R. (2000) New perspectives on hydrophobic effects, *Chem. Phys.*, **258**, 349–370.

17. Bratko, D., Curtis, R. A., Blanch, H. W., and Prausnitz, J. M. (2001) Interaction between hydrophobic surfaces with metastable intervening liquid, *J. Chem. Phys.*, **115**, 3873–3877.

18. Tsao, Y.-H., Evans, D. F., and Wennerstöm, H. (1993) Long-range attractive force between hydrophobic surfaces observed by atomic force microscopy, *Science*, **262**, 547–550, and references therein.

19. Considine, R. F., and Drummond, C. J. (2000) Long-range force of attraction between solvophobic surfaces in water and organic liquids containing dissolved air, *Langmuir*, **16**, 631–635.

20. Tyrrell, J. W. G., and Attard, P. (2001) Images of nanobubbles on hydrophobic surfaces and their interactions, *Phys. Rev. Lett.*, **87**, 176104.

21. Lee, C. Y., McCammon. J. A., and Rossky, P. J. (1984) The structure of liquid water at an extended hydrophobic surface, *J. Chem. Phys.*, **80**, 4448–4455.

22. Israelachvili, J. N., and Pashley, R. M. (1982) The hydrophobic interaction is long-range, decaying exponentially with distance, *Nature*, **300**, 341–342; (1984) Measure of the hydrophobic interaction between 2 hydrophobic surfaces in aqueous–electrolyte solutions, *J. Colloid Interface Sci.*, **98**, 500–514.

23. Pashley, R. M., McGuiggan, P. M., Ninham, B. W., and Evans, D. F. (1995) Attractive forces between uncharged hydrophobic surfaces-direct measurement in aqueous-solution, *Science*, **229**, 1088–1089.

24. Claesson, P. M., Blom, C. E., Herder, P. C., and Ninham, B. W. (1986) Interactions between water-stable hydrophobic Langmuir–Blodgett monolayers on mica, *J. Colloid Interface Sci.*, **114**, 234–242.

25. Claesson, P. M., and Christenson, H. K. (1988) Very long-range attractive forces between uncharged hydrocarbon and fluorocarbon surfaces in water, *J. Phys. Chem.*, **92**, 1650–1655.

26. Christenson, H. K., Claesson, P. M., Berg, J., and Herder, P. C. (1989) Forces between fluorocarbon surfactant monolayers: salt effects on the hydrophobic interaction, *J. Phys. Chem.*, **93**, 1472–1478.

27. Spalla, O. (2000) Long-range attraction between surfaces: existence and amplitude? *Curr. Opin. Colloid Interface Sci.*, **5**, 5–12, and references therein.

28. Wood, J., and Sharma, R. (1995) How long is the long-range hydrophobic attraction? *Langmuir*, **11**, 4797–4802.

29. Parker, J. L., Claesson, P.M., and Attard, P. (1994) Bubbles, cavities, and the long-range attraction between hydrophobic surfaces, *J. Phys. Chem.*, **98**, 8468–8480.

30. Carambassis, A., Jonker, L. C.,. Attard, Rutland, P M. W. (1998) Forces measured between hydrophobic surfaces due to a submicroscopic bridging bubble, *Phys. Rev. Lett.*, **80**, 5357–5360.

31. Craig, V. S. J. Ninham, B. W., Pashley, R. M. (1999) Direct measurement of hydrophobic forces: a study of dissolved gas, approach rate, and neutron irradiation, *Langmuir*, **15**, 1562–1569.

32. Considine, R. F., Hayes, R. A., and Horn, R. G.(1999) Forces measured between latex spheres in aqueous electrolyte: non-DLVO behavior and sensitivity to dissolved gas, *Langmuir*, **15**, 1657–1659.

33. Spikes, H. A. (2003) The half-wetted bearing. Part 2: Potential application to low load contacts, *Proc. Inst. Mech. Eng. Part J.*, **217**, 15–26.

34. Massey, B. S. (1989) *Mechanics of Fluids*, 6th ed., section 5.6, Chapman and Hall, London.

35. Tholen, S. M., and Parpia, J. M. (1991) Slip and the effect of He-4 at the He-3–silicon interface, *Phys. Rev. Lett.*, **67**, 334–337.

36. Huh, C., and Scriven, L. E. (1971) Hydrodynamic model of steady movement of a solid/liquid/fluid contact line, *J. Colloid Interface Sci.*, **35**, 85–101.

37. Reiter, G., Demirel, A. L., Granick, S. (1994) From static to kinetic friction in confined liquid-films, *Science*, **263**, 1741– 1744;

38. Reiter, G., Demirel A. L., Peanasky, J. S., Cai, L., and Granick, S. (1995) Stick to slip transition and adhesion of lubricated surfaces in moving contact, *J. Chem. Phys.*, **101**, 2606–2615.

39. Churaev, N. V., Sobolev, V. D., and. Somov, A. N. (1984) Slippage of liquids over lyophobic solid surfaces, *J. Colloid Interface Sci.*, **97**, 574–581.

40. Chan, D. Y. C., and Horn, R. G. (1985) The drainage of thin liquid films between solid surfaces, *J. Chem. Phys.*, **83**, 5311–5324.

41. Drake, J. M., *et. al.* (2000) *Dynamics in Small Confining Systems V*, Materials Research Society, Warrendale, PA.

42. Israelachvili, J. N. (1986) Measurement of the viscosity of liquids in very thin films, *J. Colloid Interface Sci.*, **110**, 263–271.

43. Nye, J. F. (1969) A calculation on the sliding of ice over a wavy surface using a Newtonian viscous approximation, *Proc. Roy. Soc. A.*, **311**, 445–467.

44. Mate, C. M., McClelland, G. M., Erlandsson, R., and Chiang, S. (1987) Atomic-scale friction of a Tungsten tip on a graphite surface, *Phys. Rev. Lett.*, **59**, 1942–1945.

45. Leger, L., Raphael, E., and Hervet, H. (1999) Surface-anchored polymer chains: their role in adhesion and friction, *Adv. Polym. Sci.*, **138**, 185–225.

46. Richardson, S. (1973) On the no-slip boundary condition, *J. Fluid Mech.*, **59**, 707–719.

47. Jansons, K. M. (1988) Determination of the macroscopic (partial) slip boundary condition for a viscous flow over a randomly rough surface with a perfect slip microscopic boundary condition, *Phys. Fluids*, **31**, 15–17.

48. Thompson, P. A., and Robbins, M. O. (1990) Shear flow near solids: epitaxial order and flow boundary condition, *Phys. Rev. A.*, **41**, 6830–6839.

49. Thompson, P.A., and Troian, S. (1997) A general boundary condition for liquid flow at solid surfaces, *Nature*, **389**, 360–362.

50. Barrat, J.-L., and Bocquet, L. (1999) Large slip effect at a nonwetting fluid–solid interface, *Phys. Rev. Lett.*, **82**, 4671–4674.

51. Pit, R., Hervet, H., and Léger, L. (2000) Direct experimental evidence of slip in hexadecane–solid interfaces, *Phys. Rev. Lett.*, **85**, 980–983.

52. Craig, V. S. J., Neto, C., and Williams, D. R. M. (2001) Shear-dependent boundary slip in aqueous Newtonian liquid, *Phys. Rev. Lett.*, **87**, 54504.

53. Kiseleva, O. A., Sobolev, V. D., Churaev, N. V. (1999) Slippage of the aqueous solutions of cetyltriimethylammonium bromide during flow in thin quartz capillaries, *Colloid J.*, **61**, 263–264.

54. Zhu, Y., and Granick, S. (2002) Apparent slip of Newtonian fluids past adsorbed polymer layers, *Macromolecules*, **36**, 4658–4663.

55. Zhu, Y., and Granick, S. (2002) The no slip boundary condition switches to partial slip when the fluid contains surfactant, *Langmuir*, **18**, 10058–10063.

Chapter 7

CHARACTERISTICS OF WATER ADJACENT TO HYDROPHILIC INTERFACES

Hyok Yoo, David R. Baker, Christopher M. Pirie,
Basil Hovakeemian, and Gerald H. Pollack

Department of Bioengineering, Box 355061,
University of Washington,
Seattle WA 98195
ghp@u.washington.edu

7.1 INTRODUCTION

Water molecules contiguous with hydrophilic interfaces are organized into ordered arrays. Such ordering is thought to project out by several molecular layers from the interface, the molecules lying beyond them undergoing thermal motion intense enough to preclude much additional ordering. Hence, beyond those few layers, water molecules are widely considered to lie within the bulk [1].

On the contrary, recent findings imply that some degree of ordering may project out from the surface by up millions of water layers or more [2, 3]. Much of the evidence and rationale for this view is presented in a recent award lecture: http://uwtv.org/programs/displayevent.aspx?rID=22222 (contents outlined and reviewed in http://www.i-sis.org.uk/liquidCrystallineWater. php), where the reader may obtain a deeper appreciation for the basis of these claims.

The core observation is a counterintuitive surprise: colloids and solutes of varied size are profoundly excluded from the region adjacent to hydrophilic materials. The EZ commonly extends up to hundreds of micrometers from the surface. An example is shown in Fig. 7.1.

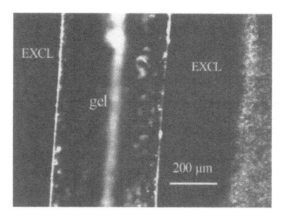

Figure 7.1 Solute exclusion (EXCL) in the vicinity of polyacrylic acid gel. The gel runs vertically. The blurred vertical white element to the right of "gel" is an optical artifact. The gel was placed on a coverslip, superfused with a suspension of 1 μm carboxylate-coated microspheres, and observed in an inverted microscope equipped with a 20× objective. The image was obtained 20 min after superfusion. Microspheres (seen on the right edge) undergo active thermal motion but do not enter the exclusion zone.

These core observations are largely detailed in two papers [2, 3]. The first deals with the question of whether the finding of long-range exclusion may have some trivial explanation. Plausible candidates were tested and ruled out by an extensive series of controls. Since then, multiple groups have informally tested and confirmed the basic finding, which is easily replicable using any of a wide range of hydrophilic surfaces, both biological and nonbiological. In fact, a similar result had been published four decades ago: in addressing the origin of the so-called "unstirred layer" adjacent to biological tissues—a region where mixing is known to be extremely slow—Green and Otori (1970) showed in both corneal tissue and contact lenses that the unstirred layer excludes microspheres; EZs several hundred micrometers wide were found, essentially the same as that in Fig. 7.1 [4].

Hence, there is little question of the existence of unexpectedly large EZs next to hydrophilic surfaces. The question is what such zones might mean, and the second paper [3] shows that the physical–chemical properties of the EZ differ from those of bulk water. Four sets of results were presented: nuclear magnetic resonance rotational relaxation time is shorter in the EZ than in bulk water, indicating a decreased degree of freedom of EZ water; infrared radiation from the EZ is less intense than from bulk water, indicating increased stability of this zone potential gradients exist in the EZ but not in bulk water; and ultraviolet–visible spectroscopy absorption spectra are markedly different in the EZ relative to bulk water.

Hence, four independent approaches show that EZ water differs from the water beyond the EZ. Collectively, these approaches imply that EZ water may well be more ordered than, and more stable than, bulk water. This difference in physicochemical properties is what appears to lead to the observed exclusion of solutes.

Here we extend these findings in four ways. We explore the possibility of ultra-long-range EZ propagation and find examples extending in length up to 1 m. We then examine the chemoelectrical features, finding charge separation that is intimately associated with the EZ. We also consider the viscosity of the EZ relative to that of bulk water and find it is much higher. And, finally, preliminary differential interference contrast data show birefringence at the Nafion–water interface extending up to hundreds of micrometers.

7.2 METHODS

For experiments on long-range EZs, a meter-long glass tube, 2.5 cm in diameter, was used. A polyacrylic acid gel (or Nafion sheet) was secured with a pin to the inside of a rubber stopper that sealed one end of the tube. The other end of the tube was sealed with another rubber stopper. The tube was filled with a suspension of 0.45 μm–diameter carboxylate-coated microspheres in deionized high-performance liquid chromatography–grade water from a Barnstead Nanopure Diamond unit (Model D11931) and held horizontally.

For the pH experiments, a universal pH indicator dye (Sigma #36803) was added to the water at the recommended concentration.

For the viscosity experiments, falling-ball viscometry was used. To track the descent of spheres, a chamber with transparent glass walls was constructed using ordinary 1 mm thick, 50 × 73 mm glass microscope slides held apart by 1 mm glass spacers, which formed the sides of the chamber. Epoxy glue (Cat. #14250, Devcon, MA) was applied around the resulting boxlike chamber to prevent water leakage. At the bottom of the chamber, we glued either a sheet of Nafion (Sigma-Aldrich #274674) or, as a control, a strip of aluminum or steel, which generated little or no EZ. Polystyrene microspheres, 5, 10, and 15 μm diameter (Bangs Laboratory, Cat. #5623, 6955, 6716, respectively) were used as falling balls. These spheres are nearly neutral, thereby minimizing any electrostatic interaction among the spheres.

To carry out experiments, the chamber was first filled with deionized water (see above). To make sure that the Nafion was fully hydrated, the microspheres were added several minutes after the chamber had been filled.

A small loop, diameter ~600 μm, was constructed out of a 100 μm–diameter copper wire for suspending microspheres into the chamber. By dipping the loop into the prepared microsphere suspension, it was possible to obtain a thin film of microsphere suspension (<0.5 µL) containing only a few microspheres. This loop was then dipped into the top of the chamber. Typically 10–20 microspheres were seen to descend.

As microspheres descended, downward displacement was tracked using a Nikon bright-field microscope with a 5× objective (Model 76023) and a Scion CCD digital camera (Model CFW-1310C). A green *light-emitting diode* (Phillips Luxeon R5JG) with a narrow localized peak at 530 nm was used as a light source. Minimizing intensity tended to minimize thermal convection. To compute the vertical component of descent (some sideways motion was often present), screens showing 1,200 µm high zones were examined using ImageJ at one frame every five seconds, and velocity was computed. This was done at a series of heights—4,800, 3,300, and 1,800, 1,000, 500, 200, 100, and 50 µm. Five runs were made at each height for microspheres of each size.

Polarizing microscopy was used to examine birefringence at the Nafion–water interface. A Nafion 117 film (Aldrich, Inc.) was placed in a 1 mm thick water chamber made of two 0.17 mm thick coverslips. The chamber was observed with a 5×/0.12NA objective lens with a maximally reduced condenser NA. Five raw polarization images were captured with an LC-polscope (CRI, Woburn, MA) and processed using a five-frame algorithm [5].

7.3 RESULTS AND DISCUSSION

7.3.1 Character of Exclusion Zones

The simplest way of envisioning a stable, ordered array is to consider it in terms of a liquid crystal, in which molecules show more alignment than a liquid but less than a solid. Like crystals of ice, liquid crystals of water could be expected to exhibit growth in any and all directions. If the EZ is liquid crystalline, then it is expected that growth would not necessarily be confined to the zone adjacent to the nucleating surface, as in Fig. 7.1. Under appropriate circumstances such as geometrical asymmetry, growth could be directed elsewhere, in odd and perhaps unpredictable directions.

Figure 7.2a shows a segment of a ribbon-shaped solute EZ that grew to extend the full length of a meter-long glass tube. A polyacrylic acid gel was positioned at one end, and the tube was filled with a microsphere suspension. Initially, a disklike EZ grew uniformly from the face of the gel surface; within 10 minutes it extended ~1 mm into the aqueous phase. The entire microsphere-free zone then shifted upward along the gel surface, creating a surge into the microsphere-containing zone. At or near the top of the surge, a thin ribbon of EZ began growing into the aqueous phase, parallel to the tube's axis (Fig. 7.2b). After several hours this ribbon extended the entire length of the meter-long tube and persisted indefinitely (i.e., for at least the maximum observation time of two weeks). During this time, microspheres undergoing rapid thermal motion just outside the ribbon failed to penetrate into the ribbon, and the ribbon retained its integrity.

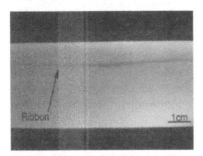 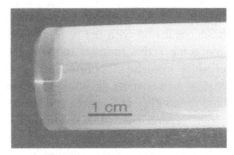

Figure 7.2 (a) Section of exclusion "ribbon" running along the length of a cylindrical tube, nucleated at the surface of the polyacrylic acid gel beyond the left side of the panel. The tube was filled with a solution containing distilled water and 15 drops of stock solution of 0.45 μm–diameter carboxylate-coated microspheres, yielding an estimated concentration of 6×10^8 microspheres per mL. (b) Top view of the chamber showing the wedge-shaped origin of the exclusion ribbon. The wedge grows out of the original disk-shaped exclusion zone adjacent to a polyacrylic acid gel surface. The prominent white object on the left is a pin used to hold the gel against the stopper. For color reference, see p. 366.

The ribbon did not grow by adding to its tip but apparently by adding from its origin near the gel's surface. This supposition came from the behavior of ribbon defects, which maintained both their shapes and their distances from the tip while translating away from the originating surface. Generally, the ribbon remained almost linear, focusing itself along a line that formed the upper edge of the cylindrical tube, although in one out of six cases, the ribbon formed elsewhere. An alternative chamber with a square cross section was made of acrylic sheets. In this case several ribbons formed simultaneously, mainly but not always along the top surface. These latter ribbons were typically wavy wisps, sometimes branched, and undergoing seemingly random fluctuations. Fluctuations notwithstanding, the structure maintained itself. Ribbon character did not appear to depend on whether it was nucleated by the polyacrylic acid gel or by a similarly positioned sheet of Nafion, a hydrophilic ionomer.

Occasionally, ribbonlike structures could also be seen in the absence of a nucleating surface. These structures were generally shorter, thinner, and less predictably evident than the ones above and could be seen only in situations in which microsphere stock had been freshly diluted immediately before use. These structures may be analogous to the solute-free "voids" seen in pure microsphere suspensions [6, 7]. Such voids were stable, and were amorphously shaped, with a characteristic dimension on the order of 100 μm. Possibly the more extended shapes seen here reflect the long chamber geometry in the present studies.

The evidence depicted in Fig. 7.2 (similar when the gel or Nafion was replaced by a hydrophilic monolayer) confirms that the EZ is a distinct phase, coexisting with the ordinary solute-containing phase. Microspheres undergoing active thermal motion just outside the ribbon phase do not penetrate into the ribbon, even for several weeks. Phase coexistence of colloidal suspensions of differing composition is well recognized [8], as is phase coexistence of mixtures of microsphere solutes of different size [9]. The current experiments show similar phase separation, one phase containing colloidal solutes, the other not.

The existence of the ribbonlike extensions is consistent with liquid crystalline behavior. Growth can be anticipated in multiple directions for extended distances; this is what was observed. On the contrary, the presence of such ribbons does not seem compatible with the alternative hypothesis that the EZ is created by some kind of electrostatic repulsive force between the gel–polymer surface and the microspheres, for meter-long extensions are difficult to envision under such a hypothesis, as electrostatic force diminishes as the square of distance. These apparently liquid crystalline structures, as mentioned, are quite stable over long periods of time, reinforcing the idea of some kind of crystal.

7.3.2 Charge Separation

A significant aspect of the EZ is its negative electrical potential [3]. The potential is most negative next to the nucleating surface and diminishes toward zero at the EZ–bulk water boundary. The implication of the negative potential is that the EZ may contain a negative charge—which would not be surprising for a liquid crystalline array of molecules. But if the EZ were negatively charged, then a positively charged counterpart ought to exist somewhere, and we considered whether the bulk water might have excess unbound protons.

Figure 7.3 shows an image of a narrow water-containing chamber with a Nafion sheet placed at the bottom. A universal pH indicator dye was added to the water in concentration as per the manufacturer's recommendation. The Nafion sheet appears dark red because of local acidity. Just above and below the Nafion are clear zones, from which the dye is excluded in much the same way as microspheres are excluded (see above). Above this EZ is a rainbow of colors, indicating a steep pH gradient. The red color indicates pH < 3; the colors above indicate progressively higher pH levels, with near neutrality at the top.

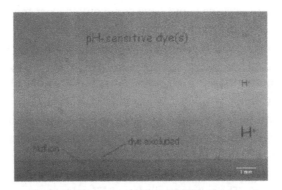

Figure 7.3 Chamber containing a Nafion tube (*bottom*) filled with water containing a pH-sensitive dye. View is normal to the wide face of a narrow chamber. Image obtained 5 min after dye solution poured into chamber. For color reference, see p. 366.

A result similar to that of Fig. 7.3, but taken at a later time, is shown in Fig. 7.4. Several features are of interest. First, the Nafion at the bottom is looped in several places; this results from expansion of the Nafion sheet as it hydrates. Second, the EZ projects nearly vertically, extending into the solution in several regions. The extensions are very much like those in Fig. 7.2. Of most interest is the fact that the red color surrounds each one of those projections, indicating a low pH even around the projecting entities. This implies that whether the EZ is closely adjacent to the nucleating surface or projects far away from it, a zone of high proton concentration surrounds it. Inevitably, a region of high proton concentration is juxtaposed next to the negatively charged EZ.

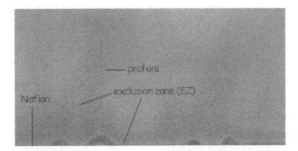

Figure 7.4 Similar to Fig. 7.3 but taken 20 min after the water with pH indicator dye is poured into the chamber. Note the exclusion zone growth perpendicular to the Nafion surface, surrounded by areas of low pH. For color reference, see p. 366.

From the latter observation, it seems likely that the free protons come from the EZ water and not from the Nafion. Nafion is an ionomer, which is expected to conduct and donate protons. Compatible with this expectation is the image of Fig. 7.3, where a region of high proton concentration lies adjacent to the Nafion. The protons could have come either from the Nafion or from the EZ water next to the Nafion, and deciding between the two options is not possible from this experiment alone. However, Fig. 7.4 implies that it may well come largely from the water, for the vertically oriented high-proton zones are far from any Nafion-proton source. Rather, those zones lie immediately adjacent to EZ water.

Another difference between Figs. 7.3 and 7.4 is that in the latter image, the pH is lower overall; the image is dominated mainly by the red color, with green (neutral pH) virtually absent. The probable reason is that the image was obtained later, after the EZ had grown substantially and included many vertical projections. The substantially larger EZ volume implies a much larger release of protons, which would diffuse out to bulk water, lowering the pH overall and accounting for the largely red-looking image.

The results imply that negative and positive zones exist side by side. It would not be surprising if the release of protons were a necessary condition for creating the negatively charged EZ. The negative–positive configuration is much like that of a capacitor or battery. Indeed, we have found (unpublished data) that when electrodes are placed in the EZ and in the region beyond, sustained currents can be obtained. Such currents provide evidence of the presence of separated charges—the EZ being negative and the region beyond being positive.

7.3.3 Time Course of Proton Discharge

To measure the time course of proton discharge, a miniature pH probe was positioned at each of a series of distances from the Nafion surface. At each distance, measurements were taken immediately after the water was added, recorded manually at 5-second intervals, for the duration of one minute or more. Representative results obtained from three sets of readings are shown in Fig. 7.5.

Consistent with the dye experiments, the pH value dropped rapidly and as a function of distance. At a distance of 1 mm, the pH drop was more than 3 pH units, which represents an H^+ increase in excess of 1,000 times. At larger distances the drop began progressively later and was of lower magnitude. Eventually, the pH at all three distances reached a similar value—reduced

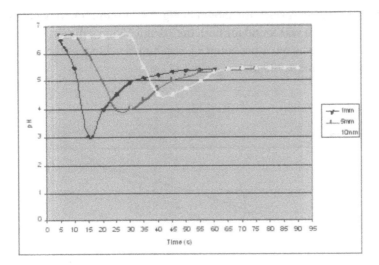

Figure 7.5 Time course of pH following addition of water to a sheet of Nafion. The pH was measured at three distances, indicated to the right, from the Nafion surface. A wave of protons appears to be generated. For color reference, see p. 367.

considerably from the original water's pH, seemingly as a result of diffusion of protons or hydronium ions released from the EZ. This result is similar to the result shown in Fig. 7.4 with the pH indicator dye: eventually, the solution pH diminishes substantially from its initial value, although the time scales may differ somewhat because of geometrical and other differences.

It appears from Fig. 7.5 that as the EZ builds, a wave of protons is generated and "pushed" outward, leaving the region beyond the EZ with a high concentration of H^+. The time course and final level would be very much dependent on both the size and the shape of the container, as well as the exposed area of the nucleating surface. Indeed, in some instances we could find pH values that on occasion fell to stable values as low as 2 or 1 with gels that occupied the majority of relatively small chambers.

7.3.4 Viscosity

The EZ's liquid-crystalline-like nature implies a cohesiveness among constituent elements, which in turn implies that the ensemble should have higher viscosity than bulk water. To test this expectation, we used falling-ball viscometry. A strip of Nafion was glued to the bottom of an experimental chamber, which was filled with water. The velocity of the descent of spheres was tracked to determine whether the spheres might slow down as they descended into the EZ.

Representative results are shown in Fig. 7.6. A reduction of velocity near the Nafion surface was found for both the 10 and 15 µm spheres. For the 5 µm spheres the fall was most pronounced, reaching a value less than 0.01 µm/s at heights less than 1,000 µm.

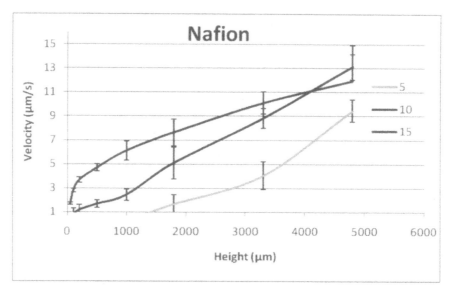

Figure. 7.6. Velocity as a function of height above the Nafion surface, for 5, 10, and 15 µm–diameter polystyrene microspheres. For color reference, see p. 367.

From the values of velocity, it was possible to compute the viscosity from the simplified version of the Navier–Stokes equation. Thus,

$$V_s(y) = \frac{2}{9} \frac{r^2 g (\rho_p - \rho_f)}{\eta}.$$

Here V_s is the particle's settling velocity (m/s), r is the Stokes radius of the particle, g is the gravitational acceleration, ρ_p is the density of the particles, ρ_f is the density of the fluid, and η is the fluid viscosity. For simplicity, nonlinear convective forces were not considered and fluid density was assumed constant. It was also assumed that the microspheres were falling at their terminal velocity.

Results are shown in Fig. 7.7. Note that computed viscosity increases markedly near the Nafion surface. By contrast, results of Fig. 7.8 show that the near-surface viscosity increase is absent for substances that have little or no EZ. Hence, the EZ is considerably more viscous compared with regions next to non-EZ-generating surfaces.

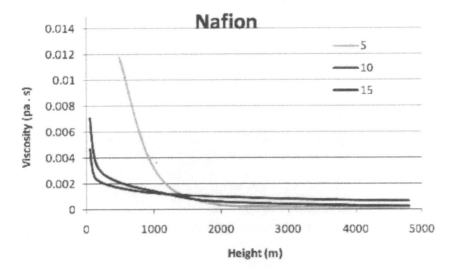

Figure. 7.7. Computed viscosity as a function of distance from the Nafion surface for 5, 10, and 15 µm–diameter polystyrene microspheres. For color reference, see p. 368.

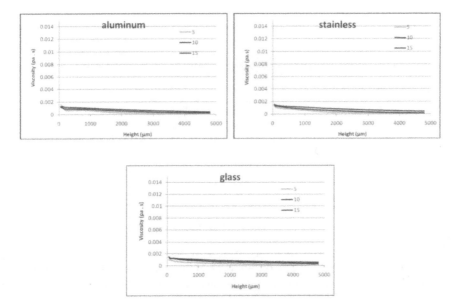

Figure 7.8 Computed viscosity as a function of distance from the surface for 5, 10, and 15 µm–diameter polystyrene microspheres for aluminum, glass, and stainless steel. For color reference, see p. 368.

7.3.5 Polarizing Microscopy

The possible liquid crystalline nature of EZ was further explored by measuring birefringence of the water near the Nafion surface using polarizing microscopy. Preliminary experiments have been carried out in the laboratory of Michael Shribak at Woods Hole, and the results are shown in Fig. 7.9a,b.

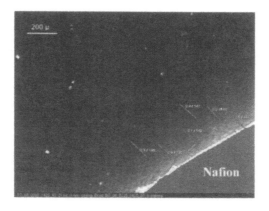

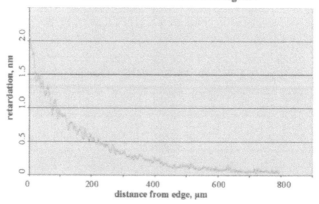

Figure 7.9 (a) Birefringent zone (bright) extends approximately 200 μm from the Nafion surface. The lines indicate the direction of the highest refractive index. Nafion sample covered by grey. (b) Optical retardation near Nafion–water interface.

A few interesting features are shown in Fig. 7.9a, which should be considered preliminary. The image brightness is linearly proportional to retardance, where the white level corresponds to 3 nm. The Nafion film itself is very birefringent; it introduces retardance of about 160 nm, which is beyond of the chosen grey scale. This is expected inasmuch as interfacial water is caged within the polymer. Perhaps, the more striking feature appears in the aqueous zone next to the Nafion. The interfacial water shows a birefringent zone with smoothly diminishing retardation extending out to about 200 μm from the edge of the Nafion (Fig. 7.9b). The dimension observed here matches nicely with the size of the EZ measured with other methods.

Another interesting finding lies in the direction of the slow axis shown in Fig. 7.9a. It was observed that the slow-axis direction within the EZ always followed the slow-axis direction of the Nafion, independent of the orientation of the Nafion edge. This implies a continuity between the water in the Nafion and the water in the EZ.

7.4 CONCLUSIONS

A near-hydrophilic aqueous zone, first identified on the basis of its proclivity to exclude solutes, shows characteristics of a liquid crystal. Previous observations have shown this zone to be more stable than bulk water [3], while current observations show that the zone not only exists contiguously with the nucleating surface, but it may also project out like stalactites, to great distances. This is one of the characteristics of crystals.

Furthermore, this liquid-crystal-like zone is negatively charged; this was shown earlier [3]. It is apparent from the present results (Figs. 7.4 and 7.5) that this negatively charged zone is contiguous with a region rich in protons and hence positively charged. This juxtaposition of negative and positive is apparent whether the EZ lies contiguously with the nucleating surface or projects out at right angles to that surface.

A possibility, then, is that the negatively charged, liquid crystalline zone arises from bulk-water molecules that are reorganized into positive and negative entities. The negatively charged species would coalesce into the stable, cohesive liquid-crystal-like structure making up the EZ, while the positive entities would remain in the bulk, free to diffuse as protons, hydronium ions, or other large-scale positively charged water-based structures. The negative entity is stable, while the positive entity is free to diffuse.

As anticipated for a liquid crystalline substance, the viscosity is considerably higher than that of bulk water. The data of Figs. 7.6–7.8 show viscosity that is an order of magnitude larger than in the vicinity of nonexcluding surfaces. The increased viscosity of EZ water may help explain the gel-like character of the cell and of various gels, where interfacial water is commonplace. If EZ water is present in abundance in these entities, then their macroscopic mechanical behavior would be more like that of a liquid crystal than that of a liquid. The implied liquid crystalline nature is further supported by preliminary polarizing microscopic data.

Although the tentative conclusions drawn here seem out of accord with generally held views, they seem the simplest interpretation of previous and current results. They imply a structural and functional aqueous framework considerably different from what is generally accepted and perhaps raise more questions than they answer. Nevertheless, they imply that aqueous behavior at interfaces may be far different than that implied by standard textbook views.

References

1. Israelachvili, J. (1992) *Intermolecular and Surface Forces*, Academic Press, San Diego.
2. Zheng, J.-M., and Pollack, G. H. (2003) Long range forces extending from polymer surfaces, *Phys. Rev. E.*, **68** (031408), 1–7.
3. Zheng, J.-M., Chin, W.-C, Khijniak, E., Khijniak, E., Jr., and Pollack, G. H. (2006) Surfaces and interfacial water: evidence that hydrophilic surfaces have long-range impact, *Adv. Colloid Interface Sci.*, **127**, 19–27.
4. Green, K., and Otori, T. (1970) Direct measurement of membrane unstirred layers, *J. Physiol. London*, **207**, 93–102.
5. Shribak, M., and Oldenbourg, R. (2003) Techniques for fast and sensitive measurements of two-dimensional birefringence distributions, *Appl. Opt.*, **42** (16), 3009–3017.
6. Ito, K., Yoshida, H., and Ise, N. (1994) Void structure in colloidal dispersions, *Science*, **263**, 66–68.
7. Yoshida, H., Ise, N., and Hashimoto, T. (1995) Void structure and vapor-liquid condensation in dilute deionized colloidal dispersions, *J. Chem. Phys.*, **103** (23), 10146–10151.
8. Albertsson, P.-A. (1960) *Partition of Cell Particles and Macromolecules*, John Wiley, New York.
9. Kaplan, P. D., Rouke, J. L., Yodh, A. G., and Pine, D. J. (2004) Entropically driven surface phase separation in binary colloidal mixtures, *Phys Rev. Lett.*, **72** (4), 582–585.

Part IV

WATER AND CELL MEMBRANES

Chapter 8

WATER AND MEMBRANES: INSIGHTS FROM MOLECULAR DYNAMICS SIMULATIONS

Mounir Tarek

UMR Structure et Réactivité des Systèmes Moléculaires Complexes,
CNRC-Nancy Université
Mounir.Tarek@edam.uhp-nancy.fr

8.1 INTRODUCTION

Membranes constitute a ubiquitous component of tissues as they hold cells together and divide them into compartments. They are formed mainly of an assembly of a wide variety of lipids, proteins, and carbohydrates that self–organize into a thin barrier that separates the interior of the cell from the outside environment [1]. The main lipid constituents of cell membranes are phospholipids—amphiphilic molecules that have a hydrophilic head and two hydrophobic tails. When lipids are exposed to water, they arrange themselves into a two-layered sheet (a bilayer) with all of their tails pointing toward the center of the sheet. Although the bilayers are only a few nanometers thick, these are impermeable to most water-soluble (hydrophilic) molecules. In order to assume a variety of biological functions necessary for the cell machinery, for example, the passive and active transport of matter, the capture and storage of energy, the control of the ionic balance, the cellular recognition and signaling, cells require the assistance of membrane proteins.

Cell membranes have a much more complex structure since they include the cytoskeleton which undertakes various functions [2]. However, several of their properties depend on the chemical composition of their lipid bilayer constituents. The length of the acyl chains and their degree of unsaturation

Water: The Forgotten Biological Molecule
Edited by Denis Le Bihan and Hidenao Fukuyama
Copyright © 2011 by Pan Stanford Publishing Pte. Ltd.
www.panstanford.com

for instance determine the thickness and the stiffness of the membrane, and often the activity of membrane proteins [3, 4]. On the other hand, the nature of polar head determines almost all physicochemical properties of the membrane interface, such as, total charge, hydration level, and interfacial ionic concentration. Also, it often modulates the interactions of the membrane with proteins and polypeptides.

While the nature of the lipid component of membranes plays a key role in their properties and function, water located in their vicinity is also of major importance. The main aim of this chapter is to shed light on the fundamental role of this water, at the microscopic level, in biophysical processes involving membranes. We will focus on two phenomena. The first one relates to the stability of membranes and their rupture upon application of an electric field, a process known as electroporation. The second is related to the function of voltage-gated ion channels, in which mounting evidence suggests that water molecules penetrating the transmembrane (TM) region of these membrane proteins are involved in the activation processes that ensure their activity.

Because of the complexity and heterogeneity of the membranes, it is often difficult to characterize, at the molecular level, many structural and dynamical processes taking place in these cell envelopes. Computer simulations in general, and molecular dynamics (MD) simulations in particular, have proven to be an effective approach to provide such a level of detail that is inaccessible to conventional experimental techniques. The molecular insight into the role of water in membranes addressed in this chapter is mainly provided by results from such simulations. We therefore describe first the commonly used methodologies and protocols in MD simulations of lipid bilayers, with an emphasis on their limitations and their strength in reproducing many experimental observables with satisfactory agreement. We then present an account of electroporation of lipid membranes resulting from the direct application of an electric field and discuss the role water plays in such a phenomenon. In the third section of the chapter, we will focus on the role that the solvent plays in the function of voltage-gated potassium channels.

8.1.1 Molecular Dynamics Simulations of Membranes

In this section, we briefly introduce some basic concepts and terminology concerning MD simulation techniques and summarize the actual state of their application to model simple lipid membranes. The reader is referred to more extensive reviews on the topic for further reading [5–10].

8.1.1.1 MD simulations: Basics

Molecular dynamics simulations refer to a family of computational methods aimed at modeling the macroscopic behavior of a given system [11, 12]. This is achieved through the numerical integration of the classical equations of motion of the corresponding microscopic many-body molecular model. Macroscopic properties are expressed as functions of particle coordinates and/or momenta, which are computed along a phase space trajectory generated by classical dynamics using forces derived from a potential energy function **U**. Common force fields used in chemistry and biophysics, for example, GROMOS [13], CHARMM [14], and AMBER [15], are based on molecular mechanics and a classical treatment of particle–particle interactions that precludes bond dissociation and, therefore, the simulation of chemical reactions. Therefore, **U** consists in a summation of bonded interactions associated with chemical bonds, bond angles, bond dihedral deformations, nonbonded interactions associated with electrostatic interactions, and van der Waals forces. The former are calculated from partial charges assigned to each atom, whereas the latter describe short-range repulsion and long-range attractions between pairs of atoms.

Molecular dynamic simulations generate a set of atomic positions and velocities of all atoms in the system as a function of time. These evolve deterministically and iteratively from an initial configuration according to the interaction potential **U**. The positions and velocities (and forces) at a given instant in time, $t + \Delta t$ are predicted from their values at a previous time, t, by integrating the classical equations of motion. Δt, the integration time step is chosen to be small enough (in the order of a femtosecond (10^{-15}s)) to avoid discretization errors, and to ensure energy conservation. When the trajectory is long enough to yield satisfactory time averages, the simulation results may be gathered to estimate observables using statistical mechanics. Hence, MD simulations can provide a detailed view of the structure and dynamics of a macromolecular system when performed under conditions corresponding to laboratory scenarios.

8.1.1.2 MD simulations of lipid bilayers

MD simulations are usually performed using a small number of molecules (few tens to few hundred thousand atoms), the system size being limited by the speed of execution of the programs and the availability of computer power. In order to simulate the macroscopic system, the sample is placed in a central cell, called the simulation box, which is replicated infinitely in three dimensions using periodic boundary conditions (PBCs) (Fig. 8.1). In membranes for instance, for computations to be tractable, the number of lipids considered is typically of the order of 100. The simulated system corresponds therefore to a tiny fragment of a liposome (cell model) or of a multilamellar lipid stack. MD trajectories for membranes simulations span in general few nanoseconds (10^{-9} s), though some have been extended to few hundreds of nanoseconds. In general, this is long enough to permit relaxation of many degrees of freedom of the system and access to local structural properties.

Bilayers formed of zwitterionic phosphatidylcholine (PC) lipids have so far been the most extensively surveyed membrane mimics. At the exception of few simulations of membranes in the gel phase [16, 17], most investigations have focused on the biologically relevant, so-called liquid crystal (L_α) phase. These included the study of saturated lipids, for example, dimiristoyl-phosphatidyl-choline (DMPC) and dipalmitoyl-phosphatidyl-choline (DPPC), polyunsaturated acyl chains such as palmitoyl-oleyl-phosphatidylcholine (POPC), and lipids based on mixtures of saturated and polyunsaturated chains, for example, stearoyl-docosahexaenoyl-glycero-phosphocholine (SDPC) [18–22]. More recently, studies have concerned lipids featuring different head groups, for example, phosphatidyl-ethanolamin, phosphatidyl-serine, and sphygomyelin [23, 24], and bilayers with mixed lipid compositions [25–29].

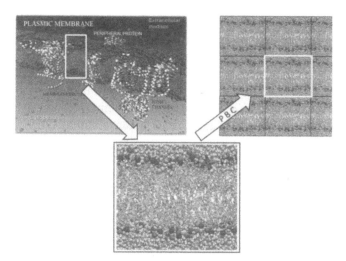

Figure 8.1 Modeling membrane bilayers. *Left*: Sketch of a plasmic membrane with incorporated membrane proteins. *Bottom*: Configuration of a palmitoyl-oleyl-phosphatidyl-choline (POPC) hydrated bilayer system from a well-equilibrated constant-pressure MD simulation performed at 300 K. Water molecules (gray), phosphate (magenta), and nitrogen atoms (blue) are represented as solid van der Waals spheres, while the acyl chains (cyan) are displayed as sticks. *Right*: During the simulation, 3D periodic boundary conditions (PBCs) are used, that is, the system is replicated infinitely in the three directions of space. For color reference, see p. 369.

Over the years, the implementation of refined MD techniques helped to elucidate microscopic events taking place in the lipid bilayer at a level of detail not always attainable by experiments alone. Bilayers built from PC lipids represented remarkable test systems to probe the simulations methodology, and the quality of these simulations (force field and MD protocol used) was often assessed by comparing calculated data to experimental results. Furthermore, MD simulations could often be used efficiently to complement and help interpret experimental data. Hence, the information provided by the density distributions of the bilayer components was confronted directly to X-ray and/or neutron diffraction [30, 31] and reflectivity [32, 33] measurements. Better refinement of NMR data has been targeted at obtaining quantities such as the average lipid chain length, the surface area per molecule [34], or the acyl chain order parameters [35] and infrared (IR) data have been reinterpreted to estimate the acyl chains *gauche-trans* conformational sequences [36].

Similar efforts were undertaken to probe the dynamics of lipids in bilayer assemblies: MD simulations were used for instance to analyze nuclear Overhauser effect spectroscopy (NOESY) cross-relaxation rates in lipid bilayers [37], to calculate relaxation rates, and to assign them to various motions of the lipid molecules. Models of individual lipid molecule dynamics have also been proposed on the basis of a thorough comparison to ^{13}C NMR T_1 relaxation rates [38] and P-31-NMR spin-lattice (R-1) relaxation rates [39] of DPPC alkyl chains. Comparison of MD simulation results to inelastic neutron scattering (INS) data [40] was used to propose detailed models for lipid dynamics in bilayers on the 100 ps time scale. MD simulations have also been used to complement inelastic X-ray data of lipid bilayers [41, 42] in probing short-wavelength collective excitations in membranes. Large systems (over 1000 lipids) studied over 10 ns led to the direct observation of bilayer undulations and thickness fluctuations of mesoscopic nature [43].

Not too surprisingly, most of the investigations concerned the structural and dynamical characterization of the lipid component of the membranes, as probing experimentally the structure and dynamics of water residing next to the lipids has been and still remains a challenge [44–47]. This stems mainly from the difficulty to single out interfacial water from bulk water in experiments. Accordingly, many of the solvent structural and dynamical properties could so far be determined exclusively from MD simulations [48–50].

The density profiles of the membrane components along the normal to the bilayer, which can be measured by neutron or X-ray diffraction techniques or calculated from MD simulations [30], indicate that the membrane–water interface is characterized by a rough lipid head group area, across which water density decays smoothly from the bulk value as the solvent penetrates deeply into the bilayer (Fig. 8.2). The interfacial region connecting the bulk water to the pure hydrocarbon region in the middle of the bilayer is very broad, in contrast to the water density oscillations observed next to flat hydrophobic surfaces [51], and the relatively narrow, noninterpenetrating air–water and pure hydrocarbon–water interfaces [52]. This lipid–water interface, that may be defined as the range over which the water density ranges from 90% to 10% of its bulk value, expands over 20 Å at each bilayer side, which is more than the total hydrophobic core thickness. Thus, counterintuitively, the membrane appears rather as a broad hydrophilic interface, with only a thin slab of pure fluid hydrocarbon in the middle.

The heterogeneous atomic distributions of species at the membrane–water interface are associated with charge and molecular dipole distributions that are at the origin of an intrinsic electrostatic profile across lipid bilayers. Indeed, the organization of the phosphate (PO_4^-) and the choline ($N(CH_3)_3^+$) groups on one hand, and of the carbonyl (C=O) groups on the other hand give rise to a permanent dipole [40]. The overall nonnegligible local electric field associated with the latter is partially compensated by a specific orientation of interfacial water molecules [53–55]. This results in a net dipole potential across each interface, for example, between the interior of the hydrocarbon layer and the aqueous phase (see Fig. 8.2).

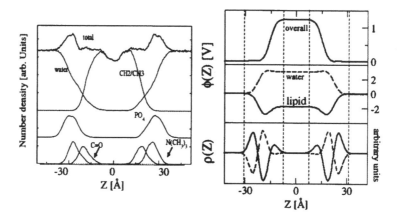

Figure 8.2 *Left*: Number density profiles (arbitrary units) along *Z*, the bilayer normal averaged over 2 ns of the MD trajectory. The total density of water and hydrocarbon chain contributions are indicated, along with those from the POPC headgroup moieties. The bilayer center is located at *Z* = 0. *Right*: Corresponding membrane electrostatic properties. *Bottom*: Water and lipid molecular charge density distributions along the lipid bilayer normal *Z* calculated from MD simulations. *Middle*: Individual contributions of water and lipids. *Top*: Overall electrostatic potential profile estimated using the

Poisson equation $\Delta\phi(Z) = \phi(Z) - \phi(0) - \frac{1}{\varepsilon_0} \int_0^Z \int_0^Z \rho(Z'') \, dZ'' \, dZ'$.

Such a dipole potential has been first introduced by Liberman and Topaly [56] who, on the basis of measurements of the partition coefficients of fat-soluble anions and cations between the membrane and the aqueous phases, hypothesized that the inner part of the bilayer membrane behaves as if it was positively charged. Because of its considerable magnitude (a few hundred millivolts), the dipole potential of a lipid bilayer is known to accounts for the much larger permeability to anions than to cations and affects the structure and function of membrane-incorporated peptides [57, 58].

The "dipole potential" of membranes has been very difficult to measure or predict (see Refs. [59–62]), and estimates obtained from various methods and various lipids range from ~200 to 1000 mV. More recent and direct measurement based on Cryo-EM imaging [61] and AFM [63] techniques show that the dipole potential can be "measured" in a noninvasive manner and estimate its value in the order or few hundred millivolts. Values ranging from 500 to 1200 mV have been reported in MD studies [7, 64–66], given the diversity of lipid studied the force fields used. Contrasting results were reported from MD simulations taking into account many-body polarization: some reported a dipole potential increase [67] while others reported a decrease [68] when the effects of polarization were included.

Similar findings were reported for charged lipid bilayers, and bilayers with added salt concentrations (see Refs. [7, 62, 69–71] and references therein). Note that in such systems, the presence of the uncompensated charges of the lipid head groups, adsorbed ions, and the counter ions that reside in the vicinity of the headgroups on both sides of the membrane contribute to the total electrostatic potential. This surface potential, which can be reasonably equated to the zeta potential, is much smaller than the intrinsic dipole potential [59, 72].

8.1.2 Membrane Electroporation

Electroporation is a phenomenon that relates to the cascade of events that follow the application of high electric fields to cells and lead to permeabilization of cell membranes [73–77]. This process, which affects the fundamental behavior of cells membranes by disturbing their integrity transiently or permanently, is efficient for the transport of drugs, oligonucleotides, antibodies, and plasmids across cell membranes [78–81]. It is nowadays used in molecular biology and biotechnology, and has also found applications in medicine [82–88].

Extensive characterization of electrical currents through planar bilayer membranes (BLM), and of transport of molecules into (or out of) cells subjected to electric field pulses tend to indicate that electroporation takes place in the following way. First, the application of electrical pulses triggers the reorganization of ions of the electrolytes at the vicinity of the membrane (charging of the membrane), which gives rise to a transient, elevated TM voltage ΔV. The local electric field [89] generated by ΔV induces a rearrangement of the membrane components (mainly water and lipids) [90–92], which ultimately leads to the formation of aqueous (hydrophilic) pores than span the membrane core. The presence of such pores increases substantially the ionic and molecular transport through the cell membrane [93].

The key features of electroporation are based on theories involving stochastic pore formation [92]. In erythrocyte membranes, large pores could be observed using electron microscopy [94], but in general, direct observation of the formation of nano-sized pores is not possible with conventional techniques. In contrast, several MD simulations have recently been conducted to model membrane electroporation [95–99], providing perhaps the most complete molecular model of the electroporation process of lipid bilayers. The following paragraphs summarize the simulation results and highlight the role water plays in the microscopic events that govern the phenomenon.

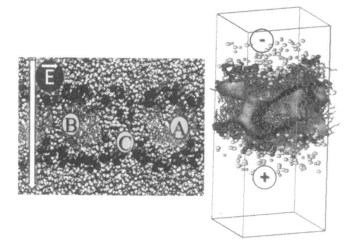

Figure 8.3. *Left*: Configuration taken from an MD simulation of a POPC bilayer subject to an electric field generating a TM potential of ~3 V, after a 2 ns run. Note that the simultaneous presence of (A) water (grey van der Waals spheres) fingers, (B) water wires, and (C) large water pores stabilized by lipid head groups (black spheres). *Right*: Configuration taken from an MD simulation of a POPC bilayer embedded in a 1 M NaCl electrolyte solution and subject to TM voltage of 2 V generated from a net charge imbalance between both sides of the membranes (unpublished data). Note here that the water baths are terminated by water–air interfaces. Na$^+$ and Cl$^-$ ions are represented as yellow and cyan spheres, respectively, and water is not shown for clarity. An isosurface of the potential is drawn in magenta to highlight the contour of the created hydrophilic pore. For color reference, see p. 369.

In membrane simulations, a TM voltage ΔV may be induced by applying a constant electric field $\vec{E}$ (implemented by applying a force $\vec{F} = q_i \vec{E}$ to all the atoms bearing a charge q_i), perpendicular to the membrane plane [100]. For electric fields of high enough magnitudes all simulations performed under this protocol led indeed to bilayer rupture according to the following sequence of events (see Fig. 8.3). First, water fingers start to penetrate deep within the lipid hydrophobic core. Ultimately these fingers join up to form water channels that span the membrane. As the electric field is maintained, few lipid head groups migrate from the membrane–water interface to the interior of the bilayer stabilizing hydrophilic pores of nanometer-sized diameters.

The time scales associated with this sequence of events taking place during electroporation were shown to depend on the strength of the applied electric field/voltage. For voltages above ~2.5–3 V, pore formation occurs in a reasonable time from the point of view of a modeler (few nanoseconds), and the lowest voltage thresholds reported in the literature are ~1.5–2 V [98, 99]. These thresholds correspond to voltage conditions where no pores formed in the membranes for MD simulation of lengths exceed 100–200 ns.

All MD studies reported pore expansion as the electric field was maintained. Interestingly enough, it was shown in one instance that hydrophilic pores could reseal within few nanoseconds when the applied field was switched off [97].

For a typical MD system size (128 lipids; 6 nm × 6 nm section) most of the simulations reported a single pore formation at high field strengths. For much larger systems, multiple pore formation was witnessed [95, 97], with diameters up to 10 nm. Such wide pores were shown to allow partial transport of large molecules, for example, a DNA plasmid across the membrane [97]. The strand considered diffused toward the interior of the bilayer and formed a stable DNA/lipid complex with support from the gene delivery model [101] in which an "anchoring step" connecting the plasmid to permeabilized cells membranes takes place during DNA transfer assisted by electric pulses.

Recently, other methods allowing simulations of realistic TM potential gradients across bilayers have been proposed [66, 102]. These methods consist of generating a net charge imbalance Qs between the two water baths located at each membrane side that induces TM voltage by explicit ion dynamics. In order to maintain the charge imbalance, the electrolytes are separated either by considering a double bilayer or by considering air–water interfaces. Overall, both protocols lead to membrane electroporation following the same scenario as in the case of the electric field method. For net charge imbalances, inducing large-enough TM voltages, such protocols allow bilayer electroporation [103, 104] (Fig. 8.3) with subsequent ionic conduction and pore resealing when the TM voltage collapses to lower values.

Hence, MD simulations have provided support to the stochastic pore formation theories. Several interesting details could further be gathered from the different investigations cited above. Perhaps the most important is the role water plays in the electroporation process. In what follows, we will briefly highlight this role by focusing on the properties of the interfacial water during electroporation, for each of the two MD protocols, namely, the electric field method and the charge imbalance method.

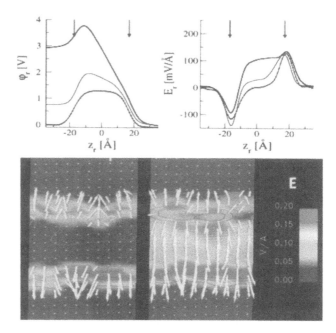

Figure 8.4 Electrostatic potential (*top left*) and electric field profiles (*top right*) along the normal to lipid membranes subject to increasing TM voltages ΔV, estimated at the initial stage of the MD simulations. The arrows indicate the approximate position of the water–membrane interface. *Bottom*: Cross-sectional view of the 3D maps of the electric field derived as the gradient of the electrostatic potential (*left*: $\Delta V = 0$ V; *right*: $\Delta V = 3$ V). The arrows indicate the direction and strength of the field. For color reference, see p. 370.

For simulations using the former method, the electrostatic potential profile across the bilayer shows that the electric field within a very short time scale—typically a few picoseconds—induces an overall TM potential difference ΔV (Fig. 8.4). Because of the use of periodic boundary conditions, $\Delta V \approx |\vec{E}|. L_z$ over the whole system, L_z being the size of the simulation box in the field direction. Decomposition of the electrostatic potential profiles in water and lipid contributions show that the electric field acts primarily on the interfacial water dipoles (small polarization of bulk water molecules), and that most of ΔV is due to the contribution of the latter. The reorientation of the lipid head groups appears not to be affected at short time scales [97, 105], and not exceeding few degrees toward the field direction at longer time scale [98]. Interestingly, simulations of a hydrated octane slab (a hydrophobic membranes mimic), confirms the main role played by reorientation of interfacial water since, even without the presence of charged lipid head groups, the field $\vec{E}$ induces a TM voltage $\Delta V \approx |\vec{E}|. L_z$.

The electrostatic properties of lipid bilayers surrounded by a 1 M NaCl electrolytes to which a net charge imbalance Qs is imposed reveal the same behavior. At moderate values, Qs is shown to reorient mainly the interfacial water molecules resulting in the creation of a TM potential ΔV, which varies linearly with Qs, since the membrane behaves as a capacitor [102].

In summary, MD simulations clearly indicate that the initial and main consequence of an applied electric field to membranes is the reorientation of interfacial water molecules to align with the field. The cascade of events following this perturbation that leads to pore formation is a direct consequence of such reorganization. Indeed, the interfacial water molecules initially forming hydrogen bonds with the lipid head groups, are now more inclined to form hydrogen bonds among themselves, which for oriented molecules may be achieved through creation of single-water files. Such fingers protrude through the hydrophobic core from both sides of the membrane. Finally, these fingers meet up to form water channels (often termed pre-pores or hydrophobic pores) that span the membrane. As the TM voltage is maintained, these water wires appear to be able to overcome the free energy barrier associated with the formation of a single file of water molecules spanning the bilayer (estimated to be ~108 kJ/mol in the absence of external electric field [52]. In agreement with experiments, these wires vanish and the hydrophilic pores reseal when the applied field is switched off [97]. If, in contrast, the electrical stress is maintained, large pores form. Note that in the second method where the voltage is induced by a charge imbalance, these hydrophilic pores allow ionic transport, which leads to a decrease in the TM potential and ultimately a resealing of the pores.

8.1.3 Activation of Voltage-Gated Channels

Potassium (K^+) channels, originally identified as the molecular entities mediating the flow of potassium ions across nerve membranes during the action potential [106] are now known in virtually all types of cells in all organisms, where they are involved in a multitude of physiological functions such as propagation of the action potential, cardiac function, and hormone regulation [107]. Several human genetic diseases, for example, pathologies involving cardiac arrhythmias, deafness, epilepsy, diabetes, and misregulation of blood pressure are caused by the alteration of K⁺ channel genes.

Many different signals can open or close the channels, a process known as gating. Potassium channels may be indeed regulated by changes in the membrane potential (voltage gated), or by transmitters and hormones (ligand gated). Voltage-gated potassium (K_v) channels constitute one of the two major subfamilies of K channels. These are complex quaternary structures in which the central hydrophilic pore results from the arrangement of four TM domains [108]. The sequence of each domain defines six TM helices (S1-S6). Segments S5 through S6 constitute the central pore domain and delineate the hydrophilic pathway for ionic current. This domain, homologous to all voltage-gated K channels [109], is surrounded by an arrangement of segments S1 to S4 identified as the voltage sensor (VS) domain [110] (Fig. 8.5).

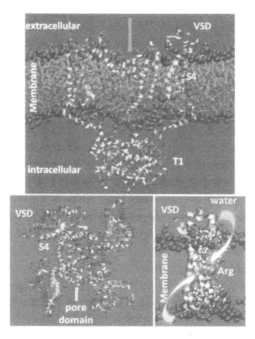

Figure 8.5 *Top*: Configuration of the Kv1.2 channel (light pink ribbons) embedded in a lipid bilayer. Only two monomers are represented for clarity. The central helices define the hydrophilic pore (direction indicated by the orange arrow) and the selectivity filter can accommodate two K⁺ ions (orange spheres). The helix S4 of the voltage sensor domain (VSD) containing highly conserved positively charged residues is highlighted in yellow. *Bottom left*: Corresponding top view. *Bottom right*: A close-up view of the VSD. MD simulations of the structure show that water (light blue) penetrates from both sides of the membrane deeply in the VSD. Atoms of the Arginine 303 depicted as yellow spheres come in contact with both water crevices. For color reference, see p. 370.

When the membrane is depolarized, this domain senses the variations of the TM voltage and triggers the conformational changes necessary for channel gating [107, 111, 112, 113]. During this voltage-sensing process, the displacement of the charges tethered to the VS gives rise to tiny transient "gating" currents. The time integral of this current is Q, the "gating charge" translocated across the membrane capacitance.

Since the seminal work of Hodgkin and Huxley [106], gating currents and gating charges have been measured for a variety of voltage-gated ion channels [114] (for K_v channels, the total gating charge $Q \approx 13e$ [115]). Three classes of models named *conventional, paddle, and transporter* have been proposed to rationalize these measurements [116], all associated with the motion of S4, the highly positively charged helix of the VSD [113].

The first two models, *conventional* and *paddle*, invoke a complete translocation of the charged S4 residues across the full thickness of the lipid membrane. The *paddle* model has been proposed on the basis of the structure of the bacterial channel KvAP [117, 118] in which S4 assumedly shifts from a conformation in which it is parallel to the membrane and exposed to the intracellular side to a perpendicular orientation. In the *conventional* model, S4 is oriented perpendicular to the membrane plane and located in the groove between the other segments of the VSD. In order to account for the gating charges, S4 is assumed to slide up and down within its protein-lined cavity environment exposing its gating charges respectively to the extra- and to the intracellular media.

While a variety of electrophysiology measurements show that during activation some S4 residues change their position from an intra- to extracellular accessible space [119–121], charge gating between transition states may involve limited TM displacements of S4. Spectroscopy [122], fluorescence [123], cysteine scanning, and accessibility measurements [124] indicate that a small displacement of S4 perpendicular to the membrane plan in the order of 2–5 Å is enough to account for current gating. Furthermore, fluorescence [125] and luminescence [126] resonance energy transfer measurements show that S4 does not translocate across the whole lipid bilayer.

This small displacement of S4 may appear not consistent with the high value of gating charges in K_v channels. However, gating currents do not measure the physical displacement of charges across the membrane, but rather the product of the charges by the fraction of the local electric field each traverses. Gating charges may be linked to the atomic scale movements of charged groups in an arbitrary structure [127]. For a voltage-gated channel undergoing a transition from a state α to a state β under an applied ΔV, the total gating charge Q can be linked to the microscopic states of the channel as follows:

$$Q.\Delta V = \Delta G(\beta, V) - \Delta G(\alpha, V). \tag{8.1}$$

Here $\Delta G(\lambda, V)$ is the conformational free energy of the channel in state λ. For a channel bearing a set of charges q_i located at r_i, $\Delta G(\lambda, V)$ may be expressed to a first approximation as [128–130]

$$\Delta G(\lambda, V) = \Delta V \cdot \sum_i^N q_i \cdot \delta_i^\lambda, \tag{8.2}$$

where, δ_i^λ is the "electrical distance." This accounts for the degree of coupling between ΔV and $\phi_i^\lambda(r_i)$ the local electrostatic potential felt by q_i located at r_i:

$$\delta_i^\lambda \equiv \frac{\partial}{\partial V} \phi_i^\lambda (r_i). \tag{8.3}$$

Hence, Q is highly dependent on the local electrostatic potential $\phi_i^\lambda(r_i)$ felt by each protein charge in both end states of the gating process.

Accordingly, Bezanilla and coworkers proposed the *transporter* model in which a large gating charge can still result from a small movement of S4, provided that the electrostatic environment around it varies [131]. The model heavily relies on the hypothesis that a specific hydration of S4 reshapes the electrostatic potential (and therefore the electric field) in the TM domain region. The variation of this electrostatic potential over short distances at the location of the S4 charges may lead to a large gating charge without large up and down movements of S4.

Even prior to the resolution of the crystal structure of a K_v channel, hydration of the VSD could be inferred from accessibility measurements of the S4 arginines residues in *Shaker* B K_v channels [132]. These results provided evidence that in both the closed and open states of the channel, several charged groups of S4 reside in aqueous crevices. It was further shown that in the closed state, mutation of Arg 362 located in S4 to His promotes proton conduction with a high turnover rate [131]. This implied that this residue located in the VSD is simultaneously accessible through water crevices extending from both sides of the membrane.

The elaborated models describing how K_v channels function at the molecular level were all derived from electrophysiology measurement as no crystal structure for the channels was available. Early molecular models of the TM-domain of K_v channels built using homology modeling and docking calculations, and satisfying a variety of experimental constraints, have provided support to the *transporter* model. Full atomistic MD simulations have indeed shown that when such a model is placed in a lipid bilayer, the solvent penetrates deeply into the voltage sensor domain (VSD) structure, and hydrates several charged residues, participating therefore in the reshaping of the local electrostatic potential [133]. Chanda *et al.* [125] used a molecular model of a *Shaker* channel embedded in a low dielectric membrane continuum that mimics a lipid bilayer. Gating charges of ~12e were measured considering a small (2 Å) vertical displacement of S4, when the membrane dielectric was distorted by protrusion of solvent crevices.

Direct and more reliable determination of the molecular properties of K_v channel became possible when the high-resolution crystal structure of a *Shaker* mammalian K_v channel (Kv1.2) was resolved in 2005 [134]. MD simulations of the channel embedded in its membrane environment could then be used to check the validity of the models of activation proposed on the basis of electrophysiology experiments [135–137] (Fig. 8.5). The crystal structure of the Kv1.2 channel [134] relaxed in a lipid bilayer shows that all the charges of S4, but the top residue (Arg 294), are protected from the lipid and are partially exposed to the solvent in agreement with EPR measurements [138]. Water protrusion deep within the VSD was noticed, and Arg 303 (362 in the *Shaker* channel) was found to bridge between intracellular and extracellular water crevices (Fig. 8.5) in agreement with its involvement in proton conduction [132].

Provided a configuration of the channel in its membrane environment is available, the 3D distributions of the electrostatic potential in the system may be generated directly from MD simulations [139]. The potential is obtained by solving Poisson's equation $\nabla^2\phi(r) = 4\pi\sum_i \rho_i(r)$, where $\rho_i(r)$ is the point charge approximated by a spherical Gaussian and the sum runs over all atoms in the system. The 3D maps of the electrostatic potential of the Kv1.2 channel hence derived (Fig. 8.6) show that specific hydration of the VSD modifies drastically the morphology of the membrane dielectric around S4, property that has been suggested to explain the exquisite electric sensitivity of voltage-gated potassium channels on the basis of fluorometric electrostatic measurements [140]. Locally, charges of S4 sense a "focused" electrical field which is a manifestation of the local abrupt variation of the local electrostatic potential over short distances in agreement with experiment [141]. For instance, the electric field in the vicinity of the top charged residues of S4, calculated as the local gradient of the 3D electrostatic potential, ranges from 80 to 100 mV Å⁻¹.

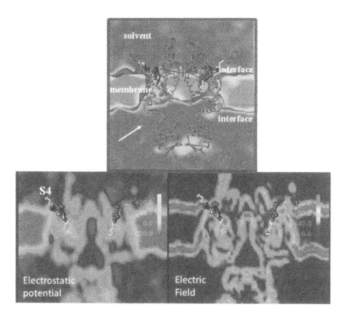

Figure 8.6 3D Electrostatic environment of the Kv1.2 channel embedded in a lipid bilayer. *Top*: An isosurface of the potential is drawn in magenta to highlight the contour of hydrophilic pore created by the water protruding inside the VSD. *Bottom*: Close-up view of the electrostatic environment of the top gating charges (Arg; black) showing the abrupt local variation of the electrostatic potential in mV (*left*) and of the electric field in mV/Å (*right*), due to the penetration of water within the TM region of the K_v channel. For color reference, see p. 371.

Therefore, the properties of the K_v channel extracted from MD simulations appear to comply with the requirements of the transporter model. It is however important to note that a complete corroboration of the latter necessitates the evaluation of the cumulative gating charge that is moved across the membrane electric field in the potassium channel during their transition between the closed and the open states. To date, the X-ray structure of the closed channel has not been resolved. Brute *force in silico* modeling of this transition have been attempted [142], but seem to necessitate an exceptionally large computational effort since the time scales involved in the transition appear to be beyond the microsecond time scale [143, 144]. Accordingly, only models of the closed channel, built in accord with electrophysiology measurements, are proposed [145]. Such models indicate that the overall motion of S4 in the order of 4–6 Å is able to account for the total gating charge during the channel transition from open to closed states. It is shown that high dielectric aqueous regions contribute to focus the TM field in both open and closed conformation, thereby controlling the magnitude of the gating charge. In the open state, the field is focused toward the extracellular half of the membrane by the large central vestibule at the intracellular pore entrance. In the closed state, the field is again focused toward the extracellular half of the membrane thanks to a wide aqueous crevice that opens up at the center of the VSD.

8.2 CONCLUSION

While water is often "forgotten" in bio-membrane images sketched in most textbooks, we have attempted in this chapter to shed light onto the importance it plays in two phenomena, the first one related to the action of an external electric field, and the second to the function of voltage-gated channels.

Because of the hydrophobic character of the acyl chains of lipids, main constituents of plasmic membranes, water is restricted to the lipid head group region. This interfacial water, whose characteristics are different from those of bulk water, plays a significant role in the electrostatic properties of the interface. It contributes in particular substantially to the membrane dipole potential. Here, it was shown that this dipole potential undergoes a drastic modification when the lipid bilayer is subject to an external electric field, for example, to a TM potential. MD simulations have provided evidence that this stems mainly from the reorientation of interfacial water molecules. Protrusion of water files through the hydrophobic core of the membrane when subject to the electrical stress seems to further initiate an electroporation process before large hydrophobic pores form and expend.

Water has been called the "lubricant of life" because of its role in promoting fast protein conformational fluctuations that may be important in globular protein-folding processes and biological function [146]. Quite interestingly, this qualification appears to hold also in the case of voltage-gated ion channels. Indeed, the presence of water deep within the TM region of these proteins appears to play a fundamental role in their activation mechanism. By reshaping the electrostatic potential near to key channel charged residues, which sense the TM voltage and trigger the conformational changes necessary for channel opening and closing, water appears to tune in the exquisite electric sensitivity of voltage-gated potassium channels.

References

1. Gennis, R. B. (1989) *Biomembranes: Molecular Structure and Function*, Springer Verlag, Heidelberg.
2. Alberts, B., Bray, D., Johnson, A., Lewis, J., Raff, M., Roberts, K., and Walter, P. (1998) *Essential Cell Biology*, Garland Publishing, New York.

3. Lee, A. G. (2003) Lipid-protein interaction in biological membranes: a structural perspective, *Biochim. Biophys. Acta*, **1612**, 1–40.

4. Lee, A. G. (2004) How lipids affects the activites of integral membrane proteins, *Biochim. Biophys. Acta*, **1666**, 62–87.

5. Anézo, C., de Vries, A. H., Höltje, H. D., Tieleman, D. P., and Marrink, S. J. (2003) Methodological issues in lipid bilayer simulations, *J. Phys. Chem. B*, **107**, 9424–9433.

6. Chipot, C., Klein, M. L., and Tarek, M. (2005) Modeling lipid membranes, in *Handbook of Materials Modeling* (ed. Yip, S.), Springer, Dordrecht, the Netherlands, pp. 929–958.

7. Berkowitz, M. L., Bostick, D. L., and Pandit, S. (2006) Aqueous solutions next to phospholipid membrane surfaces: insights from simulations, *Chem. Rev.*, **106**, 1527–1539.

8. Feller, S. E. (2008) Computational modeling of membrane bilayers, in *Current Topics in Membranes* (ed. Benos, D. J., and Simon, S.A.), Elsevier, San Diego, CA.

9. Lindahl, E., and Sansom, M. S. P. (2008) Membrane proteins: Molecular dynamics simulations, *Curr. Opin. Struct. Biol.*, **18**, 425–431.

10. Marrink, S. J., de Vries, A. H., and Tieleman, D. P. (2009) Lipids on the move: simulations of membrane pores, domains, stalks and curves, *Biochim. Biophys. Acta Biomembr.*, **1788**, 149–168.

11. Allen, M. P., and Tildesley, D. J. (1987) *Computer Simulation of Liquids*, Clarendon Press, Oxford.

12. Leach, A. R. (1996) *Molecular Modelling: Principles and Applications*, Addison Wesley, Harlow, England.

13. Schuler, L. D., Daura, X., and van Gunsteren, W. F. (2001) An improved GROMOS96 force field for aliphatic hydrocarbons in the condensed phase, *J. Comp. Chem.*, **22**, 1205–1218.

14. MacKerell, A. D., Jr., Bashford, D., Bellott, M., Dunbrack, R. L., Jr., Evanseck, J., *et al.* (1998) All-atom empirical potential for molecular modeling and dynamics studies of proteins, *J. Phys. Chem. B*, **102**, 3586–3616.

15. Case, D. A., Pearlman, D. A., Caldwell, J. W., Cheatham, T. E., III, Ross, W. S., *et al.* (1999) AMBER6, University of California, San Francisco.

16. Leekumjorn, S., and Sum, A. K. (2007) Molecular studies of the gel to liquid-crystalline phase transition for fully hydrated DPPC and DPPE bilayers, *Biochim. Biophys. Acta Biomembr.*, **1768**, 354–365.

17. Qin, S. S., Yu, Z. W., and Yu, Y. X. (2009) Structural characterization of the gel to liquid-crystal phase transition of fully hydrated DSPC and DSPE bilayers, *J. Phys. Chem. B*, **113**, 8114–8123.

18. Tu, K., Tobias, D. J., and Klein, M. L. (1995) Constant pressure and temperature molecular dynamics simulation of a fully hydrated liquid crystal phase DPPC bilayer, *Biophys. J.*, **69**, 2558–2562.

19. Berger, O., Edholm, O., and Jahnig, F. (1997) Molecular dynamics simulations of a fluid bilayer of dipalmitoylphosphatidylcholine at full hydration, constant pressure, and constant temperature, *Biophys. J.*, **72**, 2002–2013.

20. Saiz, L., and Klein, M. L. (2001) Structural properties of a highly polyunsaturated lipid bilayer from molecular dynamics simulations, *Biophys. J.*, **81**, 204–216.

21. Feller, S. E., Gawrisch, K., and MacKerell, A. D. (2002) Polyunsaturated fatty acids in lipid bilayers: intrinsic and environmental contributions to their unique physical properties, *J. Am. Chem. Soc.*, **124**, 318–326.

22. Rög, T., Murzyn, K., and Pasenkiewicz-Gierula, M. (2002) The dynamics of water at the phospholipid bilayer: a molecular dynamics study, *Chem. Phys. Lett.*, **352**, 323–327.

23. Chiu, S. W., Vasudevan, S., Jakobsson, E., Mashl, R. J., and Scott, H. L. (2003) Structure of sphingomyelin bilayers: a simulation study, *Biophys. J.* **85**, 3624–3635.

24. Mukhopadhyay, P., Monticelli, L., and Tieleman, D. P. (2004) Molecular dynamics simulation of a palmitoyl-oleoyl phosphatidylserine bilayer with Na+ Counterions and NaCl, *Biophys. J.*, **86**, 1601–1609.

25. Dahlberg, M., and Maliniak, A. (2008) Molecular dynamics simulations of cardiolipin bilayers, *J. Phys. Chem. B*, **112**, 11655–11663.

26. Patel, R. Y., and Balaji, P. V. (2008) Characterization of symmetric and asymmetric lipid bilayers composed of varying concentrations of ganglioside GM1 and DPPC, *J. Phys. Chem. B*, **112**, 3346–3356.

27. Li, Z., Venable, R. M., Rogers, L. A., Murray, D., and Pastor, R. W. (2009) Molecular dynamics simulations of PIP2 and PIP3 in lipid bilayers: determination of ring orientation, and the effects of surface roughness on a poisson-boltzmann description, *Biophys. J.*, **97**, 155–163.

28. Rog, T., Martinez-Seara, H., Munck, N., Oresic, M., Karttunen, M., *et al.* (2009) Role of cardiolipins in the inner mitochondrial membrane: insight gained through atom-scale simulations, *J. Phys. Chem. B.* **113**, 3413–3422

29. Vacha, R., Berkowitz, M. L., and Jungwirth, P. (2009) Molecular model of a cell plasma membrane with an asymmetric multicomponent composition: water permeation and ion effects, *Biophys. J.*, **96**, 4493–4501.

30. Benz, R. W., Nanda, H., Castro-Roman, F., White, S. H., and Tobias, D. J. (2006) Diffraction-based density restraints for membrane and membrane-peptide molecular dynamics simulations, *Biophys. J.*, **91**, 3617–3629.

31. Klauda, J. B., Kucerka, N., Brooks, B. R., Pastor, R. W., and Nagle, J. F. (2006) Simulation-based methods for interpreting X-ray data from lipid bilayers, *Biophys. J.*, **90**, 2796–2807.

32. Tarek, M., Tu, K., Klein, M. L., and Tobias, D. J. (1999) Molecular dynamics simulations of supported phospholipid/alkanethiol bilayers on a gold(111) surface, *Biophys. J.*, **77**, 464–472.

33. Majkrzak, C. F., Berk, N. F., Krueger, S., Dura, J. A., Tarek, M., *et al.* (2000) First principle determination of hybrid bilayer membrane structure by phase-sensitive neutron reflectometry, *Biophys. J.*, **79**, 3330–3340.

34. Petrache, H. I., Tu, K., and Nagle, J. F. (1999) Analysis of simulated NMR order parameters for lipid bilayer structure determination, *Biophys. J.*, **76**, 2479–2487.

35. Vermeer, L. S., de Groot, B. L., Reat, V., Milon, A., and Czaplicki, J. (2007) Acyl chain order parameter profiles in phospholipid bilayers: computation from molecular dynamics simulations and comparison with H-2 NMR experiments, *Eur. Biophys. J.*, **36**, 919–931.

36. Snyder, R. G., Tu, K., Klein, M. L., Mendelssohn, R., Strauss, H. L., *et al.* (2002) Acyl chain conformation and packing in dipalmitoylphosphatidylcholine bilayers from MD simulations and IR spectroscopy, *J. Chem. Phys. B.* **106**, 6273–6288.

37. Feller, S. E., Huster, D., and Gawrisch, K. (1999) Interpretation of NOESY cross-relaxation rates from molecular dynamics simulation of a lipid bilayer, *J. Am. Chem. Soc.*, **121**, 8963–8964.

38. Pastor, R. W., Venable, R. M., and Feller, S. E. (2002) Lipid bilayers, NMR relaxation, and computer simulations, *Acc. Chem. Res.*, **35**, 438–446.

39. Klauda, J. B., Roberts, M. F., Redfield, A. G., Brooks, B. R., and Pastor, R. W. (2008) Rotation of lipids in membranes: molecular dynamics simulation, P-31 spin-lattice relaxation, and rigid-body dynamics, *Biophys. J.*, **94**, 3074–3083.

40. Tobias, D. J. (2001) Membrane simulations, in *Computational Biochemistry and Biophysics* (ed. Becker, O. H., Mackerell, A. D., Roux, B., and Watanabe, M.), Marcel Dekker, New York.

41. Tarek, M., Tobias, D. J., Chen, S. H., and Klein, M. L. (2001) Short waverlength collective dynamics in phospholipid bilayers: a molecular dynamics study, *Phys. Rev. Lett.*, **87**, 238101.

42. Hub, J. S., Salditt, T., Rheinstader, M. C., and de Groot, B. L. (2007) Short-range order and collective dynamics of DMPC bilayers: a comparison between molecular dynamics simulations, X-ray, and neutron scattering experiments, *Biophys. J.*, **93**, 3156–3168

43. Lindahl, E., and Edholm, O. (2000) Mesoscopic undulations and thickness fluctuations in lipid bilayers from molecular dynamics simulations, *Biophys. J.*, **79**, 426–433.

44. Binder, H. (2007) Water near lipid membranes as seen by infrared spectroscopy, *Eur. Biophys. J.*, **36**, 265–279.

45. Gawrisch, K., Gaede, C., Mihailescu, M., and White, S. H. (2007) Hydration of POPC bilayers studied by 1H-PFG-MAS-NOESY and neutron diffraction, *Eur. Biophys. J.*, **36**, 281–291.

46. Disalvo, E. A., Lairion, F., Martini, F., Tymczyszyn, E., Frías, M., *et al.* (2008) Structural and functional properties of hydration and confined water in membrane interfaces, *Biochim. Biophys. Acta*, **1778**, 2655–2670.

47. Zhao, W., Moilanen, D. E., Fenn, E. E., and Fayer, M. D. (2008) Water at the surfaces of aligned phospholipid multibilayer model membranes probed with ultrafast vibrational spectroscopy, *J. Am. Chem. Soc.*, **130**, 13927–13937.

48. Tobias, D. J. (1999) Water and membranes: molecular details from MD simulations, in *Hydration Processes in Biology* (ed. Bellissent-Funel, M. C.), IOM Press, New York, pp. 293–310.

49. Bhide, S. Y., and Berkowitz, M. L. (2005) Structure and dynamics of water at the interface with phospholipid bilayers, *J. Chem. Phys.*, **123**, 224702.

50. Tobias, D. J., Sengupta, N., and Tarek, M. (2009) Hydration dynamics of purple membranes, *Faraday Discuss.*, **141**, 99–116.

51. Lee, C. Y., McCammon, J. A., and Rossky, P. J. (1984) The structure of liquid water at an extended hydrophobic surface, *J. Chem. Phys.*, **80**, 4448–4455.

52. Marrink, S. J., Jähniga, F., and Berendsen, H. J. (1996) Proton transport across transient single-file water pores in a lipid membrane studied by molecular dynamics simulations, *Biophys. J.*, **71**, 632–647.

53. Gawrisch, K., Ruston, D., Zimmerberg, J., Parsegian, V., Rand, R., *et al.* (1992) Membrane dipole potentials, hydration forces, and the ordering of water at membrane surfaces, *Biophys. J.*, **61**, 1213–1223.

54. Shinoda, W., Shimizu, M., and Okazaki, S. (1998) Molecular dynamics study on electrostatic properties of a lipid bilayer: polarization, electrostatic potential, and the effects on structure and dynamics of water near the interface, *J. Phys. Chem. B*, **102**, 6647–6654.

55. Cheng, J. X., Pautot, S., Weitz, D. A., and Xie, X. S. (2003) Ordering of water molecules between phospholipid bilayers visualized by coherent anti-Stokes Raman scattering microscopy, *Proc. Natl. Acad. Sci. USA*, **100**, 9826–9830.

56. Cladera, J., and O'Shea, P. (1998) Intramembrane molecular dipoles affect the membrane insertion and folding of a model amphiphilic peptide, *Biophys. J.*, **74**, 2434–2442.

57. Liberman, Y. A., and Topaly, V. P. (1969) Permeability of biomolecular phospholipid membranes for fat-soluble ions, *Biophysics. USSR*, **14**, 477.

58. Thompson, N., Thompson, G., Cole, C. D., Cotten, M., Cross, T. A., *et al.* (2001) Noncontact dipole effects on channel permeation. IV. Kinetic model of 5F-Trp(13) gramicidin A currents, *Biophys. J.*, **81**, 1245–1254

59. Mc Laughlin, S. (1989) The electrostatic properties of membranes, *Annu. Rev. Biophys. Biophys. Chem.*, **18**, 113–136.

60. Clarke, R. J. (2001) The dipole potential of phospholipid membranes and methods for its detection, *Adv. Colloid Interface Sci.*, 89-**90**, 263–281.

61. Wang, L., Bose, P. S., and Sigworth, F. J. (2006) Using cryo-EM to measure the dipole potential of a lipid membrane, *Proc. Nat. Acad. Sci. USA*, **103**, 18528–18533.

62. Demchenko, A. P., and Yesylevskyy, S. E. (2009) Nanoscopic description of biomembrane electrostatics: results of molecular dynamics simulations and fluorescence probing, *Chem. Phys. Lipids*, **160**, 63–84.

63. Yang, Y., Mayer, K. M., Wickremasinghe, N. S., and Hafner, J. H. (2008) Probing the lipid membrane dipole potential by atomic force microscopy, *Biophys. J.*, **95**, 5193–5199.

64. Smondyrev, A. M., and Berkowitz, M. L. (1999) Structure of dipalmitoylphosphatidylcholine/cholesterol bilayer at low and high cholesterol concentrations: molecular dynamics simulation, *Biophys. J.*, **77**, 2075–2089.

65. Mashl, R. J., Scott, H. L., Subramaniam, S., and Jakobsson, E. (2001) Molecular simulation of dioleylphosphatidylcholine bilayers at differing levels of hydration, *Biophys. J.*, **81**, 3005–3015.

66. Sachs, J. N., Crozier, P. S., and Woolf, T. B. (2004) Atomistic simulations of biologically realistic transmembrane potential gradients, *J. Chem. Phys.*, **121**, 10847–10851.

67. Davis, J. E., Rahaman, O., and Patel, S. (2009) Molecular dynamics simulations of a DMPC bilayer using nonadditive interaction models, *Biophys. J.*, **96**, 385–402.

68. Harder, E., MacKerell, A. D., and Roux, B. (2009) Many-body polarization effects and the membrane dipole potential, *J. Am. Chem. Soc.*, **131**, 27760–22761.

69. Bockmann, R. A., Hac, A., Heimburg, T., and Grubmuller, H. (2003) Effect of sodium chloride on a lipid bilayer, *Biophys. J.*, **85**, 1647–1655.

70. Gurtovenko, A. A., and Vattulainen, I. (2008) Effect of NaCl and KCl on phosphatidylcholine and phosphatidylethanolamine lipid membranes: insight from atomic-scale simulations for understanding salt-induced effects in the plasma membrane, *J. Phys. Chem. B*, **112**, 1953–1962.

71. Lee, S. J., Song, Y., and Baker, N. A. (2008) Molecular dynamics simulations of asymmetric NaCl and KCl solutions separated by phosphatidylcholine bilayers: potential drops and structural changes induced by strong Na+ lipid interactions and finite size effects, *Biophys. J.*, **94**, 3565–3576.

72. Xu, C., and Loew, L. M. (2003) The effect of asymmetric surface potentials on the intramembrane electric field measured with voltage-sensitive dyes, *Biophys. J.*, **84**, 2768–2780.

73. Eberhard, N., Sowers, A. E., and Jordan, C. A. (1989) *Electroporation and Electrofusion in Cell Lbiology*, Plenum Press, New York.

74. Nickoloff, J. A. (1995) *Animal Cell Electroporation and Electrofusion Protocols.* Humana Press, Totowa, NJ.

75. Weaver, J. C. (1995) Electroporation theory, in *Methods in Molecular Biology*, Humana Press, Clifton, NJ, pp. 3–29.

76. Zimmerman, U. (1996) The effect of high intensity electric field pulses on eukaryotic cell membranes: fundamentals and applications, in *Electromanipulation of Cells*, CRC Press, Boca Raton, FL, pp. 1–106.

77. Li, S. (2008) *Electroporation Protocols: Preclinical and Clinical Gene Medecine.* Humana Press, Totowa, NJ.

78. Tsong, T. Y. (1983) Voltage modulation of membrane permeability and energy utilization in cells, *Biosci. Rep.*, **3**, 487–505.

79. Tsong, T. Y. (1987) Electric modification of membrane permeability for drug loading into living cells, *Methods Enzymol.*, **149**, 248–259.

80. Prausnitz, M. R., Bose, V. G., Langer, R., and Weaver, J. C. (1993) Electroporation of mammalian skin: a mechanism to enhance transdermal drug delivery, *Proc. Natl. Acad. Sci. USA*, **90**, 10504–10508.

81. Suzuki, T., Shin, B. C., Fujikura, K., Matsuzaki, T., and Takata, K. (1998) Direct gene transfer into rat liver cells by in vivo electroporation, *FEBS Lett.*, **425**, 436–440.

82. Cemazar, M., and Sersa, G. (2007) Electrotransfer of therapeutic molecules into tissues, *Curr. Opin. Mol. Ther.*, **9**, 554–562.

83. Cemazar, M., Tamzali, Y., Sersa, G., Tozon, N., Mir, L. M., *et al.* (2008) Electrochemotherapy in veterinary oncology, *J. Vet. Int. Med.*, **22**, 826–831.

84. Teissié, J., Escoffre, J. M., Rols, M. P., and Golzio, M. (2008) Time dependence of electric field effects on cell membranes. A review for a critical selection of pulse duration for therapeutical applications, *Radiol. Oncol.*, **42**, 196–206.

85. Testori, A., J. Soteldo, A. Di Pietro, F. Verrecchia, M. Rastrelli, *et al.* (2008) The treatment of cutaneous and subcutaneous lesions with electrochemotherapy with bleomycin, *Eur. Dermatol.*, **3**, 1–3.

86. Bodles-Brakhop, A. M., Heller, R., and Draghia-Akli, R. (2009) Electroporation for the delivery of DNA-based vaccines and immunotherapeutics: current clinical developments, *Mol. Ther.*, **17**, 585–592.

87. Campana, L. G., Mocellin, S., Basso, M., Puccetti, O., De Salvo, G. L., *et al.* (2009) Bleomycin-based electrochemotherapy: clinical outcome from a single institution's experience with 52 Patients, *Ann. Surg. Oncol.*, **16**, 191–199.

88. Villemejane, J., and Mir, L. M. (2009) Physical methods of nucleic acid transfer: general concepts and applications, *Br. J. Pharmacol.*, **157**, 207–219.

89. Pucihar, G., Kotnik, T., Valic, B., and Miklavcic, D. (2006) Numerical determination of transmembrane voltage induced on irregularly shaped cells, *Ann. Biomed Eng.*, **34**, 642–652.

90. Weaver, J. C., and Chizmadzhev, Y. A. (1996) Theory of electroporation. A review, *Bioelectrochem. Bioenerg.*, **41**, 135–160.

91. Weaver, J. C. (2003) Electroporation of biological membranes from multicellular to nano scales, *IEEE Trans. Dielectr. Electr. Insul.*, **10**, 754–768.

92. Chen, C., Smye, S. W., Robinson, M. P., and Evans, J. A. (2006) Membrane electroporation theories: a review, *Med. Biol. Eng. Comput.*, **44**, 5–14.

93. Pucihar, G., Kotnik, T., Miklavcic, D., and J. T. (2008). Kinetics of transmembrane transport of small molecules into electropermeabilized cells, *Biophys. J.*, **95**, 2837–2848.

94. Chang, D. C. (1992) Structure and dynamics of electric field-induced membrane pores as revealed by rapid-freezing electron microscopy, in *Guide to Electroporation and Electrofusion* (ed. Chang, D. C., Chassy, B. M., Saunders, J. A., and Sowers, A. E.), Academic Press, Orlando, FL, pp. 9–27.

95. Tieleman, D. P. (2004) The molecular basis of electroporation, *BMC Biochem.*, **5**, 10.

96. Hu, Q., Viswanadham, S., Joshi, R. P., Schoenbach, K. H., Beebe, S. J., *et al.* (2005) Simulations of transient membrane behavior in cells subjected to a high-intensity ultrashort electric pulse, *Phys. Rev. E*, **71**, 031914.

97. Tarek, M. (2005) Membrane electroporation: a molecular dynamics simulation, *Biophys. J.*, **88**, 4045–4053.

98. Bockmann, R. A., de Groot, B. L., Kakorin, S., Neumann, E., and Grubmuller, H. (2008) Kinetics, statistics, and energetics of lipid membrane electroporation studied by molecular dynamics simulations, *Biophys. J.*, **95**, 1837–1850.

99. Ziegler, M. J., and Vernier, P. T. (2008) Interface water dynamics and porating electric fields for phospholipid bilayers, *J. Phys. Chem. B*, **112**, 13588–13596.

100. Roux, B. (2008) The membrane potential and its representation by a constant electric field in computer simulations, *Biophys. J.*, **95**, 4205–4216.

101. Golzio, M., Teissie, J., and Rols, M.-P. (2002) Direct visualization at the single-cell level of electriclly mediated gene delivery, *Proc. Natl. Acad. Sci. USA*, **99**, 1292–1297.

102. Delemotte, L., Dehez, F., Treptow, W., and Tarek, M. (2008) Modeling membranes under a transmembrane potential, *J. Phys. Chem. B*, **112**, 5547–5550.

103. Gurtovenko, A. A., and Vattulainen, I. (2005) Pore formation coupled to ion transport through lipid membranes as induced by transmembrane ionic charge imbalance: atomistic molecular dynamics study, *J. Am. Chem. Soc.*, **127**, 17570–17571.

104. Kandasamy, S. K., and Larson, R. G. (2006) Cation and anion transport through hydrophilic pores in lipid bilayers, *J. Chem. Phys.*, **125**, 074901.

105. Vernier, P. T., and Ziegler, M. J. (2007) Nanosecond field alignment of head group and water dipoles in electroporating phospholipid bilayers, *J. Phys. Chem. B*, **111**, 12993–12996.

106. Hodgkin, A. L., and Huxley, A. F. (1952) A quantitative description of membrane current and its application to conduction and excitation in nerve, *J. Physiol. (London)*, **117**, 500–544.

107. Hille, B. (1992) *Ionic Channels of Excitable Membranes*, 2nd ed., Sinauer, Sunderland, MA.

108. Choe, S. (2002) Potassium channel structures, *Nat. Rev. Neurosci..*, **3**, 115–121.

109. Doyle, D. A., Cabral, J. M., Pfuetzner, R. A., Kuo, A., Gulbis, J. M., *et al.* (1998) The structure of the potassium channel: molecular basis of K⁺ conduction and selectivity, *Science*, **280**, 69–77.

110. Lu, Z., Klem, A. M., and Ramu, Y. (2001) Ion conduction pore is conserved among potassium channels, *Nature*, **413**, 809–813.

111. Bezanilla, F. (2000) The voltage sensor in voltage-dependent ion channels, *Physiol. Rev.*, **80**, 555–592.

112. Horn, R. (2000) Conversation between voltage sensors and gates of ion channels, *Biochemistry*, **39**, 15653–15658.

113. Bezanilla, F. (2002) Voltage sensor movements, *J. Gen. Physiol.*, **120**, 465–473.

114. Fedida, D., and Hesketh, J. C. (2001) Gating of voltage-dependent potassium channels, *Prog. Biophys. Mol. Biol.*, **75**, 165–199.

115. Schoppa, N. E., and Sigworth, F. J. (1998) Activation of Shaker potassium channels. III. An activation gating model for wild-type and V2 mutant channel, *J. Gen. Physiol.*, **111**, 313–342.

116. Blaustein, R. O., and Miller, C. (2004) Shake, rattle or roll, *Nature*, **427**, 499–500.

117. Jiang, Y., Lee, A., Chen, J., Ruta, V., Cadene, M., *et al.* (2003) X-ray structure of a voltage-dependent K+ channel, *Nature*, **423**, 33–41.

118. Jiang, Y., Ruta, V., Chen, J., Lee, A., and MacKinnon, R. (2003) The principle of gating charge movement in a voltage-dependent K+ channel, *Nature*, **423**, 42–48.

119. Yang, N., George, A. L.J, and Horn, R. (1996) Molecular basis of charge movement in voltage-gated sodium channels, *Neuron*, **16**, 113–122.

120. Cha, A., and Bezanilla, F. (1997) Characterizing voltage-dependant conformational change in Shaker K channel with fluorescence, *Neuron*, **19**, 1127–1140.

121. Mannuzzu, L. M., and Isacoff, E. Y. (2000) Independence and cooperativity in rearrangements of a potassium channel voltage sensor revealed by single subunit fluorescence, *J. Gen. Physiol.*, **115**, 257–268.

122. Cha, A., Ruben, P. C., George, A. L., Fujimoto, E., and Bezanilla, F. (1999) Atomic scale movement of the voltage-sensing region in a potassium channel measured via spectroscopy, *Nature*, **402**, 813–817.

123. Asamoah, O. K., Wuskell, J. P., Loew, L. M., and Bezanilla, F. (2003) A fluorometric approach to local electric field measurements in a voltage-gated ion channel, *Neuron*, **37**, 85–97.

124. Bell, D., Yao, H., Saenger, R., Riley, J., and Sielgelbaum, S. (2004) Changes in local S4 environment provide a voltage-sensing mechanism for mammalian hyperpolarization-activated HCN channels, *J. Gen. Physiol.*, **123**, 5–19.

125. Chanda, B., Asamoah, O. K., Blunck, R., Roux, B., and Bezanilla, F. (2005) Gating charge displacement in voltage-gated ion channels involves limited transmembrane movement, *Nature*, **436**, 852–856.

126. Posson, D. J., Ge, P., Miller, C., Bezanilla, F., and Selvin, P. R. (2005) Small vertical movement of a K⁺ channel voltage sensor measured with luminescence energy transfer, *Nature*, **436**, 848–851.

127. Nonner, W., Peyser, A., Gillespie, D., and Eisenberg, B. (2004) Relating microscopic charge movement to macroscopic currents: the Ramo-Schockley Theorem applied to ion channels, *Biophys. J.*, **87**, 3716–3722.

128. Sigworth, F. J. (1994) Voltage gating of ion channels, *Q. Rev. Biophys.*, **27**, 1–40.

129. Kreusch, A., Pfaffinger, P. J., Stevens, C. F., and Choe, S. (1998) Crystal structure of the tetramerization domain of the *Shaker* potassium channel, *Nature*, **392**, 945–948.

130. Grabe, M., Lecar, H., Jan, Y. N., and Jan, L. Y. (2004) A quantitative assessment of models for voltage-dependent gating ion channels, *Proc. Natl. Acad. Sci. USA*, **101**, 17640–17645.

131. Starace, D. M., and Bezanilla, F. (2004) A proton pore in a potassium channel voltage sensor reveals a focused electric field, *Nature*, **427**, 548–553.

132. Starace, D. M., and Bezanilla, F. (2001) Histidine scanning mutagenesis of basic residues of the S4 segment of the shaker potassium channel, *J Gen Physiol.*, **117**, 469–490.

133. Treptow, W., Maigret, B., Chipot, C., and Tarek, M. (2004) Coupled motions between pore and voltage-sensor domains: a model for *Shaker* B, a voltage-gated potassium channel, *Biophys. J.*, **87**, 2365–2379.

134. Long, B. S., Campbell, E. B., and MacKinnon, R. (2005) Crystal structure of a mammalian voltage-dependent *Shaker* family K⁺ channel, *Science*, **309**, 897–903.

135. Freites, J. A., Tobias, D. J., and White, S. H. (2006) A voltage-sensor water pore, *Biophys. J.*, **91**, L90–L92.

136. Treptow, W., and Tarek, M. (2006) Environment of the gating charges in the Kv1.2 *Shaker* potassium channel, *Biophys. J.*, **90**, L64–L66.

137. Jogini, V., and Roux, B. (2007) Dynamics of the Kv1.2 voltage-gated K channel in a membrane environment, *Biophys. J.*, **93**, 3070–3082.

138. Cuello, L. G., Cortes, D. M., and Perozo, E. (2004) Molecular architecture of the KvAP voltage-dependent K⁺ channel in a lipid bilayer, *Science*, **306**, 491–495.

139. Aksimentiev, A., and Schulten, K. (2005) Imaging α–hemolysin with molecular dynamics: ionic conductance, osmotic permeability, and the electrostatic potential map, *Biophys. J.*, **88**, 3745–3761.

140. Bezanilla, F. (2005) The voltage-sensor structure in a voltage-gated ion channel, *Trends Biochem. Sci.*, **30**, 166–168.

141. Ahern, C. A., and Horn, R. (2005) Focused electric field across the voltage sensor of potassium channels, *Neuron*, **48**, 25–29.

142. Treptow, W., Tarek, M., and Klein, M. L. (2009) Initial response of the potassium channel voltage sensor to a transmembrane potential, *J. Am. Chem. Soc.*, **131**, 2107–2110.

143. Nishizawa, M., and Nishizawa, K. (2008) Molecular dynamics simulation of Kᵥ channel voltage sensor helix in a lipid membrane with applied electric field, *Biophys. J.*, **95**, 1729–1744.

144. Bjelkmar, P., Niemela, P. S., Vattulainen, I., and Lindahl, E. (2009) Conformational changes and slow dynamics through microsecond polarized atomistic molecular simulation of an Integral Kv1.2 ion channel, *PLoS Comput. Biol.*, **5**, e1000289.

145. Pathak, M. M., Yarov-Yarovoy, V., Agarwal, G., Roux, B., Barth, P., *et al.* (2007) Closing in on the resting state of the shaker K⁺ channel, *Neuron*, **56**, 124–140.

146. Barron, L. D., Hecht, L., and Wilson, G. (1997) The lubricant of life: a proposal that solvent water promotes extremely fast conformational fluctuations in mobile heteropolypeptide structure, *Biochemistry*, **36**, 13143–13147.

Chapter 9

WATER IN MEMBRANES, WATER IN CELLS: NEUTRONS REVEAL DYNAMICS AND INTERACTIONS

M. Jasnin,[a] **A. M. Stadler,**[b,c] **and G. Zaccai**[b]

[a] *Max Planck Institute of Biochemistry, Department of Molecular Structural Biology, Martinsried, Germany*
[b] *Institut Laue-Langevin, 6 rue Jules Horowitz, BP 156, Grenoble Cedex 9, France*
[c] *Forschungszentrum Jülich, Jülich, Germany*
jasnin@biochem.mpg.de

Neutron diffraction and neutron spectroscopy are particularly powerful experimental methods for the study of water in biology. They have contributed important data on the atomic-scale role played by water, not only in structuring membranes and soluble proteins, but also in permitting them to have the dynamical features necessary for biological activity. Neutron diffraction experiments localized the water in a natural intact membrane (the purple membrane [PM] of *Halobacterium salinarum* [*H. salinarum*]) under different conditions of temperature and hydration. They led to a hypothesis linking water, membrane protein dynamics, and activity, which was subsequently confirmed and refined by neutron spectroscopy. The relationship between hydration water and protein dynamics is well established from various *in vitro* experiments. In this context and following controversial reports about the dynamic state of cell water, it was important to measure, directly and *in vivo*, the atomic-scale dynamics of intracellular water. This was done by neutron scattering on *Escherichia coli* (*E. coli*) and human red blood cells (RBCs) The results clearly showed that intracellular water, other than in hydration shells, flows as freely as bulk water. In the case of RBCs, the neutron scattering experiments also showed how hemoglobin dynamical properties respond to body temperature. Moreover, the *E. coli in vivo* experiments provided a unique view of internal and global macromolecular motions within the cell environment as well as of solvent isotope effects on this dynamics. Macromolecular dynamics measured in *E. coli* and in RBCs clearly displayed significant differences with results from *in vitro* experiments on hydrated powders and solutions. Solvent isotope effects relate directly to the thermodynamics of macromolecular stabilization, in which water plays a crucial role. Comparison of results obtained from *in vivo* and *in vitro* experiments further supported the conclusion that intracellular water behaves similarly to bulk water.

9.1 BACKGROUND

Functional motions in biological macromolecules occur on various length scales, from a fraction of an Ångstrom unit to several nanometres, and on a range of time scales, from the femtosecond to the second. Proteins and nucleic acids are chain molecules, whose primary structures are determined by genes. They fulfill their specific biological activity by assuming secondary, tertiary, and quaternary structures under physiological conditions. These occur invariably *via* interactions with water molecules and other solvent components, in competition with interactions within the macromolecules themselves, with lipids in the case of membrane proteins, and sometimes with other macromolecules or smaller ligands. Folding energies are comparable with *kT*, making macromolecular structures very sensitive to temperature and solvent composition. The sensitivity is reflected in molecular dynamics; it is a necessary condition that a macromolecular structure be animated by appropriate motions in order to be biologically active. Water plays a *vital* role in macromolecular folding, stabilization, dynamics, and activity. Its physiological state and dynamics in membranes and cells are, therefore, of fundamental interest in the context of understanding biological function at the molecular level. Measurement and characterization, *in vivo*, of such molecular-scale structures and dynamics, however, is particularly challenging. Because of uniquely favorable properties of the methods, neutron diffraction and spectroscopy have been particularly powerful for such studies.

Neutrons are scattered with different *coherent* amplitudes by hydrogen (H) and its isotope deuterium (^{2}H or D), and by using standard crystallographic techniques H atoms and water molecules can be identified in complex structures by neutron diffraction using H-D labeling or water/*deuterium oxide* exchange, respectively. The *incoherent* scattering cross section of H is much larger than that of other atoms found in biological material and of D. In energy-resolved neutron scattering (spectroscopy), this strong incoherent scattering of H atoms is observed and analyzed. In the time scales examined, H atom motions reflect those of the chemical groups to which they are bound. They are, therefore, excellent indicators of internal and global dynamics in macromolecules, as well as of water dynamics. The incoherent scattering cross section of D or ^{2}H is much lower than that of H, so that H-D isotope labeling and water/deuterium oxide exchange provide powerful tools to focus on the dynamics of different components of a complex system.

The cytoplasm of all cells contains at least 70% of water [1, 2]. It is a crowded environment with the free distance between macromolecules in the order of Ångstrom units [3]. There has been much controversy about similarities and differences between the structure and dynamics of cytoplasmic and bulk water [4]. Membranes are often considered structures that exclude water, but the effects of hydration water around lipid head groups and on membrane surfaces are significant, especially with respect to membrane protein dynamics [5–7]. Hydration-shell water displays structural and dynamical differences with bulk water [8, 9]. Does the cytoplasm environment modify the properties of water beyond the immediate hydration shells of macromolecules? It is an important question because most biochemical studies are performed in fractionated systems in a dilute solution, with the assumption that the observations are significant for the corresponding cellular process. *In vitro* neutron spectroscopy on proteins in membranes [7, 10], in solution, and in hydrated powder samples [11, 12] established that protein dynamics is strongly hydration dependent. It was of paramount interest, therefore, to measure protein dynamics *in vivo*.

9.1.1 Experimental

Neutron diffraction experiments to characterize the water structure in the PM were performed on the D16 diffractometer at the Institut Laue-Langevin (ILL), Grenoble (http://www.ill.eu/instruments-support/instruments-groups/instruments/d16/) The instrument had been built and optimized for the measurements and became a model for *membrane diffractometers* at other neutron-scattering centers.

Energy-resolved neutron-scattering experiments provide information on both the time scale and amplitudes of atomic fluctuations. The energy resolution of a spectrometer defines the time scale, while its scattering vector range defines the observable length scale for amplitude determination. In experiments on membrane and cell water dynamics, energy-resolved neutron-scattering data were measured on spectrometers with various energy resolution values, at various neutron-scattering centers in Europe. The corresponding time scales covered the range from that of bulk water to that of internal macromolecular motions and reduced mobility interfacial water motions.

H-D isotope labeling was performed by water/deuterium oxide exchange in cell culture media and by *in vivo* adaptation to the D of bacterial cultures at the D-LAB at ILL (http://www.ill.fr/YellowBook/deuteration/).

9.1.2 Membrane Water

The PM of *H. salinarum* is currently the natural membrane whose structure and dynamics are best characterized. It functions as a light-driven proton pump and is composed of specific lipids, the retinal binding protein, bacteriorhodopsin (BR), and water molecules all organized on a well-ordered two-dimensional lattice. A photocycle in which the retinal binding protein goes through a series of different absorption states is associated with proton pump activity [13]. Neutron diffraction experiments localized the water in the membrane under different conditions of temperature and hydration. They led to a hypothesis linking water, membrane protein dynamics, and activity, which was subsequently confirmed and refined by neutron spectroscopy. The unit cell dimension of the BR lattice in PM decreases by the same amount (2%) upon drying the membranes at room temperature as when they are cooled to liquid-nitrogen temperatures. Neutron diffraction experiments with water/ deuterium oxide exchange, however, show that whereas in dry membranes the lipid head groups are dehydrated and the decrease in dimension is due to a smaller area occupied by the lipid molecules, the water of hydration remains

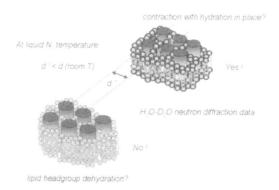

Figure 9.1. A schematic illustration of the results of Zaccai (1987). The decrease in crystallographic cell dimension is similar for dehydration of the membranes at room temperature and for when the membranes are cooled to liquid-nitrogen temperature. The neutron diffraction results, from experiments using H_2O/D_2O exchange, showed unambiguously that cooled membranes were not dehydrated but had contracted with the head group hydration still in place. For color reference, see p. 371.

in place in cooled membranes and the decrease in dimension is due to thermal contraction only (Fig. 9.1) The data suggested a hypothesis that functional BR, in the wet state at room temperature, has a relatively soft environment that would allow large-amplitude motions of the protein; in dry membranes at room temperature (which are inactive), the amplitudes of protein motions would be inhibited by a more close-packed environment (Fig. 9.2) as they are reduced, due to thermal contraction, in cold membranes.

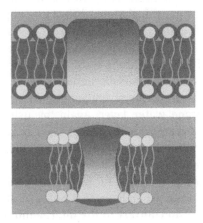

Figure 9.2 A side view of BR in the membrane, showing how dehydration of lipid head groups could lead to tightening of the protein environment, which could interfere with its internal dynamics. For color reference, see p. 372.

The hydration-temperature-dynamics hypothesis for PM was verified in the first neutron spectroscopy experiments on a natural membrane [10]. Small-amplitude atomic fluctuations (mean square displacements [MSDs]) in the 100 ps time scale were plotted against temperature for samples of dry and hydrated PM, respectively (such plots are called *elastic temperature scans*) In the time-length window examined, the dry PM MSDs increased linearly with temperature for the entire range, as would be expected for a harmonic solid. The wet PM MSDs, on the other hand, displayed a *dynamical transition* at about 200 K (−73°C) above which they increased more steeply with temperature, indicating the onset of a "softer" membrane state [14], which is apparently necessary for BR photocycle activity. A similar dynamical transition had been observed in myoglobin [15]; the larger amplitudes of fluctuations above the transition temperature are necessary for ligand access to the heme pocket.

PM samples are made up of stacks of regular spacing in which membranes and water layers alternate (Fig. 9.3a) By using elastic temperature scans, specific deuteration, and molecular dynamics simulations, Wood *et al.*

examined the correlation between water layer and membrane dynamics [7]. Examining completely deuterated PM, hydrated in water, allowed the direct experimental exploration of water dynamics. The study of natural abundance PM in deuterium oxide focused on membrane dynamics. They concluded from elastic temperature scans that ps–ns hydration water dynamics are not directly coupled to membrane motions on the same time scale at temperatures less than 260 K. In agreement with the experimental results, molecular dynamics simulations of hydrated PM in the temperature range of 100–296 K revealed an onset of hydration water translational diffusion at approximately 200 K but no transition in the PM at the same temperature. The results suggested that, in contrast to soluble proteins [12], the dynamics of the membrane protein is not controlled by that of intermembrane water layer at temperatures less than 260 K. And indeed, the experiments discussed in Figs. 9.1 and 9.2 indicate that lipid dynamics may have a stronger impact on membrane protein dynamics than the intermembrane water layer. Again, by using specific deuterium labeling, the lipid and membrane dynamics were measured separately and plotted as elastic temperature scans [16]. The plot in Fig. 9.3b shows that the lipid and the membrane display dynamical transitions at about 250 K. Since lipids require head group hydration for increased mobility, a corollary of the result is that the lipid head group hydration water (dark blue in Fig. 9.2) also "melts" at about 250 K, in contrast to the intermembrane water, which was shown to become mobile at about 200 K.

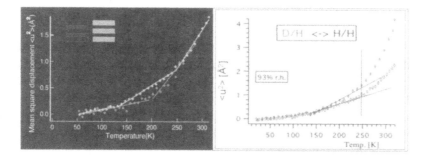

Figure 9.3 (a) Elastic temperature scans of the intermembrane water layer (red) and the membrane (green) obtained by specific H/D labeling of the components. Note that dynamical transitions occur at different temperatures for the red and green lines: ~200 K for the water and ~250 K for the membrane (the inflection in the membrane data at ~150 K is due to methyl group rotations) [7]. (b) Elastic temperature scans of the lipid (red) and membrane (black) fractions in PM, obtained by H/D labeling and membrane reconstitution [16]. The inflection in both plots at ~150 K is due to methyl group rotations (the density of methyl groups is higher in the lipid than in the membrane fraction) Note that dynamical transitions occur at the same temperature, ~250 K, for the membrane and lipid fractions, suggesting that they are coupled. For color reference, see p. 372.

9.1.3 Intracellular Water

There are claims that the cell interior is a gel or colloidal-like structure in which confined water is interfacial with properties that are significantly different at the molecular level from those of pure water [17]. These claims were derived from indirect experimental observations and remain controversial [4]. In this context, we note that a recent neutron spectroscopy study showed that water diffusion in clay gels was similar to bulk water [18]. The apparent diffusion coefficient of water in various biological tissues, including the brain, has been measured by *nuclear magnetic resonance* (NMR) methods. On a micrometric scale, water diffusion appears to be reduced by a factor of between 2 and 10 compared to pure water. The reduction could be explained partly by tortuosity effects, macromolecular crowding, and confinement effects [19].

On the atomic length and time scales, water diffusive motions were measured by neutron spectroscopy and specific H/D labeling, *in vivo*, in *E. coli* [20], in a halophilic archaeon from the Dead Sea [21], and in RBCs [22]. The observed length and time scale range covered that of motions from interfacial to bulk water. The translational and rotational diffusion of the major water fraction in the cytoplasms of *E. coli* and RBCs was found to be similar to that of bulk water. In contrast, a large fraction of water with a slowed-down translational diffusion was found in the extreme halophile.

The data from RBCs indicated two dynamics populations of water: a major fraction (~90%) has dynamical properties similar to those of bulk water (time scale of a few picoseconds), and a minor fraction (~10%) was interpreted as bound hydration water with significantly slower dynamics (time scale slower than ~40 ps) [22]. The temperature dependence of the hydration fraction followed Arrhenius behavior with an activation energy of 2.15 ± 0.10 kcal/mol. Hydrodynamic experiments on hemoglobin required the assumption of bound water coverage of around 50% of the first hydration layer around the protein in order to predict quantitatively the measured parameters [23, 24]. From the known hemoglobin concentration inside RBCs, Stadler *et al.* calculated that 50% of the hydration layers do in fact correspond to about 10% of the total cytoplasmic water, in agreement, with the neutron data analysis [22].

The neutron results on *E. coli* and RBCs established firmly that water diffusion, beyond the narrow hydration shells around macromolecules, is neither confined nor significantly slowed down compared with pure water. Contrary to the widespread belief that water is "tamed" by macromolecular confinement, cell water diffusion beyond the hydration shell is similar to that of pure water at physiological temperature. The same conclusion has been reached using NMR spectroscopy by Persson and Halle [25]. They examined the rotational spin relaxation rate of *E. coli* water, over a wide time range, from the millisecond to the picosecond time scale, and found that about 85% of *E. coli* water presented rotational relaxation times similar to those of pure water.

The behavior of water in the Dead Sea archaeon (*Haloarcula marismortui* [*H. marismortui*]) was found to be significantly different from that in *E. coli* and RBCs [21]. *In vivo* neutron spectroscopy on deuterated cells in water, measured in the time scale from the picosecond to the nanosecond, showed that a significant part of cell water presented a translational diffusion slowed down by two orders of magnitude compared with pure water. NMR work, on the other hand, established that the rotational spin relaxation rate of *H. marismortui* water was very close to that found in pure water [25]. These results can be interpreted in terms of the known exceptional salt and water-binding interactions of halophilic proteins and the almost saturated salt environment in the cell cytoplasm [26]. Water molecules involved in carboxyl group interactions with hydrated potassium ions would have their translational diffusion affected but not their ability to rotate by H-bond formation and reformation.

9.1.4 Intracellular Macromolecular Dynamics

Internal macromolecular dynamics occurs on the picosecond to nanosecond time scale; the dynamics lubricates larger conformational changes on slower, millisecond time scales, which are necessary for important biological processes, such as enzyme catalysis and intermolecular recognition [27, 28] So-called quasi-elastic incoherent neutron scattering (QENS) provides information on internal diffusive motions in macromolecules [29]. The technique has been used to study internal molecular motions in hydrated protein powders [15, 30–36] and in the integral membrane protein BR in PM stacks [37–41]. A QENS study of *E. coli*, using specific H/D labeling, provided a dynamical mapping of macromolecular dynamics *in vivo* [42]. The average global macromolecular diffusion in the bacterial cytoplasm was found to be similar to that of hemoglobin in RBCs [43]. Comparison with previous *in vitro* work revealed significant differences between physiological internal motions and motions extracted from experiments on hydrated powders and in solution. In the picosecond time domain, both internal molecular flexibility and diffusion rates are increased in the cell interior compared to parameters extracted from fully hydrated powders. The result showed that the large amount of cell water contributes to fast internal macromolecular motions. In contrast, picosecond internal flexibility measured in *E. coli* was reduced significantly, compared with that measured in solution, suggesting that weak forces due to the vicinity of macromolecules may attenuate the lubricating effect of cell water.

The dynamics of hemoglobin in whole human RBCs was measured by QENS and compared to that of hemoglobin-hydrated powders and solutions [44]. Global protein diffusion in the cells could be separated from internal hemoglobin motions in the data and was analyzed by using a model based on diffusion at infinite dilution (measured by dynamic light scattering) and interactions in a concentrated suspension of noncharged hard-sphere colloids. The value found for the hemoglobin diffusion coefficients *in vivo* was similar to that expected from the model calculation for short-time self-diffusion and larger than the long-time self-diffusion coefficient reported by Doster and Longeville [43], as expected. The geometry of internal protein motions was also extracted from the data and interpreted by using the model for "diffusion in a sphere" proposed by Volino and Dianoux [45] with a Gaussian distribution of sphere radii. An interesting transition in motion geometry for hemoglobin internal dynamics was found at human body temperature from the measurements on human RBCs (Fig. 9.4). Similar to the *E. coli* results, it was revealed that hemoglobin dynamics under physiological conditions in RBCs was not accurately represented by *in vitro* sample preparations, such as hydrated powders.

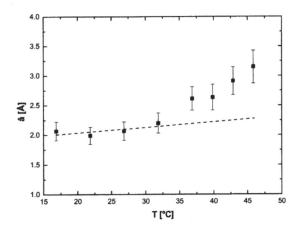

Figure 9.4 Average sphere size of internal hemoglobin motions in human RBCs as a function of temperature [44]. The dashed line indicates low-temperature linear behavior. At ~37°C, which is close to human body temperature, a break in the dynamics is visible, which was interpreted as due to partial unfolding of hemoglobin. Hemoglobin appears to be in a softer state above body temperature than at temperatures below.

Solvent interactions are essential for macromolecular stability, and protein folding is the result of a delicate balance between hydration, intramolecular interactions, ligand binding, etc., as discussed in section 9.1. The difference between water and deuterium oxide molecules and their H-bonding properties is known to influence macromolecular stabilization and dynamics. The slope of an elastic temperature scan is proportional to the apparent "softness" of the structure, or inversely proportional to its *resilience* [14]. It has been shown by neutron-scattering experiments that deuterium oxide increased the resilience of a halophilic protein but reduced it (made it appear softer) in the case of mesophilic bovine serum albumin (BSA) [46]. Deuterium oxide increases protein stability in both cases. The result then indicated that stabilization is entropy driven in the case of BSA (the apparent increased softness being due to increased sampling of conformational substates) and enthalpy driven in the case of the halophile (stronger hydration interactions leading to higher resilience). Jasnin *et al.* explored the water/deuterium oxide solvent isotope effect on the average macromolecular dynamics in the *E. coli* interior [47]. Measurements were performed on living *E. coli* containing water and deuterium oxide, respectively, close to physiological conditions of temperature. Flexibility (as expressed in the MSD) and resilience were both found to be smaller in deuterium oxide. The smaller resilience value suggested larger entropy content in the case of deuterium oxide, as for BSA *in vitro*. The solvent isotope effect is expected to depend critically on the state of water, and the agreement between the *in vivo* and *in vitro* results on mesophile stabilization is a further indication that the state of water in the cell is similar to that of bulk water.

References

1. Minton, A. P. (2001) The influence of macromolecular crowding and macromolecular confinement on biochemical reactions in physiological media, *J. Biol. Chem.*, **276**, 10577–10580.

2. Hall, D., and Minton, A. P. (2003) Macromolecular crowding: qualitative and semiquantitive successes, quantitative challenges, *Biochim. Biophys. Acta*, **1649**, 127–139.

3. Krueger, S., and Nossal, R. (1998) SANS studies of interacting hemoglobin in intact erythrocytes, *Biophys. J.*, **53**, 97–105.

4. Ball, P. (2008) Water as an active constitutent in cell biology, *Chem. Rev.*, **108**, 74–108.

5. Zaccai, G. and Gilmore, D. J. (1979) Areas of hydration in the purple membrane of *Halobacterium halobium*: a neutron diffraction study, *J. Mol. Biol.*, **132**, 181–191.

6. Zaccai, G. (1987) Structure and hydration of purple membranes in different conditions, *J. Mol. Biol.*, **194**, 569–572.

7. Wood, K., Plazanet, M., Gabel, F., Kessler, B., Oesterhelt, D., Tobias, D. J., Zaccai, G., and Weik, M. (2007) Coupling of protein and hydration-water dynamics in biological membranes, *Proc. Natl. Acad. Sci. USA*, **104**, 18049–18054.

8. Svergun, D. I., Richard, S., Koch, M. H., Sayers, Z., Kuprin, S. and Zaccai, G. (1998) Protein hydration in solution: experimental observation by X-ray and neutron scattering, *Proc. Natl. Acad. Sci. USA*, **95**, 2267–2272.

9. Bellissent-Funel, M.-C. (2001) Structure of confined water, *J. Phys. Condens. Matter*, **13**, 9165–9177.

10. Ferrand, M., Dianoux, A. J., Petry, W. and Zaccai, G. (1993) Thermal motions and function of bacteriorhodopsin in purple membranes: effects of temperature and hydration studied by neutron scattering, *Proc. Natl. Acad. Sci. USA*, **90**, 9668–9672.

11. Stadler, A. M., Digel, I., Embs, J. P., Unruh, T., Tehei, M., Zaccai, G., Büldt, G., and Artmann, G. M. (2009) From powder to solution: hydration dependence of human hemoglobin dynamics correlated to body temperature, *Biophys. J.*, **96**, 5073–5081.

12. Wood, K., Frölich, A., Paciaroni, A., Moulin, M., Härtlein, M., Zaccai, G., Tobias, D. J., and Weik, M. (2008) Coincidence of dynamical transitions in a soluble protein and its hydration water: direct measurements by neutron scattering and MD simulations, *J. Am. Chem. Soc.*, **130**, 4586–4587.

13. Subramaniam, S., Lindahl, M., Bullough, P., Faruqi, A. R., Tittor, J., Oesterhelt, D., Brown, L., Lanyi, J. and Henderson, R. (1999) Protein conformational changes in the bacteriorhodopsin photocycle, *J. Mol. Biol.*, **287**, 145–161.

14. Zaccai, G. (2000) How soft is a protein? A protein dynamics force constant measured by neutron scattering, *Science*, **288**, 1604–1607.

15. Doster, W., Cusack, S. and Petry, W. (1989) Dynamical transition of myoglobin revealed by inelastic neutron scattering, *Nature*, **337**, 754–756.

16. Frölich, A., Gabel, F., Jasnin, M., Lehnert, U., Oesterhelt, D., Stadler, A. M., Tehei, M., Weik, M., Wood, K. and Zaccai, G. (2009) From shell to cell: neutron scattering studies of biological water dynamics and coupling to activity, *Faraday Discuss.*, **141**, 117–130; discussion 175–207.

17. Pollack, G. H., Cameron, I. L., and Wheatley, D. N. (2006) *Water and the Cell*, Springer, Dordrecht, the Netherlands.

18. Seydel, T., Wiegart, L., Juranyi, F., Struth, B., Schober, H (2008) Unaffected microscopic dynamics of macroscopically arrested water in dilute clay gels, *Phys. Rev. E.*, **78**, 061403.

19. Le Bihan, D. (2007) The "wet mind": water and functional neuroimaging, *Phys. Med. Biol.*, **52**, R57–R90.

20. Jasnin, M., Moulin, M., Haertlein, M., Zaccai, G., and Tehei, M. (2008a) Down to atomic-scale intracellular water dynamics, *EMBO Rep.*, **9**, 543–547.

21. Tehei, M., Franzetti, B., Wood, K., Gabel, F., Fabiani, E., Jasnin, M., Zamponi, M., Oesterhelt, D., Zaccai, G., Ginzburg, M., and Ginzburg, B. Z. (2007) Neutron scattering reveals extremely slow cell water in a Dead Sea organism, *Proc. Natl. Acad. Sci. USA*, **104**, 766–771.

22. Stadler, A. M., Embs, J. P., Digel, I., Artmann, G. M., Unruh, T., Büldt, G., and Zaccai, G. (2008a) Cytoplasmic water and hydration layer dynamics in human red blood cells, *J. Am. Chem. Soc.*, **130**, 16852–16853.

23. Garcia de la Torre, J. (2001) Hydration from hydrodynamics. General considerations and applications of bead modelling to globular proteins, *Biophys. Chem.*, **93**, 159–170.

24. Garcia de la Torre, J., Huertas, M. L., and Carrasco, B. (2000) Calculation of hydrodynamic properties of globular proteins from their atomic-level structure, *Biophys. J.*, **78**, 719–730.

25. Persson, E., and Halle, B. (2008) Cell water dynamics on multiple time scales, *Proc. Natl. Acad. Sci. USA*, **105**, 6266–6271.

26. Madern, D., Ebel, C., and Zaccai, G. (2000) Halophilic adaptation of enzymes, *Extremophiles*, **4**, 91–98.

27. Jimenez, R., Salazar, G., Yin, J., Joo, T., and Romesberg, F. E. (2004) Protein dynamics and the immunological evolution of molecular recognition, *Proc. Natl. Acad. Sci. USA*, **101**, 3803–3808.

28. Tousignant, A., and Pelletier, J. N. (2004) Protein motions promote catalysis, *Chem. Biol.*, **11**, 1037–1042.

29. Bée, M. (1988) *Quasielastic Neutron Scattering: Principles and Applications in Solid State Chemistry, Biology and Materials Science*, Adam Hilger, Philadelphia.

30. Andreani, C., Filabozzi, A., Menzinger, F., Desideri, A., Deriu, A., and Di Cola, D. (1995) Dynamics of hydrogen atoms in superoxide dismutase by quasielastic neutron scattering, *Biophys. J.*, **68**, 2519–2523.

31. Dellerue, S., Petrescu, A. J., Smith, J. C., and Bellissent-Funel, M.-C. (2001) Radially softening diffusive motions in a globular protein, *Biophys. J.*, **81**, 1666–1676.

32. Fitter, J. (1999) The temperature dependence of internal molecular motions in hydrated and dry α-amylase: the role of hydration water in the dynamical transition of proteins, *Biophys. J.*, **76**, 1034–1042.

33. Paciaroni, A., Orecchini, A., Cinelli, S., Onori, G., Lechner, R. E., and Pieper, J. (2003) Protein dynamics on the picosecond timescale as affected by the environment: a quasielastic neutron scattering study, *Chem. Phys.*, **292**, 397–404.

34. Pérez, J., Zanotti, J. M., and Durand, D. (1999) Evolution of the internal dynamics of two globular proteins from dry powder to solution, *Biophys. J.*, **77**, 454–469.

35. Roh, J. H., Curtis, J. E., Azzam, S., Novikov, V. N., Peral, I., Chowdhuri, Z., Gregory, R. B., and Sokolov, A. P. (2006) Influence of hydration on the dynamics of lysozyme, *Biophys. J.*, **91**, 2573–2588.

36. Zanotti, J.-M., Bellissent-Funel, M.-C., and Parello, J. (1997) Dynamics of a globular protein as studied by neutron scattering and solid-state NMR, *Phys. B*, **234–236**, 228–230.

37. Fitter, J., Ernst, O. P., Hauß, T., Lechner, R. E., Hofmann, K. P., and Dencher, N. A. (1998) Molecular motions and hydration of purple membranes and disk membranes studied by neutron scattering, *Eur. Biophys. J.*, **27**, 638–645.

38. Fitter, J., Lechner, R. E., and Dencher, N. A. (1997) Picosecond molecular motions in bacteriorhodopsin from neutron scattering, *Biophys. J.*, **73**, 2126–2137.

39. Fitter, J., Lechner, R. E., Büldt, G., and Dencher, N. A. (1996) Internal molecular motions of bacteriorhodopsin: hydration-induced flexibility studied by quasielastic incoherent neutron scattering using oriented purple membranes, *Proc. Natl. Acad. Sci. USA*, **93**, 7600–7605.

40. Fitter, J., Lechner, R. E., Büldt, G., and Dencher, N. A. (1996) Temperature dependence of molecular motions in the membrane protein bacteriorhodopsin from QINS, *Phys. B*, **226**, 61–65.

41. Fitter, J., Lechner, R. E., and Dencher, N. A. (1999) Interactions of hydration water and biological membranes studied by neutron scattering, *J. Phys. Chem. B*, **103**, 8036–8050.

42. Jasnin, M., Moulin, M., Haertlein, M., Zaccai, G., and Tehei, M. (2008) *In vivo* measurement of internal and global macromolecular motions in *Escherichia coli*, *Biophys. J.*, **95**, 857–864.

43. Doster, W. and Longeville, S. (2007) Microscopic diffusion and hydrodynamic interactions of hemoglobin in red blood cells, *Biophys. J.*, **93**, 1360–1368.

44. Stadler, A. M., Digel, I., Artmann, G. M., Embs, J. P., Zaccai, G., and Büldt, G. (2008) Hemoglobin dynamics in red blood cells: correlation to body temperature, *Biophys. J.*, **95**, 5449–5461.

45. Volino, F., and Dianoux, A. J. (1980) Neutron incoherent-scattering law for diffusion in a potential of spherical symmetry: general formalism and application to diffusion inside a sphere, *Mol. Phys.*, **41**, 271–279.

46. Tehei, M., Madern, D., Pfister, C., and Zaccai, G. (2001) Fast dynamics of halophilic malate dehydrogenase and BSA measured by neutron scattering under various solvent conditions influencing protein stability, *Proc. Natl. Acad. Sci. USA*, **98**, 14356–14361.

47. Jasnin, M., Moulin, M., Haertlein, M., Zaccai, G., and Tehei, M. (2008) Solvent isotope effect on macromolecular dynamics in *E. coli*, *Eur. Biophys. J.*, **37**, 613–617.

Chapter 10

STRUCTURE AND INHIBITOR OF WATER CHANNEL IN BRAIN

Yukihiro Tanimura, Kazutoshi Tani, Hiroshi Suzuki, Kouki Nishikawa, Akiko Kamegawa, Yoko Hiroaki, and Yoshinori Fujiyoshi

Department of Biophysics, Faculty of Science, Kyoto University,
Oiwake, Kitashirakawa, Sakyo-ku, Kyoto 606-8502, Japan
yoshi@em.biophys.kyoto-u.ac.jp

Water channel aquaporin-4 (AQP4) is predominantly expressed in glial cells and has been implicated in brain edema resulting from water intoxication, brain ischemia, or meningitis. An AQP4-specific inhibitor suitable for clinical use would thus be expected to help protect against brain edema. We confirmed acetazolamide (AZA) specifically inhibits water permeation of the rat AQP4 (rAQP4) channel, by measurement of water conduction through AQP4-reconstituted liposomes. The structural study of this channel is also important for understanding complex functions of the water channel in the brain. The *aqp4* gene is known to express two different isoforms, and only the shorter isoforms assemble into orthogonal arrays, while the longer isoforms interfere with array formation. Two-dimensional (2D) crystals of AQP4 were revealed to be the same AQP4 packing as orthogonal arrays observed *in vivo*. By sodium dodecyl sulfate (SDS)-digested freeze-fracture replica labeling (FRL), palmitoylation of the N-terminal cysteine was also identified to be involved in array inhibition of the longer isoform. The structure of the double-layered 2D crystal of AQP4 suggested weak but specific interactions between tetramers in adjoining membranes, which may provide us insights into an unknown and unexpected role for AQP4 in the adhesion of membrane layers. Knowledge about multiplicity of AQP4 observed in structural studies will provide an important building block for understanding the functional complexity in our brain, while these functions remain to be elucidated.

Water: The Forgotten Biological Molecule
Edited by Denis Le Bihan and Hidenao Fukuyama
Copyright © 2011 by Pan Stanford Publishing Pte. Ltd.
www.panstanford.com

10.1 INTRODUCTION

Water-permeable channels are named AQPs, which provide the main route for water movement across the membranes in many cells [1, 2]. The AQP family proteins fall into two subfamilies, aquaporins and aquaglyceroporins. The AQPs are specifically selective for water, whereas aquaglyceroporins can also conduct some other small, neutral solutes such as glycerol. Some of their structures have been analyzed, revealing the structural determinants for their functions [3–8]. The AQP monomer consists of six membrane-spanning α-helices that assume an unusual fold, named the AQP fold [3]. The monomers assemble into a homotetramer, with each subunit forming an independent pore that can conduct water in both directions. Thirteen AQPs have been identified in humans, and more than 300 AQPs have so far been identified in other organisms, ranging from plants to animals and procaryotes to eucaryotes. AQP4 [9] is the predominant water channel in the mammalian brain [10], where it is known to be mainly expressed in glial cells. Its distribution pattern suggests that it controls water fluxes in the brain and that it plays an essential role in brain homeostasis [11]. Specifically, AQP4 is expressed in the end feet of astrocytes, in glial lamellae of the hypothalamus, and in ependymal cells and subependymal astrocytes [12, 13]. Interestingly, characteristic orthogonal arrays were observed in the end feet of astrocytes [14] and were later found to be formed by AQP4 [15]. The size of the arrays depends on the expression ratio of the two AQP4-splicing variants named AQP4M1 and AQP4M23 [16]. The structure of AQP4M23 was analyzed by electron crystallography and revealed the mechanism of orthogonal array formation [7]. The mechanism of array disruption by the longer isoform AQP4M1 could, however, not be accurately deduced from the high-resolution structure of AQP4M23. Subsequent freeze-fracture studies combined with chemical analyses eventually revealed that the disruption of AQP4 arrays by AQP4M1 could be attributed to N-terminal palmitoylation of either Cys13 or Cys17, residues that are not present in AQP4M23 [17].

The structural analysis of AQP4 revealed another important feature of the channel, its cell-adhesive function. This function was confirmed by experiment that L-cells without endogenous cell-adhesive molecules start to form clusters upon expressing AQP4 [7]. A recent study suggested that AQP4 is related to higher-order brain functions [18], but the molecular mechanism for this function remains to be elucidated. The study of AQP4 knockout mice indicated that AQP4 deletion reduces cytotoxic brain edema resulting from water intoxication, brain ischemia, or meningitis [19, 20]. Therefore, AQP4 inhibitors suitable for clinical use would be expected to help protect against cytotoxic brain edema.

Considering the importance of AQP4 in the pathophysiology of numerous clinical conditions, specific and reversible AQP4 inhibitors are highly desirable. Mercury, silver, and gold have been shown to inhibit several AQPs [21, 22], but these metals are not specific for AQPs and are highly toxic to living cells. AZA, a sulfonamide carbonic anhydrase (CA) inhibitor, has recently been reported to inhibit AQP1-mediated water permeability in oocytes [23] as well as in mammalian cells [24]. AZA has also been reported to inhibit AQP4-mediated water permeability in oocytes [25]. In conflicting studies, AZA was suggested to show no inhibition of water conduction by AQP1 or AQP4 [26, 27] and was also reported to have no effect on water permeability of oocytes expressing AQP1 [28]. Because AZA is an organic molecule and can be chemically modified, it is a potential lead compound for the development of specific and selective inhibitors for AQP4. It is therefore important to fully characterize the effect of AZA on AQP-mediated water conduction.

We reconstituted purified recombinant rAQP4 and human AQP1 (hAQP1) into lipid vesicles and used the proteoliposomes in conjunction with stopped-flow light-scattering measurements to study the effect of AZA on the water permeability of these two AQPs [29]. To establish the efficacy of AZA, we also tested two additional compounds, methazolamide (MZA) and valproic acid (VPA). MZA belongs to the same family of sulfonamide CA inhibitors as AZA, and their chemical structures are very similar. The chemical structure of VPA differs from those of AZA and MZA, but it was reported to also inhibit AQP4-mediated water transport in oocytes [30]. Since our studies were performed with pure proteins reconstituted into liposomes, our results should be more accurate than those previously obtained with less well-defined systems such as oocytes and mammalian cells.

The importance of AQP4 as the predominant water channel in the brain and its propensity to form ordered arrays encouraged us to study the structure of AQP4 by electron crystallography, the technique used to solve the first AQP structure of AQP1 [3]. On the basis of the structure of AQP1, we proposed the hydrogen bond (H-bond) isolation mechanism, which explains how AQPs block proton permeation. Since the accuracy of permeability measurement in a water channel is poor than that in an ion channel, even a simple result that the water permeation speed of AQP4 is as fast as that of AQP1 had been controversial. On the basis of the structural determination of AQP1, the H-bond isolation mechanism was proposed. After then, some other mechanisms were proposed to explain proton exclusion in the water channel, namely, *global orientation tuning* theory [31] and the *desolvation effect of ions* model [32–34]. Global orientation tuning theory poses that water molecules in the cytoplasmic and extracellular halves of the channel adopt opposite orientations and that this bipolar orientation of water molecules impedes proton conduction. On the other hand, the desolvation effect argues that proton permeation is prevented by the electrostatic field created by the dipole moment of the two short pore helices. Recent molecular dynamics simulations state that proton exclusion is dependent mostly on the desolvation effect and less on the arginine (Arg) residue of the ar/R site or on the asparagine (Asn) residues of the asparagine-proline-alanine (NPA) motifs [35, 36]. The mechanism of proton exclusion keeping fast water permeation is thus still a major subject to be discussed on the basis of structural analysis of water channel molecules in a membrane layer. Here we show that the structural view of AQP4 at high resolution is indispensable for entering a deep understanding of the functions of AQP4 in the brain.

10.2 EXPRESSION, PURIFICATION, AND RECONSTITUTION OF WATER CHANNELS INTO LIPOSOMES

Expression of histidine (His)-tagged rAQP4 in Sf9 insect cells yielded more than 3 mg of pure protein from 1 L of cell culture [7]. His-tagged hAQP1 expressed equally well in insect cells and also produced yields of more than 3 mg of pure protein from 1 L of cell culture. The water channels were solubilized in n-octyl-β,D-glucopyranoside (OG), purified by nickel affinity chromatography and reconstituted into liposomes using the rapid detergent dilution method [37]. The proper integration of AQPs into the liposomes was verified by sucrose-gradient centrifugation [38]. Liposomes were prepared in the same way but without AQP.

10.3 MEASUREMENT OF WATER CHANNEL INHIBITION BY AZA, MZA, AND VPA

The chemical structures of the two sulfonamide CA inhibitors, AZA and MZA, and VPA are shown in Fig. 10.1a. Stopped-flow measurements were carried out on proteoliposomes in the presence and absence of one of the three compounds, namely, AZA, MZA, and VPA. As control, parallel experiments were also carried out with empty liposomes. Vesicles were subjected to a 100 mM inwardly directed sucrose gradient at 22°C. The kinetics of vesicle shrinkage were measured by recording over time the change in intensity of 90° scattered light at a wavelength of 450 nm. Osmotic water permeability coefficients (P_f) were calculated as described [29]. While water permeation is in general difficult to measure precisely, our measurements were reproducible with only a small standard deviation.

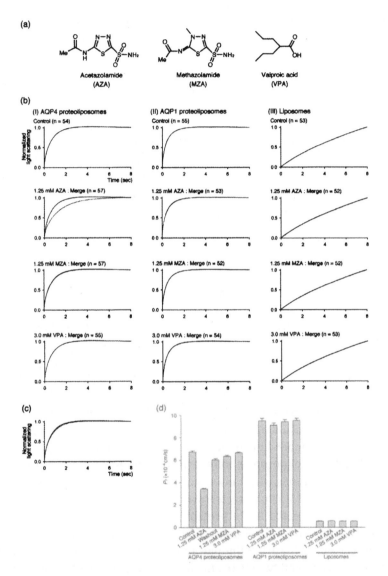

Figure 10.1 Water permeability of AQP proteoliposomes and effect of AZA, MZA, and VPA. (a) Chemical structures of AZA, MZA, and VPA. (b) Stopped-flow measurements of AQP proteoliposomes and control liposomes performed under control conditions and in the presence of the indicated compounds. Time courses of measurements were normalized at 8 sec and fitted to single-exponential curves. Graphs show averaged time courses. (c) Reversibility of the inhibition of AQP4 water permeability by AZA. After incubating AQP4 proteoliposomes for 30 min with 1.25 mM AZA, the proteoliposomes were washed two times and subjected to stopped-flow analyses in the absence of AZA. The graph shows averaged time courses (pink, $n = 50$). (d) Averaged osmotic water permeability coefficients (P_f) for the experiments shown in (b) and (c) (mean ± SD).

When AQP4 proteoliposomes were pretreated for 5 minutes with 1.25 mM AZA, we observed a pronounced reduction in water flow. Determination of P_f revealed that the water permeability of AQP4 in the presence of AZA is only about 46.7 ± 0.7% compared with that of untreated AQP4 proteoliposomes (Fig. 10.1b,d). Similar results were obtained with AZA even if the AQP4 proteoliposomes were not first preincubated with the compound. In contrast, 1.25 mM MZA had no significant inhibitory effect on the water permeability of AQP4 in proteoliposomes; 3 mM VPA also showed no inhibition of AQP4-mediated water permeation even if the AQP4 proteoliposomes were preincubated with VPA for up to 30 minutes (Fig. 10.1b,d). Furthermore, all three compounds had no effect on the water permeability of AQP1 proteoliposomes or on the water permeability of control liposomes lacking AQPs (Fig. 10.1b, d). The data obtained by stopped-flow measurements are summarized in Table 1 in the paper by Tanimura *et al.* [29].

10.4 AZA INHIBITS AQP4-MEDIATED WATER PERMEATION

The water permeability of AQP4 was reduced to 46.7% of untreated AQP4 by 1.25 mM AZA (Fig. 10.1d). The water permeability of AQP4 was also measured at different concentrations of AZA. Since the solubility of AZA in water is limited, it was not possible to measure the inhibitory effect of AZA at concentrations significantly higher than 1.25 mM. It was thus difficult to directly measure the IC_{50} value of AZA, but using even lower AZA concentrations we could demonstrate that the inhibition of AQP4-mediated water conduction depended on the AZA concentration (Fig. 10.2).

The AQP4 proteoliposomes were incubated for 30 minutes with 1.25 mM AZA, washed two times with buffer, and subjected to stopped-flow analysis in the absence of AZA. The water permeability was recovered to 88.6 ± 1.8% of the level of untreated AQP4 proteoliposomes (Fig. 10.1c,d). This result revealed that AZA reversibly inhibits AQP4 water permeability.

In agreement with earlier findings [25], we confirmed that AZA inhibited AQP4 water permeability. Our results also showed that AZA inhibits AQP4-mediated water conduction in a concentration-dependent manner (Fig. 10.2) and that an AZA concentration of 1.25 mM reduces the water permeability of AQP4 proteoliposomes to 46.7% (Fig. 10.1d). Our results thus seem to contradict the conclusion that AZA does not inhibit AQP4 water permeability [27], but this could be attributed to different experimental conditions, namely, the measurement temperatures of 10°C and our 22°C. When we repeated our stopped-flow measurements at 4°C, we also found that 1.25 mM AZA had no significant effect on the water permeability of AQP4 proteoliposomes, even after incubation of the proteoliposomes with AZA overnight at 4°C. Efficient inhibition of AQP4-mediated water conduction by AZA thus seems to depend critically on the temperature, with higher temperatures being required for inhibition of AQP4.

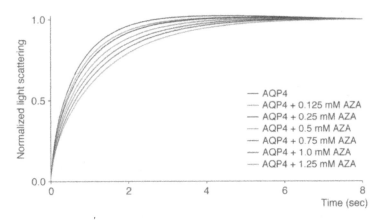

Figure 10.2 Dose dependence of the inhibition of AQP4 water permeability by AZA. The graph shows averaged time courses: 0.125 mM AZA, n = 16; 0.25 mM AZA, n = 15; 0.5 mM AZA, n = 17; 0.75 mM AZA, n = 17; 1.0 mM AZA, n = 19. For color reference, see p. 373.

By MZA and VPA, we observed no significant inhibition of the water permeability of AQP4 proteoliposomes. This finding is different from the result that 20 µM VPA causes a 40% inhibition of hAQP4 expressed in oocytes [30]. We cannot explain this difference at this moment, and more careful experiments are required for conclusive results, although we conclude that VPA is not an inhibitor of rAQP4. It is surprising and interesting that MZA, the chemical structure of which is remarkably similar to that of AZA, does not inhibit the water permeability of AQP4. This result suggests that the methyl group attached to the thiadiazole group prevents MZA from binding to AQP4. Together with the results reported by Huber *et al.* [25], it seems that 1,3,4-thiadiazole-2-sulfonamide would be the key chemical structure of compounds that can specifically bind to rAQP4. This information should be useful for screening and designing new inhibitors of AQP4. We found that none of the three compounds we tested affected the water permeability of AQP1. AZA does not inhibit AQP1 even with a 30-minute preincubation. Because AQP1 is the closest homolog of AQP4, the finding that AZA efficiently inhibits rAQP4 but not hAQP1 suggests that AZA is specific for AQP4.

In conclusion, our data confirm that AZA inhibits the water permeability of rAQP4 but has no effect on hAQP1. Because AZA binds reversibly to rAQP4, it is potentially a promising lead compound to develop AQP4 inhibitors suitable for clinical use. The structural analysis of AQP4 with bound AZA should reveal the mechanism how AZA inhibits water conduction by AQP4 and may help to design new inhibitors against AQP4.

10.5 CRYSTALLIZATION AND DATA COLLECTION FOR STRUCTURAL ANALYSIS OF AQP4

The AQP4M23 gene was obtained from the rat brain QUICK-Clone cDNA library (Clontech). For mimicking phosphorylation of serine 180 in loop D, the point mutation of the residue to aspartic acid was introduced using the QuikChange II site-directed mutagenesis kit (Stratagene) and confirmed by DNA sequencing. The AQP4M23S180D construct was subcloned into pBlueBacHis2B (Invitrogen), and plasmid DNA was purified with the QIAGEN plasmid midi kit. We succeeded in obtaining more than 3 mg of purified 6xHis-tagged rAQP4M23 and AQP4M23S180D from 1 L of cultured Sf9 cells.

Reconstitution of AQP4M23 as well as AQP4M23S180D produced large double-layered 2D crystals, which gave rise to sharp high-resolution diffraction spots and were therefore suitable for electron crystallographic structural analysis. Structural analysis was, however, complicated because of variations in the lateral alignment of the two membranes in the double-layered 2D crystals, making it necessary to classify the crystals. Electron diffraction data, even when high-resolution information was included, proved to be not sensitive enough to distinguish crystal variants. We subsequently took an image of each crystal that produced a high-resolution diffraction pattern. The images provided phase information to a resolution better than 6 Å, sufficient to distinguish among different crystal variants. The need for a corresponding image for each diffraction pattern made the structural analysis very challenging, but our effective cryoelectron microscope enabled us to collect all data for high-resolution structural analyses [39]. Classification based on image data identified one predominant crystal type, with the distance between each membrane centre 45 Å, with a specific molecular interaction between adjoining layers. This type of crystal comprised about 70% of all analyzed crystals and therefore was selected for structural analysis. The final 3.2 Å and 2.8 Å resolution amplitude data sets were used to determine the AQP4 structure and its mutant one, respectively, by the molecular replacement method.

10.6 STRUCTURE OF THE AQP4 WATER CHANNEL

Both structures of AQP4M23 and AQP4M23S180D adopt the typical AQP fold but feature a short 3_{10} helix, named HC, as shown in Fig. 10.3, which is not seen in AQP1. The narrowest region in the AQP1 pore, previously termed ar/

R [40], is located close to the extracellular entrance of the pore. The higher-resolution structure of bovine AQP1 was analyzed by Sui *et al.* [41]. The pore diameter of the ar/R region in AQP4 is similar to that in AQP1. At the NPA site in the center of the pore, the water molecule (W5 in Fig. 10.4b) forms hydrogen bonds with Asn97 and Asn213 of rAQP4 (Asn78 and Asn194 in bAQP1, Asn76 and Asn192 in hAQP1) (Fig. 10.3). The water molecule is thus isolated from the other water molecules in the intra- and extracellular halves of the pore. The diameter is just fit to water permeation because the diameter of the water molecule is 2.8 Å, which is too small for permeation of hydrated ions and solutes (Fig. 10.3). The H-bond isolation mechanism was proposed to be responsible for blocking proton permeation by keeping fast water permeation in AQPs [3].

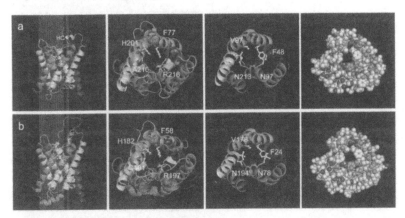

Figure 10.3 Structures of (a) rAQP4M23 and (b) bAQP1 showing the typical AQP fold. AQP4 features a short 3_{10} helix, named, HC, which is important for the adhesive function, but AQP1 has no such HC (first and second figures from the left in both [a] and [b]). The narrowest region in the AQP4 and AQP1 pores, previously termed ar/R [40], is located close to the extracellular entrance of the pore (figures in the second row). The pore diameter at NPA motifs in AQP4 is similar to that in AQP1, although the main chain of Ala210 is in a different conformation and His201 is slightly shifted away from the pore centre. At the NPA sites in the centre of the pores of AQP4 and AQP1 (third figure), the pore diameters of both channels are about 3 Å of Van der Waals distances (fourth figure). For color reference, see p. 373.

To compare wild-type AQP4 with the S180D mutant, we calculated the root-mean-square deviation (RMSD) between their backbone atoms. The RMSD was 0.77 Å overall and 1.26 Å for loop D, which contains the S180D mutation, demonstrating that the structures are essentially identical. Our structural studies thus show that the S180D mutation, mimicking the dopamine-dependent phosphorylation of Ser180 proposed by Zelenina *et al.* [42], does not induce a significant conformational change in AQP4. We also calculated the RMSD between the rAQPM23S180D structure at 2.8 Å resolution [43] and hAQP4 at 1.8 Å resolution by X-ray crystallography [42], which resulted in the value of 0.61 Å, suggesting that both structures are highly identical.

10.7 WATER MOLECULES IN THE AQP4 CHANNEL

The structure of AQP4M23S180D clearly resolved seven water molecules in the channel (Fig. 10.4).

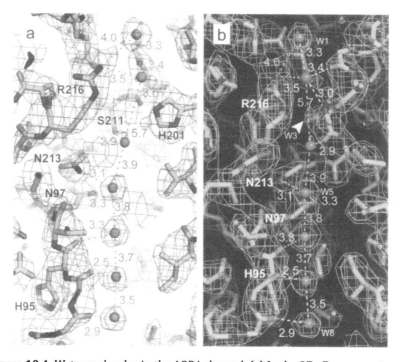

Figure 10.4 Water molecules in the AQP4 channel. (a) In the 2Fo-Fc map contoured at 1.2 σ (marine mesh), water molecules in the channel (red spheres) are clearly resolved as spherical densities. (b) In the Fo-Fc map contoured at 3.0 σ (orange-yellow mesh), a small density (marked by a white arrowhead and labeled W3) appeared in the channel, which we interpreted as the quasi-stable position of a water molecule. Distances (Å) between water molecules and their closest protein atoms are depicted as dotted lines. Distances (Å) between neighboring water molecules are depicted as dotted lines if in hydrogen-bonding distance or as dotted cyan lines if the distance is too far for hydrogen bonding. For color reference, see p. 374.

In addition, the Fo–Fc map showed an additional spherical density at the ar/R constriction site (white arrow head in Fig. 10.4b). Since the side chains of AQP4 around the ar/R region were represented by clear density and the atoms had low-temperature factors in this region, we assigned an eighth water molecule (W3 in Fig. 10.5a) to the spherical density at the ar/R site. The narrow diameter at this constriction would make it an unfavorable position for a water molecule, potentially explaining the weak density for water at this position. The two neighboring water molecules (W2 and W4) on either side of the ar/R constriction, which formed hydrogen bonds with the unstable water molecule (W3) in the constriction, showed higher-temperature factors compared with those of all the other water molecules in the channel (2.9 Å^2 to 13.2 Å^2 [2]) (Fig. 10.5a,b).

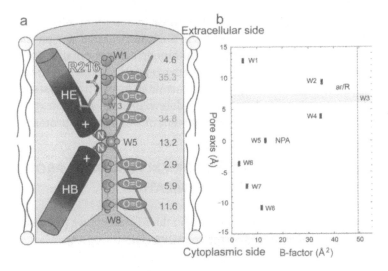

Figure 10.5 Schematic diagram of the water arrangement in the AQP4 channel and the temperature factors of water molecules. (a) The two short pore helices, HB and HE, which are shown as rods with the blue and red colors indicating their electrostatic dipoles, form an electrostatic field that causes water molecules in the two half channels to adopt opposite orientations. All water molecules, except the one at the channel entrances, form two or three hydrogen bonds with the neighboring water, main chain carbonyl oxygen (C=O in the red oval), and/or nitrogen of the NPA motifs (N in the cyan circle). The residue Arg216, part of the ar/R constriction site, is shown in a ball-and-stick representation. The water molecules in the pore are designated W1–W8. The temperature factors of the water molecules are denoted to the right of the channel. (b) plot of the B-factors of the water molecules in the channel. All eight water molecules in the pore are represented. The labeling of the water molecules is the same as in (a). The channel region occupied by W3 is depicted as a wheat-colored band because the B-factor of W3 could not be determined. The red dashed line indicates the average B-factor of all the atoms in the AQP4 molecule. For color reference, see p. 375.

The B-factors of the clearly observed seven water molecules in the pore are lower than 40 Å^2 [2] (Fig. 10.5b). Our density map of water molecules in the AQP4 structure analyzed at 2.8 Å resolution by electron crystallography shows better quality than that of water molecules in the hAQP4 structure analyzed at 1.8 Å resolution by X-ray crystallography (compare with Fig. 10.4 in this review and Fig. 3 in the paper by Ho *et al.* [44]). The comparison gave us a counterintuitive notion that the lower-resolution map determined by electron crystallography gave better quality than the map by X-ray. The reason is not clear, but the difference of the observed map qualities for water molecules in the channel could possibly be attributed to environmental effects. The dipole moments formed by helices, especially the two short helices HB and HE (Fig. 10.5a), may be influenced by the nonglobular environment of the lipid bilayer, as discussed by Sengupta *et al.* (2005) [45]. These dipole moments in the membrane layer may be larger than those in detergent micelles and thus have a force to orient water molecules in the water channel at a physiological condition of the membrane inside. Although experimental proof will be required to verify the effect of different environments on membrane proteins, the structure of a membrane protein determined by electron crystallography would be more adequate because it is determined in the context of a lipid bilayer, very close to the native environment of a membrane protein.

10.8 WATER ARRANGEMENT FOR EXCLUSION OF PROTON KEEPING FAST WATER PERMEATION

The water positions determined to date in the AQP structure [6, 8, 41, 46–48] were superimposed on the structure of AQP4 (Fig. 10.6) [49].

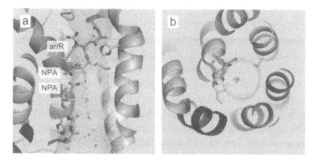

Figure 10.6 AQP4 channel with positions of all water molecules seen in AQP structures determined to date. The pore in AQP4 as calculated by "HOLE" [49] is shown as a transparent surface. The color indicates the pore radius from red (narrow; ar/R) to green (NPA) to purple (wide). Residues interacting with water molecules in the channel are represented as ball-and-stick models (carbon, oxygen, and nitrogen atoms are shown in green, red, and blue, respectively). Red crosses indicate the positions of water molecules seen in other AQP structures. Side view (a) and slice around (b) the NPA region of the AQP4 channel with water molecules seen in the structures of highly water-selective AQPs. Reproduced from Ref. [43] by permission. For color reference, see p. 375.

The water molecules clearly seem to pass through the channel in a single-file arrangement, following a clearly defined route, as seen in the end-on view in Fig. 10.6b. The positions of the water molecules are not coincident with the centre line of the channel cavity calculated by "HOLE" [49], but the position shifts close to hydrogen-bonding partners. The deviation from the centre line was observed especially close to the position near NPA motifs at the membrane centre because the water molecules make specific hydrogen bonds with the amide groups of asparagine residues of the NPA motifs (Fig. 10.6).

The eight water molecules in the AQP4 channel are also in a single-file arrangement but tend to be arranged in a favorable position like stepping stones (Fig. 10.5). The inner surface of water channels is largely hydrophobic, except the oxygen atoms of the main chain carbonyl groups as well as the nitrogen atoms of the side chain amide groups, namely, Asn213 and Asn97 of the NPA motifs (Fig. 10.6a). The eight water positions in the channel are stabilized by carbonyl groups and amide groups both of which are hydrogen-bonding partners for the hydrogen atoms and the oxygen atoms, respectively, of the permeating water molecules. These partners for water molecules also formed a guide rail on the hydrophobic wall of the channel, fitting to the water orientation forced by dipole moments of two short helices B and E (Fig. 10.5).

The guide rail is presumed to be an important molecular machinery for fast water permeation of AQP4, whose water permeation speed was actually measured and revealed to be about 3 billion water molecules per second. The speed approaches that of AQP1, which is known to be one of the fast water channels. From the measured distances between successive water molecules in the channel (Fig. 10.4), all water molecules appear to form hydrogen bonds with their neighbors (red dotted lines in Fig. 10.5a), except the water at the NPA site and the subjacent one. The two hydrogen atoms of the water molecule, which is hydrogen-bonding with amide groups of the asparagines (red dotted lines in Fig. 10.4a), are importantly forcing to orient them perpendicular to the channel axis because of the arrangement of the amide groups and the molecular orbital for water. The hydrogen bonds of the water molecule (W5 in Fig. 10.5) at the NPA site thus are separated from the other water molecules in the channel, lending support to the H-bond isolation mechanism [3]. Each water molecule of the single file in the channel can thus form two or three hydrogen bonds. Since water in bulk solution usually forms three or four hydrogen bonds with neighboring water molecules, water molecules entering the channel only have to sacrifice a single hydrogen bond, an energy cost of about 3 kcal. The arrangement of carbonyl and amide groups in the AQP4 channel thus dramatically lowers the energy barrier for water molecules entering the narrow AQP channel and allows for very fast water permeation through the otherwise hydrophobic channel (Fig. 10.5). The ar/R constriction site is important for blocking H_3O^+ but not for the separation of hydrogen bonds [50], as shown by the arrangement of water molecules in the channel. Finally, if a narrow, positively charged channel region were sufficient to block proton permeation, water channels would not have had to evolve the complex AQP fold seen in all analyzed structures of water channels to date.

10.9 REGULATION OF ORTHOGONAL ARRAY FORMATION

Studies by freeze-fracture electron microscopy revealed prominent arrays of orthogonally arranged intramembrane particles in the perivascular membranes of astrocyte endfeet [14]. Such orthogonal arrays have also been found in the sarcolemmas of myofibers [51] and the basolateral membranes of renal collecting ducts [52]. Using immunogold labeling of freeze-fracture replicas, direct evidence was provided that AQP4 is the major constituent of the arrays [15]. While the function of these arrays in the brain remains unclear, orthogonal arrays were reported to rapidly shrink or even disappear after circulatory arrest [53], suggesting the assembly of AQP4 tetramers into orthogonal array changes in response to the physiological conditions of the water homeostasis in the brain.

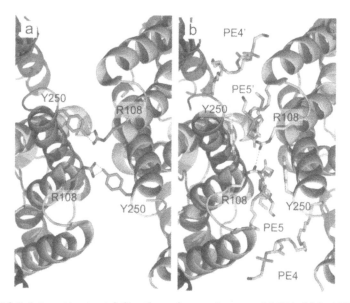

Figure 10.7 Interaction to stabilize the orthogonal arrays of AQP4. (a) In AQP4 wild-type crystals, AQP4 tetramers are stabilized by a specific interaction, such as tyrosine and arginine. (b) Structure of the AQP4S180D mutant analyzed at a resolution of 2.8 Å revealed the insertion of lipid molecules between tetramers. In this crystal, the arrays are mediated by hydrophobic interaction with lipid molecules. For color reference, see p. 376.

On the basis of structural analysis by 2D crystals of AQP4, the characteristic intermolecular interactions in AQP4 crystals were demonstrated to be mediated by Arg108 and Tyr250 and also by hydrophobic interactions of Gly157, Trp231, and Ile239 (Fig. 10.7a). However, the structure of the AQP4S180D mutant analyzed at an improved resolution of 2.8 Å revealed the insertion of lipid molecules between the tetramers, and thus the array could be mediated by hydrophobic interactions with lipid molecules (Fig. 10.7b). This lipid insertion is consistent with experimental data that crystallization of AQP4M23S180D required a larger amount of lipid molecules and amazingly gave better quality of resolution than that of the AQP4 wild type, while a mechanism was unclear, except the result that mutation of S180D stabilized the AQP4 tetramer.

AQP4 has two alternative splice variants resulting from differential translation initiation either at the first methionine (AQP4M1) or at the second methionine (AQP4M23) [10]. As a result, AQP4M1 has a 22-residue-longer N-terminus than AQP4M23. Endogenous AQP4 is considered to form heterotetramers containing both isoforms *in vivo*, but SDS–polyacrylamide gel (SDS-PAGE) analysis showed that in rat brain and other tissues AQP4M23 is much more abundant than AQP4M1 [54]. While the water permeabilities of AQP4M1 and AQP4M23 are likely to be similar [10, 54], freeze-fracture analyses of Chinese hamster ovary (CHO) cells expressing AQP4M1 or AQP4M23 revealed that the two isoforms have different morphological properties. AQP4M23 can form large orthogonal arrays, whereas AQP4M1 restricts the formation of orthogonal arrays [16, 55]. Co-expression of AQP4M1 and AQP4M23 leads to orthogonal arrays that are similar in size to those observed in astrocytes, implying that regulation mechanisms exist that may define the size and disassembly efficiency of orthogonal arrays. To elucidate the mechanism that regulates array formation and disruption, we constructed systematic N-terminal deletion mutants of AQP4 and examined their abilities to form orthogonal arrays by using SDS-digested FRL (SDS-FRL) [56] to directly visualize the distribution of AQP4 molecules in transiently transfected CHO cells. When deletion constructs shorter than 16 amino acid residues were expressed in CHO cells, no orthogonal arrays were observed and the gold labels were evenly dispersed over the P-faces of the plasma membranes. In the case of deletion to C17, orthogonal arrays labeled by many of immunogold particles appeared on the P-face images. Furthermore, mutagenic substitution of the two cysteine residues at positions 13 and 17 in the N-terminus of AQP4M1 resulted in orthogonal array formation, while no such arrays were observed when one of the two cysteine residues remained in the N-terminus of AQP4 (Fig. 10.8). Biochemical analysis and metabolic labeling of transfected CHO cells revealed that the two N-terminal cysteines of AQP4M1 are palmitoylated. These results suggest that palmitoylation of the N-terminal cysteines is involved in the mechanism of the inability of AQP4M1 to form orthogonal arrays (Fig. 10.8).

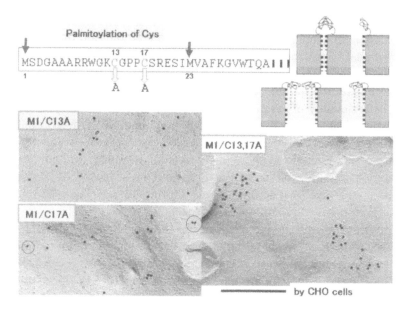

Figure 10.8 The array formation on CHO cell membranes was observed by deleting amino acids one by one from the N-terminus on the basis of the freeze-fracture method. When the 17th amino acid cysteine residue was deleted, suddenly large arrays were observed. Cysteine residues 13 and 17 were revealed to be essential for array disruption by palmitoylation. The lipid modification even on one cysteine was also confirmed to disrupt the arrays of AQP4. Without any lipid modification, interaction of intertetramers has no hindrance and forms orthogonal arrays, but lipid-modified amino terminus hinders the array-forming interaction. The scale bar indicates 4,000 Å. For color reference, see p. 376.

10.10 CELL ADHESIVE FUNCTION OF AQP4

AQP4 forms double-layered 2D crystals due to interactions between extracellular sides of the AQP4 molecules (Fig. 10.9). The interactions between adjoining AQP4 molecules are formed by two hydrophobic contacts. One is mediated by a proline (Pro) at the end of membrane-spanning helix 3 (Fig. 10.9), which is part of a Pro–Pro motif found in AQP4 and AQP0. Val142 mediates the second interaction, which is not seen in AQP0. This residue is part of a short 3_{10} helix in the extracellular loop C (formed by residues Ser140 to Gly143), which is orienting parallel to the membrane surface (Fig. 10.9).

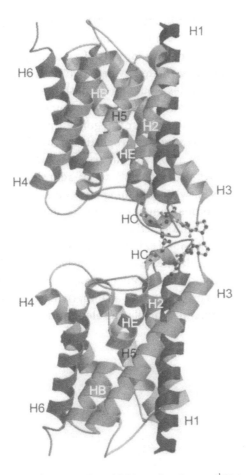

Figure 10.9 One pair of interacting AQP4 molecules stabilizing the membrane junction is shown in a ribbon representation. A ribbon diagram of a pair of interacting AQP4 molecules highlighting residues Pro139 and Val142 of 3_{10} helices responsible for junction interactions is shown. For color reference, see p. 377.

The AQP4 molecule conducts water very rapidly, but membrane junction formation may increase the resistance for water conductance in AQP4, because the tetramers in the two interacting membranes are shifted with respect to each other by half a unit cell ($p42_12$ symmetry) (Fig. 10.10). This packing results in a tight tongue-into-grooves interaction between the two crystalline layers. The interaction of the tetramers in the two adjoining membranes would interfere with AQP4 water conductance due to partial blocking of the extracellular pore openings. Consequently, different from the adhesive structure of AQP0, the force generated by rapid water flow through the channels in the double-layered membranes would destabilize the interaction between the adjoining membranes (Fig. 10.10). Only a single subunit in an AQP4 tetramer can interact with a subunit in the tetramer in the adjoining membrane when the tetramer is not part of an orthogonal array. Since this interaction is weak and is mediated by only two residues, it is unlikely that a single pair of AQP4 molecules would promote stable membrane adhesion. In our crystals, each AQP4 tetramer interacts with four tetramers in the adjoining membrane, so that formation of an orthogonal array, such as that observed *in vivo*, would enhance the adhesive properties of AQP4. The variations in the relative positions of the two layers in our crystals, however, suggest that even crystalline AQP4 arrays promote only weak adhesion. This indicates that while the interactions between adjoining tetramers are specific (Fig. 10.10), AQP4-mediated membrane junctions may be dynamic. Partly separated AQP4-containing membrane junctions have already been observed in glial laminae of the hypothalamus [12], a finding we corroborated with similar images of separated membranes within extensive membrane junction areas [7]. Since junction formation may reduce water conductance, formation and separation of junctions may indeed be an underlying mechanism to modulate AQP4 water conductance and/or play a role in osmo-, thermo-, and glucose sensing.

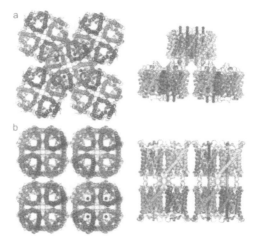

Figure 10.10 Two types of double-layered interactions, which were observed in the 2D crystals of (a) AQP4 and (b) AQP0. All adjoining molecules in AQP0 crystals can adhere even without array formation, but only one pair of molecules of AQP4 can adhere when orthogonal arrays were disrupted, because the tetramer of AQP4 at the upper layer was shifted from the tetramer at the lower layer. As shown by blue rods, adjoining channels of AQP4 are shifted and water permeation is blocked when these two membrane layers are strongly adhered, which is different from the case of AQP0. For color reference, see p. 378.

Our structure of the AQP4M23S180D mutant was revealed to directly interact with three lipid, phosphatidylethanolamine (PE) molecules, by mainly forming negatively charged surface areas (Fig. 10.11). The interactions seen in the crystals suggest that AQP4 may also interact with PE lipids in the outer leaflet of membranes of adjacent cells even *in vivo*. However, the extracellular leaflet of cell membranes usually does not contain PE lipids but instead contains mainly phosphatidylcholine (PC) lipids, which is responsible for asymmetry in eukaryotic lipid bilayers and is also considered to be maintained by several lipid flippases [57]. The PC molecule cannot bind with AQP4 because it has no amine group, in which PE forms the hydrogen bond with residues in loops A and C of AQP4 (Fig. 10.11).

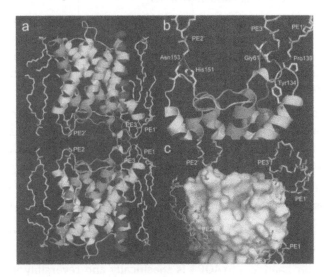

Figure 10.11 AQP4 and lipid interactions stabilizing the membrane junction. (a) Two interacting AQP4 molecules (ribbon diagram) with surrounding lipid molecules (ball-and-sticks representation) viewed parallel to the membrane plane. (b) Close-up view of the interactions between the extracellular AQP4 surfaces with lipids of the adjoining membrane. Lipids (cyan) and their interacting AQP4 residues (yellow) are shown in a ball-and-stick representation. Yellow dotted lines depict hydrogen bonds between lipid head groups and AQP4 residues. (c) Electrostatic potential of the AQP4 surface, ranging from −4kT (red) to +4kT (blue), and a ball-and-stick representation of the associated PE molecules, showing that the lipid head groups interact with neutral (grey) or negatively charged surface areas of AQP4. Reproduced from Ref. [43] by permission. For color reference, see p. 379.

10.11 CONCLUSION

There are complex but important structural components of AQP4 that could carry molecular mechanisms related to the complexity of brain functions.

The known or presumed functions of AQP4 are as follows:

1. AQP4 specifically conducts water but excludes protons, ions, and solutes.

2. AQP4 exhibits fast water permeation at a rate of 3 billion water molecules per second, per channel.

3. Possible gating mechanisms are suggested for AQP4, in which the dopamine signal or protein kinase C is mediated.

4. Orthogonal arrays are regulated by at least two alternative splice variants and also lipid modification on cysteine residues of C13 and C17 at the N-terminus of the long isoform AQPM1.

5. The cell-adhesive function of AQP should be dynamic because the adhesive force is rather weak and interferes with water conductance due to packing into a tight tongue-into-grooves interaction. The AQP molecule can directly bind with PE molecules when they come out by a lipid flippase.

6. AQP4 co-localizes with various proteins, including membrane proteins. For example, AQP4 has a binding site for the postsynaptic density protein PSD-95, the *Drosophila* discs-large tumor suppressor protein and zonula occludens-1 protein (PDZ) domain.

7. The autoantibody to AQP4 is known to cause diseases such as multiple sclerosis and neuromyelitis optica.

8. AQP4 is implicated in brain edema resulting from water intoxication, brain ischemia, or meningitis.

9. The water channel of rAQP4 is specifically and reversibly inhibited by AZA.

The water channel is responsible for changes of the local volume of extracellular space in conjunction with the released amount of K^+ and glutamate and also with increased amounts of CO_2 [58]. Thus, high water selectivity of AQP4 might act as an important rolelike foundation for supporting signal transduction mechanisms regulated mainly by ion channels. However, there is a gap between the understanding of water channels and high-order brain functions.

The brain is protected by the cranial bone from outer mechanical pressures and forces. Conversely, osmotic pressure affects neural cells in the brain because the cranial bone has no flexibility to release the pressure, and thus an osmo-sensor in the brain is important. The osmo-, thermo-, and glucose-sensory functions were known to be related to glial lamellae in the hypothalamus, where AQP4 molecules were localized. The major molecule responsible for these sensory functions is not defined yet, but AQP4's dual role as a water channel and a weak adhesion molecule might be good for osmo-sensing in the glial lamellae because we confirmed AQP4 expression at the adhesive area of glial lamellae as well as at the partly separated area. For sensing osmotic pressure, gushing water through adhesive AQP4 molecules may separate the glial lamellae layers, as shown in Fig. 10.12a, because of the packing of a tight tongue-into-grooves interaction of AQP4 tetramers between the two adjoining membranes with AQP4. This could be named *gushing water model 1*. From an early stage, the proton was known to migrate along the membrane surface faster and at a longer range thanin the bulk-water phase [59]. Although no experimental evidence was revealed, we think that the proton is a good candidate to transfer a signal through cell surfaces and AQP4 could modify or block the signal transfer with the proton by gushing water without the proton. Although this is highly speculative, we named this proton signal transfer and regulation mechanism by the water channel *gushing water model 2*, as schematically shown in Fig. 10.12b.

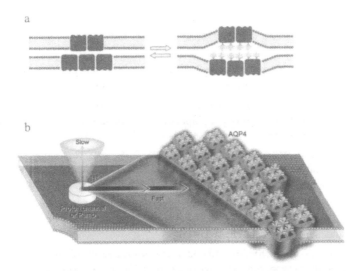

Figure 10.12 (a) Schematic for explaining a possible function of a cell-adhesive channel (adhennel [7]). Adhesive AQP4 molecules may separate membrane layers by pressure of gushing water, as shown in this figure, because the packing of AQP4 tetramers is not end-on but a staggered arrangement between the two adjoining channels. (b) Schematic for explaining proton signal transfer and regulation mechanism by the water channel as gushing water model 2. For color reference, see p. 379.

Most importantly, the fast water permeation function may influence cell adhesion and vice versa. Therefore change of strength in the adhesion might be important for widening the dynamic range of these functions. The adhesive force could be changed by the array size, which is regulated by alternative splice variants and lipid modification on the two cysteine residues at the N-terminus of AQP4 and also by flipping out of inner PE molecules.

The variety of localization in the cells as well as co-localization with other protein molecules, for example, through binding with PDZ domains, may give a chance for AQP4 molecules to take up complex functions in the brain. Knowledge about the multiplicity of AQP4 conducted by structural studies as well as by analysis of genetically engineered mice may provide important building blocks for understanding the functional complexity of the brain.

Acknowledgments

We acknowledge Dr. Mari Dezawa for her critical reading. The research was supported by Grants-in Aid for Specially Promoted Research, and the Japan New Energy and Industrial Technology Development Organization (NEDO).

References

1. Borgnia, M., Nielsen, S., Engel, A., and Agre, P. (1999) Cellular and molecular biology of the aquaporin water channels, *Annu. Rev. Biochem.*, **68**, 425–458.

2. Agre, P., King, L. S., Yasui, M., Guggino, W. B., Ottersen, O. P., Fujiyoshi, Y., Engel, A., and Nielsen, S. (2002) Aquaporin water channels: from atomic structure to clinical medicine, *J. Physiol.*, **542**, 3–16.

3. Murata, K., Mitsuoka, K., Hirai, T., Walz, T., Agre, P., Heymann, J. B., Engel, A., and Fujiyoshi, Y. (2000) Structural determinants of water permeation through aquaporin-1, *Nature*, **407**, 599–605.

4. Fu, D., Libson, A., Miercke, L. J., Weitzman, C., Nollert, P., Krucinski, J., and Stroud, R. M. (2000) Structure of a glycerol-conducting channel and the basis for its selectivity, *Science*, **290**, 481–486.

5. Gonen, T., Sliz, P., Kistler, J., Cheng, Y., and Walz, T. (2004) Aquaporin-0 membrane junctions reveal the structure of a closed water pore, *Nature*, **429**, 193–197.

6. Gonen, T., Cheng, Y., Sliz, P., Hiroaki, Y., Fujiyoshi, Y., Harrison, S. C., and Walz, T. (2005) Lipid-protein interactions in double-layered two-dimensional AQP0 crystals, *Nature*, **438**, 633–638.

7. Hiroaki, Y., Tani, K., Kamegawa, A., Gyobu, N., Nishikawa, K., Suzuki, H., Walz, T., Sasaki, S., Mitsuoka, K., Kimura, K., Mizoguchi, A., and Fujiyoshi, Y. (2006) Implications of the aquaporin-4 structure on array formation and cell adhesion, *J. Mol. Biol.*, **355**, 628–639.

8. Törnroth-Horsefield, S., Wang, Y., Hedfalk, K., Johanson, U., Karlsson, M., Tajkhorshid, E., Neutze, R., and Kjellbom, P. (2006) Structural mechanism of plant aquaporin gating, *Nature*, **439**, 688–694.

9. Hasegawa, H., Ma, T., Skach, W., Matthay, M. A., and Verkman, A. S. (1994) Molecular cloning of a mercurial-insensitive water channel expressed in selected water-transporting tissues, *J. Biol. Chem.*, **269**, 5497–5500.

10. Jung, J. S., Bhat, R. V., Preston, G. M., Guggino, W. B., Baraban, J. M., and Agre, P. (1994) Molecular characterization of an aquaporin cDNA from brain: candidate osmoreceptor and regulator of water balance, *Proc. Natl. Acad. Sci. USA*, **91**, 13052–13056.

11. Amiry-Moghaddam, M., and Ottersen, O. P. (2003) The molecular basis of water transport in the brain, *Nat. Rev. Neurosci.*, **4**, 991–1001.

12. Nielsen, S., Nagelhus, E. A., Amiry-Moghaddam, M., Bourque, C., Agre, P., and Ottersen, O. P. (1997) Specialized membrane domains for water transport in glial cells: high-resolution immunogold cytochemistry of aquaporin-4 in rat brain, *J. Neurosci.*, **17**, 171–180.

13. Takata, K., Matsuzaki, T., and Tajika, Y. (2004) Aquaporins: water channel proteins of the cell membrane, *Prog. Histochem. Cytochem.*, **39**, 1–83.

14. Landis, D. M., and Reese, T. S. (1974) Arrays of particles in freeze-fractured astrocytic membranes, *J. Cell Biol.*, **60**, 316–320.

15. Rash, J. E., Yasumura, T., Hudson, C. S., Agre, P., and Nielsen, S. (1998) Direct immunogold labeling of aquaporin-4 in square arrays of astrocyte and ependymocyte plasma membranes in rat brain and spinal cord, *Proc. Natl. Acad. Sci. USA*, **95**, 11981–11986.

16. Furman, C. S., Gorelick-Feldman, D. A., Davidson, K. G., Yasumura, T., Neely, J. D., Agre, P., and Rash, J. E. (2003) Aquaporin-4 square array assembly: opposing actions of M1 and M23 isoforms, *Proc. Natl. Acad. Sci. USA*, **100**, 13609–13614.

17. Suzuki, H., Nishikawa, K., Hiroaki, Y., and Fujiyoshi, Y. (2008) Formation of aquaporin-4 arrays is inhibited by palmitoylation of N-terminal cysteine residues, *Biochim. Biophys. Acta*, **1778**, 1181–1189.

18. Iwamoto, K., Kakiuchi, C., Bundo, M., Ikeda, K., and Kato, T. (2004) Molecular characterization of bipolar disorder by comparing gene expression profiles of postmortem brains of major mental disorders, *Mol. Psychiatry.*, **9**, 406–416.

19. Manley, G. T., Fujimura, M., Ma, T., Noshita, N., Filiz, F., Bollen, A. W., Chan, P., and Verkman, A. S. (2000) Aquaporin-4 deletion in mice reduces brain edema after acute water intoxication and ischemic stroke, *Nat. Med.*, **6**, 159–163.

20. Papadopoulos, M. C., and Verkman, A. S. (2005) Aquaporin-4 gene disruption in mice reduces brain swelling and mortality in pneumococcal meningitis, *J. Biol. Chem.*, **280**, 13906–13912.

21. Preston, G. M., Jung, J. S., Guggino, W. B., and Agre, P. (1993) The mercury-sensitive residue at cysteine 189 in the CHIP28 water channel, *J. Biol, Chem.*, **268**, 17–20.

22. Niemietz, C. M., and Tyerman, S. D. (2002) New potent inhibitors of aquaporins: silver and gold compounds inhibit aquaporins of plant and human origin, *FEBS Lett.*, **531**, 443–447.

23. Ma, B., Xiang, Y., Mu, S. M., Li, T., Yu, H. M., and Li, X. J. (2004) Effects of acetazolamide and anordiol on osmotic water permeability in AQP1-cRNA injected Xenopus oocyte, *Acta Pharmacol. Sin.*, **25**, 90–97.

24. Gao, J., Wang, X., Chang, Y., Zhang, J., Song, Q., Yu, H., and Li, X. (2006) Acetazolamide inhibits osmotic water permeability by interaction with aquaporin-1, *Anal. Biochem.*, **350**, 165–170.

25. Huber, V. J., Tsujita, M., Yamazaki, M., Sakimura, K., and Nakada, T. (2007) Identification of arylsulfonamides as Aquaporin 4 inhibitors, *Bioorg. Med. Chem. Lett.*, **17**, 1270–1273.

26. Yang, B., Kim, J. K., and Verkman, A. S. (2006) Comparative efficacy of HgCl$_2$ with candidate aquaporin-1 inhibitors DMSO, gold, TEA$^+$ and acetazolamide, *FEBS Lett.*, **580**, 6679–6684.

27. Yang, B., Zhang, H., and Verkman, A. S. (2008) Lack of aquaporin-4 water transport inhibition by antiepileptics and arylsulfonamides, *Bioorg. Med. Chem.*, **16**, 489–493.

28. Søgaard, R., and Zeuthen, T. (2008) Test of blockers of AQP1 water permeability by a high-resolution method: no effects of tetraethylammonium ions or acetazolamide, *Pflugers Arch.*, **456**, 285–292.

29. Tanimura, Y., Hiroaki, Y., and Fujiyoshi, Y. (2009) Acetazolamide reversibly inhibits water conduction by aquaporin-4, *J. Struct. Biol.*, **166**, 16–21.

30. Huber, V. J., Tsujita, M., Kwee, I. L., and Nakada, T. (2009) Inhibition of Aquaporin 4 by antiepileptic drugs, *Bioorg. Med. Chem.*, **17**, 418–424.

31. Tajkhorshid, E., Nollert, P., Jensen, M. Ø., Miercke, L. J. W., O'Connell, J., Stroud, R. M., and Schulten, K. (2002) Control of the selectivity of the aquaporin water channel family by global orientational tuning, *Science*, **296**, 525–530.

32. de Groot, B. L., Frigato, T., Helms, V., and Grubmüller, H. (2003) The mechanism of proton exclusion in the aquaporin-1 water channel, *J. Mol. Biol.*, **333**, 279–293.

33. Chakrabarti, N., Tajkhorshid, E., Roux, B., and Pomès, R. (2004a) Molecular basis of proton blockage in aquaporins, *Structure*, **12**, 65–74.

34. Chakrabarti, N., Roux, B., and Pomès, R. (2004b) Structural determinants of proton blockage in aquaporins, *J. Mol. Biol.*, **343**, 493–510.

35. de Groot, B. L., and Grubmüller, H. (2005) The dynamics and energetics of water permeation and proton exclusion in aquaporins, *Curr. Opin. Struct. Biol.*, **15**, 176–183.

36. Hub, J. S., and de Groot, B. L. (2008) Mechanism of selectivity in aquaporins and aquaglyceroporins, *Proc. Natl. Acad. Sci. USA*, **105**, 1198–1203.

37. Zeidel, M. L., Ambudkar, S. V., Smith, B. L., and Agre, P. (1992) Reconstitution of functional water channels in liposomes containing purified red cell CHIP28 protein, *Biochemistry*, **31**, 7436–7440.

38. Yakata, K., Hiroaki, Y., Ishibashi, K., Sohara, E., Sasaki, S., Mitsuoka, K., and Fujiyoshi, Y. (2007) Aquaporin-11 containing a divergent NPA motif has normal water channel activity, *Biochim. Biophys. Acta*, **1768**, 688–693.

39. Fujiyoshi, Y. (1998) The structural study of membrane proteins by electron crystallography, *Adv. Biophys.*, **35**, 25–80.

40. de Groot, B. L., and Grubmüller, H. (2001) Water permeation across biological membranes: mechanism and dynamics of aquaporin-1 and GlpF, *Science*, **294**, 2353–2357.

41. Sui, H., Han, B.-G., Lee, J. K., Walian, P., and Jap, B. K. (2001) Structural basis of water-specific transport through the AQP1 water channel, *Nature*, **414**, 872–878.

42. Zelenina, M., Zelenin, S., Bondar, A. A., Brismar, H., and Aperia, A. (2002) Water permeability of aquaporin-4 is decreased by protein kinase C and dopamine, *Am. J. Physiol. Renal Physiol.*, **283**, F309–F318.

43. Tani, K., Mitsuma, T., Hiroaki, Y., Kamegawa, A., Nishikawa, K., Tanimura, Y., and Fujiyoshi, Y. (2009) Mechanism of aquaporin-4's fast and highly selective water conduction and proton exclusion, *J. Mol. Biol.*, **389**, 694–706.

44. Ho, J. D., Yeh, R., Sandstrom, A., Chorny, I., Harries, W. E. C., Robbins, R. A., Miercke, L. J. W., and Stroud, R. M. (2009) Crystal structure of human aquaporin 4 at 1.8 Å and its mechanism of conductance, *Proc. Natl. Acad. Sci. USA*, **106**, 7437–7442.

45. Sengupta, D., Behera, R. N., Smith, J. C., and Ullmann, G. M. (2005) The α helix dipole: screened out? *Structure*, **13**, 849–855.

46. Savage, D. F., Egea, P. F., Robles-Colmenares, Y., O'Connell, J. D., and Stroud, R. M. (2003) Architecture and selectivity in aquaporins: 2.5 Å X-ray structure of aquaporin Z, *PLoS Biol.*, **1**, 334–340.

47. Harries, W. E. C., Akhavan, D., Miercke, L. J. W., Khademi, S., and Stroud, R. M. (2004) The channel architecture of aquaporin 0 at a 2.2-Å resolution, *Proc. Natl. Acad. Sci. USA*, **101**, 14045–14050.

48. Horsefield, R., Norden, K., Fellert, M., Backmark, A., Törnroth-Horsefield, S., Terwisscha van Scheltinga, A. C., Kcassman, J., Kjellbom, P., Johanson, U., and Neutze, R. (2008) High-resolution x-ray structure of human aquaporin 5, *Proc. Natl. Acad. Sci. USA*, **105**, 13327–13332.

49. Smart, O. S., Neduvelil, J. G., Wang, X., Wallace, B. A., and Sansom, M. S. (1996) HOLE: a program for the analysis of the pore dimensions of ion channel structural models, *J. Mol. Graph.*, **14**, 354–360.

50. Beitz, E., Wu, B., Holm, L. M., Schultz, J. E., and Zeuthen, T. (2006) Point mutations in the aromatic/arginine region in aquaporin 1 allow passage of urea, glycerol, ammonia, and protons, *Proc. Natl. Acad. Sci. USA*, **103**, 269–274.

51. Rash, J. E., Staehelin, L. A., and Ellisman, M. H. (1974) Rectangular arrays of particles on freeze-cleaved plasma membranes are not gap junctions, *Exp. Cell. Res.*, **86**, 187–190.

52. Orci, L., Humbert, F., Brown, D., and Perrelet, A. (1981) Membrane ultrastructure in urinary tubules, *Int. Rev. Cytol.*, **73**, 183–242.

53. Landis, D. M., and Reese, T. S. (1981) Membrane structure in mammalian astrocytes: a review of freeze-fracture studies on adult, developing, reactive and cultured astrocytes, *J. Exp. Biol.*, **95**, 35–48.

54. Neely, J. D., Christensen, B. M., Nielsen, S., and Agre, P. (1999) Heterotetrameric composition of aquaporin-4 water channels, *Biochemistry*, **38**, 11156–11163.

55. Silberstein, C., Bouley, R., Huang, Y., Fang, P., Pastor-Soler, N., Brown, D., and Van Hoek, A. N. (2004) Membrane organization and function of M1 and M23 isoforms of aquaporin-4 in epithelial cells, *Am. J. Physiol. Renal. Physiol.*, **287**, F501–F511.

56. Fujimoto, K. (1995) Freeze-fracture replica electron microscopy combined with SDS digestion for cytochemical labeling of integral membrane proteins. Application to the immunogold labeling of intercellular junctional complexes, *J. Cell. Sci.*, **108**, 3443–3449.

57. Devaux, P. F. (1992) Protein involvement in transmembrane lipid asymmetry, *Annu. Rev. Biophys. Biomol. Struct.*, **21**, 417–439.

58. Nagelhus, E. A., Mathiisen, T. M., and Otterson, O. P. (2004) Aquaporin-4 the central nervous system: cellular and subcellular distribution and coexpression with Kir4.1, *Neuroscience*, **129**, 905–913.

59. Heberle, J., Riesle, J., Thiedemann, G., Oesterhelt, D., and Dencher, N. A. (1994) Proton migration along the membrane surface and retarded surface to bulk transfer, *Nature*, **370**, 379–382.

Chapter 11

WATER, MEMBRANES, AND LIFE WITHOUT WATER

John H. Crowe

Department of Molecular and Cellular Biology,
University of California, Davis,
CA 95616, USA
jhcrowe@ucdavis.edu

11.1 WATER BUILDS STRUCTURE IN MEMBRANES

11.1.1 The Hydrophobic Effect and Spontaneous Formation of Bilayers

Like other amphiphiles, membrane phospholipids self-assemble in aqueous solution into aggregates, provided their density exceeds a certain critical micelle concentration (cmc), which depends on their chemical structure and the ions present [1–4]. In the case of phospholipids, the assembly takes the form of bilayers, which aggregate into multilamellar structures (Fig. 11.1). The consensus view of the driving force behind this self-assembly is based on the poor solubility of hydrocarbons in water, or what is known as the hydrophobic effect [5]. The presence of hydrocarbon residues induces the formation of a cavity in the water structure, leading to an increased order in the water near the hydrocarbons (known as *clathrate* water) and thus to a decrease in the entropy of the system [6, 7]. When hydrocarbon residues aggregate [8, 9], the cavities fuse and expel water from the interface, thereby increasing the entropy of the system. This entropically driven process leads to the spontaneous formation of stable aggregates [10].

Water: The Forgotten Biological Molecule
Edited by Denis Le Bihan and Hidenao Fukuyama
Copyright © 2011 by Pan Stanford Publishing Pte. Ltd.
www.panstanford.com

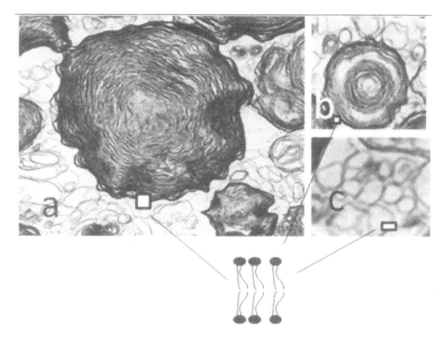

Figure 11.1 When dry phospholipids are placed in water, they spontaneously assemble into (a) multilamellar vesicles consisting of multiple bilayers stacked against each other. Application of physical forces such as sonication or filtration can convert these multilamellar vesicles into (b) plurilamellar ones consisting of a few layers of bilayers or (c) unilamellar ones consisting of a single bilayer.

Although it does not drive the aggregation, the hydrophilic head group forms an interface with water and contributes to the size, the shape of the aggregates through the interactions between the molecules, and the physical properties of the assembly. For example, the phase state of the assembled bilayer is profoundly influenced by the state of water surrounding the head group, a subject that will occupy much of this review. Geometric packing considerations predict the conformation of the aggregate, given some elementary structural information about the amphiphiles, as clearly presented by Israelachvili *et al.* [11]. To assess the conformation, simple geometry can be used: the dimensionless packing parameter p, where $p = v/a_0 l_c$ (v = the hydrocarbon volume, a_0 = the optimal head group area, and l_c = the critical chain length beyond which the hydrocarbon chain is no longer fluid) [3]. This parameter determines whether the amphiphiles will form spherical micelles ($p < 1/3$), nonspherical micelles ($1/3 < p < 1/2$), vesicles or bilayers ($1/2 < p < 1$), or inverted structures ($p > 1$). So long as the system contains only one amphiphile and water, this prediction holds reasonably well. With more complex systems interactions between components—such as electrostatic interactions, van der Waals forces, or hydrogen bonding—reorganize the system, following a complex phase diagram. For example, adding salt, changing the pH, or altering the phase state of the components by raising or lowering the temperature can lead to segregation of the amphiphiles and gives rise to unexpected aggregates such as nanodiscs, punctured planes, and facetted icosahedra, depending on stoichiometry [12–14]. Such phase-separated complex structures, often formed under the influence of water and inorganic ions, are the subject of the next section.

11.1.2 The Hydrophobic Effect and Membrane Organization

Biological membranes possess a functional lipid domain structure, often spoken of as "rafts" [15–20]. The domains are usually small in cell membranes at physiological temperatures—on the scale of nanometers—and consist of phase-separated ordered lipids surrounded by fluid phase domains (hence the term "rafts"). Although the discussion concerning the origins and biological significance of rafts in native membranes continues, there is abundant evidence that they are involved in important biological processes such as cell signaling, lipid and protein trafficking, cell–cell adhesion, membrane transport, and protein targeting to the cell surface. The latter may have special significance in that evidence is emerging that rafts are involved in infection of cells by viruses and bacteria [18]. Indeed, there is growing evidence that rafts are intimately involved in disease states such as Alzheimer's disease [22], human immunodeficiency virus [23], and certain types of cancer [24], apoptosis [25] and in binding of prions to cell surfaces [26]. Clearly, it is important that we understand the origins of rafts and the mechanisms by which they are stabilized in native membranes, and it appears that the dominant force driving their formation is water.

An extensive amount of work has been done on the phase behavior of lipid mixtures to obtain a clearer picture of the possible interactions between different lipid species in cell membranes that might lead to the formation of rafts. Several binary lipid mixtures have been analyzed, using a variety of techniques, including differential scanning calorimetry (DSC) [21], Fourier-transformed infrared spectroscopy (FTIR) [27], nuclear magnetic resonance (NMR) spectroscopy [28, 29], and computer simulations [30, 31]. Phase diagrams for a large variety of mixtures have thus been generated [21]. In general, in a binary mixture in which the lipid species have similar phase transition temperatures, there is a high degree of mixing and the phase transitions merge. In the case of lipid mixtures that have lipids with phase transitions that differ by more than 10°C, the system behaves as a nonideal mixture, in which case the phase transitions of the lipid species do not merge completely. This is observed by DSC as two cooperative events representing a fraction enriched in the lower-melting-point lipid and a fraction enriched in the higher-melting-point lipid, respectively [32]. The lower the miscibility between the lipid species, the farther the phase transitions from each other, and the closer the phase transitions are to the individual lipid components' transition temperatures. This implies that a multicomponent bilayer with a degree of immiscibility between lipid species can show multiple cooperative phase transitions.

Phase separation is segregation of membrane components in the plane of the bilayer. Although several forces are involved, one of the main driving forces for phase separation is the hydrophobic mismatch, which arises from a difference in membrane thickness between two species within a bilayer, such as a protein and a lipid or a lipid and a lipid [33, 34]. The differences in thickness lead to exposure of hydrophobic residues to water (Fig. 11.2) and, consequently, to a decrease in entropy of the system, resulting from ordering of the water. Thus, the assembly of components of similar thickness into relatively homogeneous domains is entropically driven. The net increase in entropy driving the process, which is contributed by water, can arise from differences in their respective phase states; for example, dipalmitoylphosphatidylcholine (DPPC) in the gel phase has fully extended acyl chains, whereas DPPC in the liquid-crystalline phase has an increased number of trans-gauche isomerizations. This induces a disordering in the acyl chains, which results in an overall reduction in the average length of the molecule. Thus, two different species in different phase states may have a marked molecular length difference and as a result will have a strong tendency to phase-separate (Fig. 11.2).

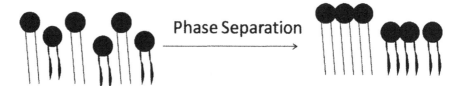

Figure 11.2 In a simple binary mixture of two phospholipids, one with short acyl chains, in the liquid-crystalline phase, the other with longer acyl chains, in the gel phase, a hydrophobic mismatch occurs, exposing portions of the acyl chains in the long-chain lipid to water. As a result, the two lipid phases separate. For color reference, see p. 380.

The main gel to liquid-crystalline phase transition is not quite an ideal first-order transition, so even for a single lipid such as DPPC, there is a small temperature range in which the DPPC gel phase and the DPPC liquid-crystalline phase coexist. Spectroscopic evidence has shown that gel-phase DPPC and liquid-crystalline-phase DPPC separate within this narrow temperature range [35]. In a nonideal mixture such as dilauroylphosphatidylcholine (DLPC)/disteaorylphosphatidylcholine (DSPC), there will be a much broader temperature range of phase coexistence [32]. The degree of phase separation in this system will vary with temperature; at low temperatures, where both species are in the gel phase, the hydrophobic mismatch is not as accentuated compared with intermediate temperatures at which DLPC is mostly in the liquid-crystalline phase and DSPC is in the gel phase. The degree of phase separation is at its maximum in the gel/liquid-crystalline coexistence temperature regime, where the hydrophobic mismatch is the largest. The same arguments apply to aggregation of membrane proteins because of hydrophobic mismatch [33].

Some of the most convincing evidence that phase separation can be macroscopic has come from fluorescence microscopy. Korlach *et al.* [36] used confocal microscopy to study the distribution of fluorescent dyes in mixtures of DLPC/DPPC, which were known from calorimetric and spectroscopic evidence to have low miscibility. Complete phase separation of the two lipids was observed, with domains in the μm range. Addition of cholesterol altered the shape of the phase boundaries, suggesting that cholesterol is active at those boundaries [37].

Epifluorescence methods with this same mixture prepared as supported bilayers [38] gave the results shown in Fig. 11.3. The membranes are labeled with a fluorescent probe that partitions into ordered domains. As the proportion of DSPC was increased, the fluorescent domains increased in size, thus indicating that this is the ordered, DSPC-enriched phase (Fig. 11.3a–c). When the sample was warmed above the DSPC transition, the fluorescence became homogeneous (Fig. 11.3d), suggesting that the two lipids mixed, as expected. Upon cooling, the fluorescent domains reformed, reaching maximal intensity in the phase coexistence region for this pair of lipids (Fig. 11.3e,f). Along the same lines, Bagatolli and Gratton [39] used two-photon fluorescence microscopy to characterize a series of binary mixtures of lipids. Their images showed clear phase separation at temperatures in the solid/fluid coexistence region for each pair of lipids. They also showed that the extent of miscibility appeared to be related to the hydrophobic mismatch of the lipid pairs, as one might expect.

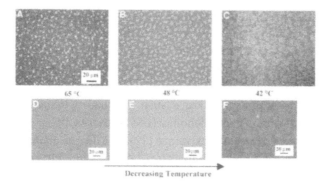

Figure 11.3 Epifluorescence images of DMPC/DSPC-supported single bilayers labeled with 0.5 mol% NBD-PE (di14:0). As the mole ratio of DMPC/DSPC is decreased (A–C) the fluorescent domain increases in size, indicating that this is the DSPC fraction. When the system was heated above the DSPC transition, the domain structure was lost (D). Cooling then resulted in reformation and growth of the domains (E, F). Modified from Ref. [27]. For color reference, see p. 380.

11.2 WATER AND THE POLAR HEAD GROUPS OF MEMBRANE PHOSPHOLIPIDS

Despite the dominating effect of the hydrocarbon chains on the lipid–lipid associations, the hydration state of the polar head group may also influence such associations. In this and the following sections we will describe the hydration state of the polar head group and the consequences of removal of that water. Most of the work to be described here has been done on multilamellar structures, and it is uncertain how much of it applies directly to single bilayers. Nevertheless, there is good evidence that at least the water associated with the polar head groups has similar properties.

11.2.1 Hydration Shells

Numerous studies have been carried out in order to understand the hydration of lipid membranes. Experiments and simulations have found that water molecules penetrate the interfacial region (phospholipid head groups) of bilayers, but water has a low probability of being found in the hydrophobic core [40–45]. Fully hydrated liquid-crystalline-phase phosphatidylcholine (PC) lipids such as DPPC can take up a maximum of ~30 water molecules per lipid, distributed in the interfacial water region and the water layer between the two head group regions of adjacent bilayers [40]. However, the amount of water that can hydrate the membrane depends on the membrane's structure.

In the more structurally ordered gel phase, only ~13 water molecules per DPPC are absorbed [40]. The number of water molecules in the interfacial water region also changes from ~4 per DPPC in the gel phase to ~9 per DPPC in the liquid-crystalline phase. There have been various reports on how the head groups associate with water molecules in bilayers. A variety of techniques including infrared (IR), NMR, neutron scattering, and electrical conductivity measurements have been employed in studies on water–head group interactions [40, 46]. In spite of this, a clear picture is still lacking for lipid bilayers at various hydration levels. Molecular dynamics (MD) simulations on DMPC bilayers suggest that the presence of the hydrophobic choline group in the lipid is responsible for the formation of a *clathrate-like* shell at the surface of the lipid bilayer [42, 44, 47] have conducted MD simulations on the hydrogen-bonding structure of water at the fully hydrated bilayer surface of DMPC. In addition to the formation of a hydration shell around the choline group, where the water molecules are able to form hydrogen bonds with each other by creating a clathrate structure, they also found that the double-bonded oxygen atoms of the phosphate group have a 74% probability of forming hydrogen bonds with two water molecules. The importance of the choline group in hydration has also been revealed by electrical conductivity measurements, which indicate that the choline group is as important as the phosphate group for absorbing water [48]. Using NMR measurements, Hsieh and Wu [49] studied the structure and dynamics of the hydration shell of DMPC bilayers at temperatures below 0°C and suggested that the hydration shell of the head group consists of two distinct regions, a clathrate-like water cluster near the hydrophobic choline group and hydration water molecules associated with the phosphate group. About one or two water molecules are associated with the phosphate group and are frozen, whereas about five or six water molecules near the choline group behave as a water cluster and remain unfrozen at temperatures as low as –70°C.

Because of the nanoscale confinement of water between bilayers in multilamellar structures, the dynamics of water in the interbilayer region are expected to be very different from those of bulk water in analogy to the substantial deviations from bulk-water dynamics observed in other nanoscopic water systems [50]. Ultrafast vibrational spectroscopy has become a powerful tool to study the dynamics of water in nanoscopic environments [45]. Experiments probing the OD stretching mode of dilute HOD in water trapped in reverse micelles [50–52] have shown that the vibrational population relaxation (vibrational lifetime) and the orientational relaxation of water both slow as the amount of water in the system decreases.

In reverse micelles, a core-shell model can be used to quantitatively explain the size-dependent IR absorption spectra and population relaxation times. The water dynamics are subdivided into two contributions, a "shell" region consisting of water closely associated with the head groups and a "core" region with bulklike properties consisting of water molecules away from the head groups.

The physical data on water associated with the polar head group can be fit using a two-state model consisting of bulk water and head group water. For example, recent IR measurements on DMPC phospholipids multilayers that are weakly hydrated with deuterium oxide (D_2O) suggest that water loses most of its bulk properties and has a high degree of structural heterogeneity in the interfacial region [53, 54].

In summary, our level of understanding of the water associated with the polar head groups is still rudimentary, but we do know that it has a profound influence on the physical properties of the lipids with which it interacts.

11.3 HYDRATION-DEPENDENT PHASE TRANSITIONS IN PHOSPHOLIPIDS

The preceding discussion of the hydration state of phospholipids in bilayers suggests that dehydration should profoundly affect the nature of lipid–lipid interactions. Because water occupies volume between the polar head groups, removal of that water would be expected to decrease the intermolecular distance between the polar groups, leading to increased van der Waals interactions between the hydrocarbon chains. Such increased interactions are characteristic of certain types of cooperative phase transitions. We will not review the enormous amount of literature that deals with phase transitions: such reviews are readily available [21, 55]. Instead, we will comment here only on those findings that are of interest in explaining the hydration-dependent phase transitions.

11.3.1 The Gel- to Liquid-Crystalline Phase Transition

Probably the most studied of all lipid phase transitions is the gel to liquid-crystalline transition first suggested for phospholipids by Dennis Chapman *et al.* This sort of transition involves ordered packing of the hydrocarbon chains, with all trans, fully extended acyl chains, thus maximizing van der Waals interactions. Upon heating, a gradual increase in mobility of the hydrocarbon chains is evident (as seen with IR spectroscopy, X-ray diffraction, and NMR) until at the phase transition temperature the chains show a sudden increase in mobility with further increases in temperature. The molecular events responsible for this effect have been extensively studied and continue to constitute the subject of numerous publications. Of particular interest here is the finding that hydration state has a profound effect on the physical status of membrane phospholipids and the transition of lipid bilayers from the liquid-crystalline to gel states. For example, Janiak and Small [reviewed in 21] showed that when dry phospholipids are rehydrated, the area occupied by a polar head group increases to a limiting value at about 10–12 moles water/mole lipid. Simultaneously, the thickness of the bilayer decreases with hydration, again reaching a limiting value at the same hydration (Fig. 11.4).

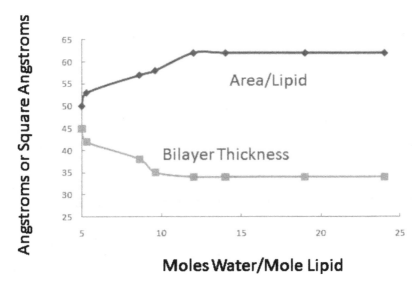

Figure 11.4 Effects of hydration on the cross-sectional area occupied by a lipid and the bilayer thickness. Note that both parameters reach a constant value at about 12 moles water/mole lipid. Data from Janiak and Small [reviewed in Ref. 21].

In other words, water is forcing the head groups apart and at the same time plasticizing the bilayer. This latter point is made by Chapman *et al.* and Kodama *et al.* [reviewed in 21], who showed that when phase transition temperatures were recorded for PC containing variable amounts of water, the transition temperature fell to a limiting value as the water content rose about 12 moles water/mole lipid. At higher water contents there was no further decrease in the transition temperature (Fig. 11.5). Thus, with dehydration PC exists in the gel phase at temperatures that would permit in the hydrated lipid to exist in the liquid-crystalline phase.

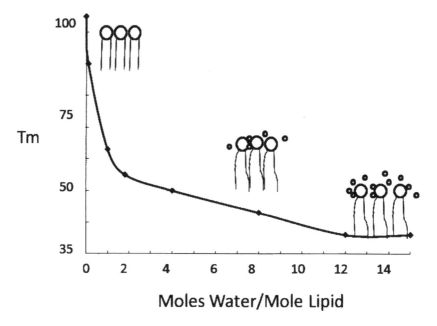

Figure 11.5 Effect of removal of water associated with the polar head group on T_m. T_m increases by about 60°C. The small dots around the polar head groups represent water molecules.

11.3.2 The Consequences of Lipid Phase Transitions During Drying

Hydration-dependent phase transitions can lead to dramatic changes in the physical state and macroscopic organization of membranes. Here are some examples:

1. *Fusion.* Membranes are separated in aqueous solution by bulk water, which is known to resist the close approach of adjacent bilayers, a phenomenon known as the *hydration pressure* [55]. Working with multilamellar vesicles, Rand and Parsegian, in a long series of studies in which an external osmotic pressure was applied to the vesicles, measured the repeat distance between bilayers, using X-ray diffraction. They found that when two hydrated lipid bilayers come within short distances (4–10 Å) of each other, they experience a strong repulsive pressure, which increases exponentially with decreasing interbilayer separation [55]. Despite the critical role of this repulsion in a variety of biological processes (such as membrane fusion), its origin still remains a controversial issue. The existing views fall roughly into two groups. One group is based on the concept that the short-range repulsion is mainly associated with

the water–lipid interaction and originates from the perturbation of the water structure by the hydrophilic lipid surface [55]. An alternative explanation of the water-mediated repulsion implies the dominant role of entropy-driven deviations of hydrated bilayers from ideal planar geometry [56]. According to this view, most important deviations are thermal undulations of the membrane surface, fluctuations in the bilayer thickness, and protrusions of individual lipid molecules and their head groups into the aqueous phase. An attempt to combine the protrusion and hydration pressures in a simplified model of supported lipid bilayers was reported by Lipowsky and Grotehans [57]. Two different regimes were found in which the water-mediated repulsion was dominated either by protrusion or by hydration forces, depending on the temperature and hydration level. Several groups of investigators have studied these effects in supported bilayer membranes deposited on a flat sheet, so that surface undulations are minimized. Most recently, Pertsin *et al.* [58] carried out molecular simulations on such supported bilayers and reported that the short-range water-mediated repulsion originates from the hydration component of the intermembrane pressure, whereas the direct interaction between the membranes remains attractive throughout the pressure range studied. In short, the origin of this phenomenon is still enigmatic, although the best evidence suggests that water is certainly involved.

2. *Leakage.* The main phase transition in pure phospholipids is quite abrupt in some phospholipids, such as DPPC, which may require as little as 1–2°C for completion. The measured transition temperature T_m and the transition temperature range depend on the composition of the phospholipid head group, the length of the acyl chains, and the degree of unsaturation. In complex mixtures of phospholipids such as those found in native biological membranes, the range over which the transition occurs is usually very broad—often tens of degrees in duration [59]—or, in the case of membranes with high cholesterol content, a discrete transition is not seen at all [60]. Since passage though the phase transition, even in pure phospholipids with the sharpest, most cooperative transitions, does not occur absolutely simultaneously in all members of the lipid population, domains of gel- and liquid-crystalline-phase lipids must coexist for the duration of the transition. The literature suggests that the increased permeability of the vesicles seen during the transition may be due to packing defects in phase boundaries between these coexisting gel and liquid-crystalline phases [61, 62]. The rate of movement of trapped solutes across the membrane during the phase transition is enhanced in a binary lipid mixture, and the more immiscible the lipids, the higher the permeability, presumably again because of packing defects at

boundaries between the lipid domains. Hays *et al.* [63] reported that the following variables affected leakage from liposomes during chilling: (i) an increase in the rate of cooling and warming resulted in decreased leakage; (ii) maximal leakage occurred at the measured phase transition temperature; (iii) addition of defect-forming additives such as a second phospholipid or a surfactant increased leakage from the liposomes during the phase transition but not above or below that temperature; and (iv) small unilamellar vesicles leaked much more rapidly than large unilamellar vesicles (a phenomenon no doubt related to the high surface to volume ratio of small vesicles). There is no doubt that such leakage occurs even in intact cells during thermotropic phase transitions. For example, Oliver *et al.* [64] showed that internal [Ca^{++}] increased in human platelets during chilling, which was ascribed to leakage of Ca^{++} across membranes coincident with a lipid phase transition. This rise in internal [Ca^{++}] appears to be directly related to the fact that human platelets are activated by chilling. Similar damage from chilling is seen in a wide variety of cells [65]. Leakage across membranes during drying, by contrast, may not occur. Inspection of the phase diagram for a phospholipid (Fig. 11.5) will indicate why: the phase transition does not occur until all the bulk water is gone from the system; thus, there is no water into which the contents of the interior of the membrane can leak. However, when water is added back to the system, the situation changes dramatically. As the membrane is rehydrated it will undergo a phase transition, during which leakage may occur. For example, when dry yeast cells are rehydrated they must be placed in warm water. This requirement had been known for many years (indeed, the instructions on packets of commercial baker's yeast are quite specific about this), but why this is so was not understood until Leslie *et al.* [66] measured lipid phase transitions in yeasts. The fully hydrated cells showed a transition around 0°C, while the dry ones had a transition between 35°C and 40°C. Thus, when the dry cells are put in water below 35°C they will undergo a phase transition during the rehydration and will leak their contents to the medium and die. If they are placed in warm water (about 40°C), the membranes will have already passed through the transition before they become hydrated and the leakage is obviated. The same mechanism seems to apply to many cells that survive drying [65]. In the following main section we will review some of what is known about adaptive mechanisms that permit these organisms to survive drying.

3. *Phase separation in membranes during drying.* As we described earlier one of the driving forces behind the phase separation of microdomains in water the hydrophobic mismatch between long-chain lipids with high T_m and shorter-chain ones with low T_m. Now consider the consequences

for the simple binary system of DLPC and DPPC used in Fig. 11.2. At full hydration, the DLPC is liquid crystalline and the DPPC gel, and the resulting hydrophobic mismatch leads to complete phase separation (Fig. 11.2). But as water is removed, the DLPC enters the gel phase as well and the driving force that maintains the phase separation disappears. As a result, the two lipids mix, and the domain structure disappears. This appealing model is satisfying but misleading; a native membrane is far more complex than this simple model, and the consequences for drying are more complex. The various lipid components in the membrane will all have discrete phase diagrams and thus will enter the gel phase at different water contents. Under these conditions, gel-phase domains will tend to exclude liquid-crystalline lipids and extensive phase separation may result. For example, when a native membrane, sarcoplasmic reticulum, was dried, massive cholesterol crystals were seen [67]. Furthermore, nonbilayer hexagonal phases were seen, due to phase separation of the phosphatidylethanolamines [67]. Such membranes were irreversibly damaged from the dehydration. Similarly, the domain structure necessary to membrane function was irreversibly violated in platelets that were dried without protection and phase separation of other components was seen [68].

If all this damage occurs due to dehydration, how is it that some organisms can survive drying?

11.4 LIFE WITHOUT WATER: OVERCOMING THE DAMAGE FROM LIPID PHASE TRANSITIONS

11.4.1 Trehalose Production and Stress

Trehalose is a sugar that accumulates at high concentrations—as much as 20% of the dry weight—in the tissues of many organisms capable of surviving complete dehydration, spread across all taxonomic kingdoms. For example, baker's yeast cells, which have been the subject of the most intensive investigation, do not survive drying in the log phase of growth and do not contain significant amounts of trehalose, but in the stationary phase they accumulate the sugar and may then be dried successfully [69]. Until relatively recently, trehalose was thought to be a storage sugar in these cells, and the correlation between survival in the dry state and its presence was believed to be related to repair functions during rehydration, providing a ready energy source. Evidence is accumulating that trehalose production may be a universal stress response in yeasts; it can even prevent damage from environmental insults such as ethanol production during fermentation

[70–72]. In fact, overproduction of trehalose in some yeast strains by increasing synthesis and decreasing degradation is being used to boost ethanol tolerance and thus to boost industrial ethanol production. Along the same lines, trehalose has been shown to inhibit toxic effects of toluene in a bacterium [70]. Trehalose production may be more widespread as a stress response than had been previously appreciated; the analog of trehalose in higher plants has been thought to be sucrose [73], but several higher plants have recently been reported to produce trehalose in response to drought stress, either by the plant itself or by a symbiont. Recent evidence suggests that the genes for trehalose biosynthesis may be widespread in the genome of higher plants; they appear to be involved in signaling [74].

11.4.2 Trehalose and Biostability

Beginning in the early 1980s we established that biomolecules and molecular assemblages such as membranes and proteins can be stabilized in the dry state in the presence of trehalose [75]. Since that time, trehalose has come to be widely applied in the stabilization of biological materials [75]. Because the present article is focused on membranes, a brief discussion of stabilization of membranes in the dry state follows.

In the initial studies [76, 77], liposomes were prepared from a lipid with low T_m, palmitoyloleoylphosphatidylcholine (POPC). A fluorescent marker, carboxyfluorescein, was trapped in the aqueous interior. When the liposomes were freeze-dried with trehalose and rehydrated, the vesicles were seen to be intact, and nearly 100% of the carboxyfluorescein was retained. It quickly emerged that stabilization of POPC liposomes, and other vesicles prepared from low-melting-point lipids, had two requirements, inhibition of fusion between the dry vesicles and depression of T_m in the dry state. In the hydrated state, T_m for POPC is about −1°C and rises to about +70°C when it is dried without trehalose. In the presence of trehalose, T_m is depressed in the dry state to −20°C. Thus, the lipid is maintained in the liquid-crystalline phase in the dry state, and phase transitions are not seen during rehydration at room temperature.

The significance of this phase transition during rehydration is that when phospholipids pass through such transitions, the bilayer becomes transiently leaky, which resembles the effects resulting from passage through a main phase transition in fully hydrated membranes during changes in temperature, as discussed earlier. (The physical basis for this leakiness has recently been investigated in some detail by Hays *et al.* [63]. Thus, the leakage that normally accompanies this transition must be avoided if the contents of membrane vesicles and whole cells are to be retained. During drying, leakage is probably not a problem because T_m is not affected until all the bulk water has been removed. But during rehydration, it is a serious problem; the membranes are placed in water and will undergo the phase transition in the presence of excess bulk water, thus allowing leakage. In addition to the damage that occurs during passage through the phase transition, and perhaps even more importantly in the present context, phase separation of membrane components can occur in the absence of trehalose during drying, as the membranes undergo the transition into the gel phase, an event that is often irreversible.

Low-temperature-melting lipids such as POPC all seem to behave as described previously; T_m is depressed to a minimal value immediately in the presence of trehalose after drying, independent of the thermal history (Fig. 11.6). Saturated lipids with high T_m, such as DPPC, behave quite differently, and effects of trehalose on T_m depend strongly on the thermal history (Fig. 11.6). When T_m in DPPC dried without trehalose is measured, it is seen to rise from 41°C in the hydrated lipid to 110°C when it is dried. In the presence of trehalose, T_m is about 60°C until the acyl chains are melted once, after which T_m is depressed to 24°C [78]. If the lipid is then incubated at temperatures <24°C, T_m rapidly reverts to about 60°C. Thus, the stable T_m renders DPPC in the gel phase at physiological temperatures, regardless of whether it is hydrated or dry. This effect may have special relevance for biological membranes, because microdomains contain lipids with elevated T_m.

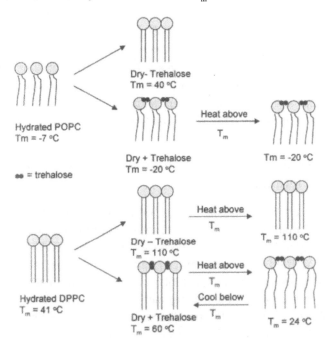

Figure 11.6 Effects of thermal history and the presence of trehalose on T_m in dry phospholipids.

11.4.3 Mechanism of Depression of T_m in Dry Phospholipids

The mechanism of depression of T_m has received a great deal of attention since the discovery of this effect [75–79]. Three main hypotheses have emerged: The water replacement hypothesis suggests that sugars can replace water molecules by forming hydrogen bonds with polar residues, thereby stabilizing the structure in the absence of water [76]. The water entrapment hypothesis suggests that sugars concentrate water near surfaces, thereby preserving its salvation [80]. The vitrification hypothesis suggests that the sugars form amorphous glasses, thus reducing structural fluctuations [81, 82].

A consensus has emerged that these three mechanisms are not mutually exclusive [76]. Vitrification may occur simultaneously with direct interactions between the sugar and polar residues. Direct interaction, on the other hand, has been demonstrated by a wide variety of physical techniques, including IR spectroscopy [79, 83], NMR [84, 85], and X-ray [86, 87]. Theoretical analyses have contributed greatly to this field in recent years. Chandrasekhar and Gaber [88] and Rudolph *et al.* [89], in the earliest studies, showed that trehalose can form energetically stable conformations with phospholipids, binding three adjacent phospholipids in the dry state. Sum *et al.* [90] showed by molecular simulations that the sugars adapt molecular conformations that permit them to fit onto the surface topology of the bilayer through hydrogen bonds. The sugars interact with up to three adjacent phospholipids. Golovina *et al.* [91] recently confirmed these results with molecular simulations and showed that trehalose increases the area per lipid in the dry state under conditions that seem to be inconsistent with any model that does not require direct interaction between the sugar and polar head group.

11.4.4 Trehalose Maintains Microdomains in Dry Model Membranes

We have carried out modeling studies in an attempt to discover whether trehalose can maintain the structural integrity of rafts in the dry state and, if so, to elucidate the mechanism. Because the phase behavior of the binary mixture DLPC/DSPC is so well characterized in the hydrated state, we started there, rather than with the more complex mixture seen in a native membrane. Using deuterated DPPC and hydrogenated DLPC, it has been possible to monitor phase behavior of the two lipids in a 1:1 mixture, using FTIR [83], with the following results: (i) the freshly prepared fully hydrated mixture is completely phase-separated in the gel/liquid-crystalline coexistence temperature regime, as expected; (ii) when this mixture is dried without trehalose, mixing occurs; (iii) if the mixture is dried with trehalose, the lipid phase separation is maintained, although a small fraction of the DLPC is mixed with the DSPC; and (iv) in the mixture dried with trehalose, most of the DLPC fraction has a transition below 0°C in the dry state, while the DSPC transition is seen at about 80°C [83].

We propose that trehalose maintains phase separation in this mixture of lipids in the dry state by the following mechanism. The DLPC fraction, with its low T_m in the hydrated state, might be expected to behave like unsaturated lipids described earlier, in that T_m in the dry state is reduced to a minimal and stable value immediately after drying with trehalose, regardless of the thermal history. That appears to be the case. The DSPC fraction, by contrast, would be expected to behave like DPPC, as described earlier. DSPC is in the gel phase in the hydrated state at room temperature, and it remains in the gel phase when it is dried with trehalose. In other words, we are proposing that by maintaining one of the lipids in the liquid-crystalline phase during drying, while the other remains in the gel phase, trehalose maintains the phase separation (Fig. 11.7). We suggest that this is the fundamental mechanism by which trehalose maintains microdomains in native membranes during drying. By maintaining the liquid-ordered/liquid-crystalline microdomain structure in platelets during drying and rehydration, trehalose could preserve the small-scale phase separation that appears to be important for membrane function and thus prevent macroscopic phase separation.

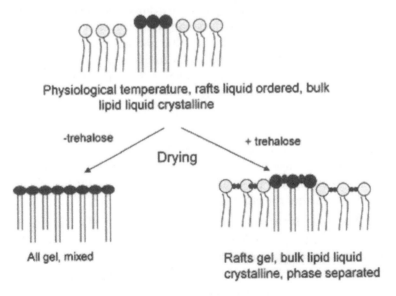

Figure 11.7 Proposed mechanism for maintaining lipid domains intact in dry membranes in the presence of trehalose.

11.4.5 There Is More Than One Way to the Same End

Although the occurrence of trehalose at high concentrations is common in anhydrobiotic animals, some such animals have only small amounts of trehalose or none at all [92–95]. It is tempting to construe these findings as evidence against a role for sugars in anhydrobiosis [96]. We suggest that it is not the sugars *per se* that are of interest in this regard, but rather the physical principles of the requirements for stabilization, as described above. There are multiple ways to achieve such stabilization:

1. Hydroxyethyl starch (HES) alone will not stabilize dry membrane vesicles composed of lipids with low T_m, but a combination of a low-molecular-weight sugar such as glucose and HES can be effective [97]. Here is the apparent mechanism: Glucose depresses T_m in the dry lipid but has little effect on inhibiting fusion, except at extremely high concentrations. On the other hand, the polymer has no effect on the phase transition, but inhibits fusion. Thus, the combination of the two meets both requirements, while neither alone does so. It seems likely that such combinations of molecules might be found in anhydrobiotes in nature.

2. In fact, a glycan isolated from the desiccation tolerant alga *Nostoc* apparently works in conjunction with oligosaccharides [98]. Similarly, certain proteins have been shown to affect the phase state of the sugars and either enhance or are required for stabilization [99].

3. Hincha *et al.* [100] have shown that fructans from desiccation-tolerant higher plants will by themselves both inhibit fusion and reduce T_m in dry phospholipids such as egg PC. The mechanism behind this effect is still unclear. The interaction is similar to that shown by sugars, but it is also specific to fructans and is not shown by other polymers [101]. In a related study, Hincha *et al.* [102] reported that a series of raffinose family oligosaccharides are all capable of stabilizing dry liposomes.

4. Hoekstra and Golovina [103] have reported that amphiphiles that are free in the cytoplasm in fully hydrated cells of anhydrobiotes apparently are inserted into membranes during dehydration. The role of this phenomenon in stabilization is uncertain, but presumably the amphiphiles alter the order of the acyl chains.

5. Goodrich *et al.* [104] reported that disaccharides tethered to the bilayer surface by a flexible linker esterified to cholesterol have an effect on membrane stability similar to that seen in free sugar. Such molecules could provide stability in anhydrobiotes, although they have not yet been reported. However, Popova and Hincha [105] found that digalactosyl diacylglycerol depressed T_m in dry phospholipids, perhaps in keeping with this suggestion.

The point is there are many ways to achieve stability. Once an understanding of the physical requirements for preservation was achieved, it became apparent that many routes can lead to the same end. Similar observations on the stability of dry proteins have been made by Carpenter *et al.*, with similar conclusions [106, 107].

Is trehalose special? No, not really, but it does have some useful properties that have made it a favorite for stabilizing biological materials, pharmaceuticals, vaccines, and the like. Those properties are summarized, along with some conclusions about the myth that has grown up about trehalose, in Crowe *et al.* [75, 76].

11.5 TREHALOSE CAN STABILIZE DRY MAMMALIAN CELLS

The use of disaccharides as protectants for mammalian cells has been a promising approach, but a major challenge has been to introduce the sugar into the cell at high concentrations. Current efforts aimed at solving this problem include (i) transfecting cells with the genes for trehalose synthesis [108], (ii) using transient pores in the membrane that permit passage of trehalose [109], and (iii) taking advantage of the transient leakiness of membranes at the phase transition temperature [110]. These studies suggest that trehalose improves freeze tolerance and survival of partial dehydration. However, conflicting results have been reported on the ability of trehalose loaded cells to survive complete dehydration [111–113].

We have reported successful freeze-drying of human platelets, which results in survival exceeding 90% [114]. We started this project at the invitation of the United States Department of Defense, where there is an obvious need for platelets for use in severe trauma cases. At present, platelets are stored in blood banks for a maximum of three to five days, by federal regulation, after which they are discarded. Furthermore, the platelets are stored at room temperature; they cannot even be refrigerated without rendering them useless therapeutically, a phenomenon for which we have provided an explanation [115–118]. There is a chronic shortage of platelets in hospitals, and field hospitals operated by the military rarely have access to platelets at all. Thus, prolonging the shelf life of platelets would be a valuable contribution. Freeze-dried platelets have the following properties:

1. The dry platelets are stable for at least two years when stored at room temperature under vacuum. During that time we have seen no loss of platelets.

2. The freeze-dried, rehydrated cells respond to normal platelet agonists, including thrombin, ADP, collagen, and ristocetin.

3. Studies on the morphology of the trehalose-loaded, freeze-dried, and rehydrated platelets show that they are affected by the drying but are morphologically similar to fresh platelets. When they were dried without trehalose, on the other hand, most of the platelets disintegrated during the rehydration event, but of the small number that were left, most had fused with adjacent cells, forming an insoluble clump.

4. We have extended the freeze-drying to mouse and pig platelets as animal models for *in vivo* testing. The rehydrated platelets are far from perfect, but they nevertheless show surprisingly good regulation of key elements of cellular physiology such as intracellular calcium [64]. For instance, when fresh platelets are challenged with thrombin they show an increase in [Cai] that is dose dependent. The rehydrated platelets show a similar response, although it is strongly attenuated. Nevertheless, the increase in [Cai] appears to be sufficient to trigger morphological and physiological changes required for coagulation.

11.6 MEMBRANE DOMAINS ARE PRESERVED IN PLATELETS FREEZE-DRIED WITH TREHALOSE

We incubated the rehydrated platelets in a fluorescent probe that partitions into ordered domains (dil) and found that they were labeled uniformly, similar to the image seen with fresh platelets [27]. When they were then chilled or exposed to thrombin, the label aggregated, again similar to the pattern seen in the fresh platelets, suggesting that microdomains are intact in the platelets freeze-dried with trehalose. Studies with FTIR appear to confirm this result; the thermal behavior in cells freeze-dried with trehalose is similar to that seen in fresh platelets. By contrast, those freeze-dried without trehalose showed extensive phase separation, with little similarity to the pattern seen in fresh platelets. We propose that the mechanism of stabilization of these membrane domains in the dry platelets is that shown in Fig. 11.7.

11.7 SUMMARY

Water is essential for the formation of the lipid bilayers that constitute the fundamental backbone of membranes. While other factors are also involved, water plays a major role in driving and maintaining the domain structure of membranes, which appears to be essential for biological functions. Nevertheless, many anhydrobiotic organisms are capable of stabilizing membranes, including their functions, in the absence of water. The mechanism by which this is accomplished is having some impact on stabilization of cells of interest in human welfare that previously have been refractory to preservation.

References

1. Hunter, R. J. (2001) *Foundations of Colloid Science*, Oxford University Press, New York.

2. Gelbart, W. M., and Ben-Shaul, A. (1996) *J. Phys. Chem.*, **100**, 13169.

3. Israelachvili, J. (1991) Intermolecular and Surface Forces, Academic Press, New York.

4. Tresset, G. (2009) *PMC Biophys.*, **2**, 1–25.

5. Tanford, C. (1980) *The Hydrophobic Effect: Formation of Micelles and Biological Membranes*, Wiley, New York.

6. Chandler, D. (2005) *Nature*, **437**, 640.

7. Willard, A. P., and Chandler, D. (2008) *J. Phys. Chem. B*, **112**, 6187.

8. Meyer, E. E., Rosenberg, K. J., and Israelachvili, J. (2006) *Proc. Natl. Acad. Sci. USA*, **103**, 15739.

9. Despa, F., and Berry, R. S. (2007) *Biophys. J.*, **92**, 373.

10. Maibaum, L., Dinner, A. R., and Chandler, D. (2004) *J. Phys. Chem. B*, **108**, 6778.

11. Israelachvili, J. N., Mitchell, D. J., and Ninham, B. W. (1976) *J. Chem. Soc. Faraday Trans.*, **272**, 1525.

12. Zemb, T., Dubois, M., Deme, B., and Gulik-Krzywicki, T. (1999) *Science*, **283**, 816.

13. Dubois, M., Demé, B., Gulik-Krzywicki, T., Dedieu, J. C., Vautrin, C., Désert, S., Perez, E., and Zemb, T. (2001) *Nature*, **411**, 672.

14. Dubois, M., Lizunov, V., Meister, A., Gulik-Krzywicki, T., Verbavatz, J. M., Perez, E., Zimmerberg, J., and Zemb, T. (2004) *Proc. Natl. Acad. Sci. USA*, **101**, 15082.

15. Brown, D. A. (2001) *Proc. Natl. Acad. Sci. USA*, **98**, 10517.

16. Kenworthy, A. (2002) *Trends Biochem. Sci.*, **27**, 435.

17. London, E. (2002) *Curr. Opin. Struct. Biol.*, **12**, 480.

18. Szöőr, A., Szöllősi, J., and Vereb, G. (2010) *Immunol. Lett.*, **130**, 2-12

19. Weisz, O. A., and Rodriguez-Boulan, E. (2009) *J. Cell Sci.*, **122**, 4253.

20. Lingwood, D., and Simons, K. (2010) *Science*, **327**, 46.

21. Marsh, D. (2010) *Biochim. Biophys Acta*, **1798**, 688-699

22. Martín, V., Fabelo, N., Santpere, G., Puig, B., Marín, R., Ferrer, I., and Díaz, M. J. (2010) *J. Alzheimers Dis.*, **19**, 489.

23. Patil, A., Gautam, A., and Bhattacharya. (2010) *J. Virol.* (in press).

24. Xu, Z. X., Ding, T., Haridas, V., Connolly, F., and Gutterman, J. U. (2009) *PLoS One*, **4**, e8532.

25. Chaigne-Delalande, B., Mahfouf, W., Daburon, S., and Moreau, J. F., and Legembre P. (2009) *Cell Death Differ.*, **16**, 1654.

26. Brouckova, A., and Holada, K. (2009) *Thromb. Haemostasis*, 5, 966.

27. Leidy, C., Gousset, K., Ricker, J. Wolkers, W. F.,Tsvetkova, N. M., Tablin, F., and Crowe, J. H. (2004) *Cell Biochem. Biophys.*, **40**, 123.

28. Thewalt, J. L., and Bloom, M. (1992) *Biophys. J.*, **63**, 1176.

29. Sankaram, M. B., and Thompson, T. E. (1991) *Proc. Natl. Acad. Sci. USA*, **8**, 8686.

30. Nudelmang G., Weigert, M., and Louzoun, Y. (2009) *Mol. Immunol.*, **15**, 3141.

31. Risselada, H. J., and Marrink, S. J. (2008) *Proc. Natl. Acad. Sci. USA*, **105**, 17367.

32. Mabrey, S., and Sturtevant, J. M. (1976) *Proc. Natl. Acad. Sci. USA*, **73**, 3862.

33. Killian, J. A. (1998) *Biochim. Biophys. Acta*, **137**, 401.

34. Gil, T., Ipsen, J. H., Mouritsen, O. G., Sabra, M. C., Sperotto, M. M., and Zuckermann, M. J. (1998) *Biochim. Biophys. Acta*, **1376**, 245.

35. Pedersen, S., Jorgensen, K., Baekmark, T. R., and Mouritsen, O. G. (1996) *Biophys. J.*, **71**, 554.

36. Korlach, J., Schwille, P., Webb, W. W., and Feigenson, G. W. (1999) *Proc. Natl. Acad. Sci. USA*, **96**, 8461.

37. Feigenson, G. W., and Buboltz, J. T. (2001) *Biophys. J.*, **80**, 2775.

38. Leidy, C., Wolkers, W. F., Jørgensen, K., Mouritsen, O. G., and Crowe, J. H. (2001) *Biophys. J.*, **80**, 1819.

39. Bagatolli, L. A., and Gratton, E. (2000) *Biophys. J.*, **79**, 434.

40. Nagle, J. F., and Tristram-Nagle, S. (2000) *Biochim. Biophys. Acta: Rev. Biomem.*, **1469**, 159.

41. Tristram-Nagle, S., and Nagle, J. F. (2004) *Chem. Phys. Lipids*, **127**, 3.

42. Lopez, C. F., Nielsen, S. O., Klein, M. L., and Moore, P. B. (2004) *J. Phys. Chem. B.*, **108**, 6603–6610.

43. Khandelia, H., and Kaznessis, Y. N. (2007) *Biochim. Biophys. Acta*, **1768**, 509.

44. Damodaran, K. V., and Merz, J. K. M. (1993) *Langmuir*, **9**, 1179.

45. Zhao, W., Moilanen, D. E., Fenn, E. E., and Fayer, M. D. (2008) *J. Am. Chem. Soc.*, **130**, 13927.

46. Milhaud, J. (2004) *Biochim. Biophys. Acta*, **1663**, 19.

47. Fitter, J., Lechner, R. E., and Dencher, N. (1999) Interactions of hyration water and biological membranes studied by neutron scattering, *J. Phys. Chem. B*, **103**, 8036–8050.

48. Jendrasiak, G. L., and Smith, R. L. (2004) *Chem. Phys. Lipids*, **131**, 183.

49. Hsieh, C. H., and Wu, W. G. (1996) *Biophys. J.*, **71**, 3278.

50. Howe-Siang, T., Piletic, I. R., and Fayerb, M. D. (2005) *J. Chem. Phys.*, **122**, 174501.

51. Piletic, I. R., Moilanen, D. E., Spry, D. B., Levinger, N. E., and Fayer, M. D. (2006) *J. Phys. Chem. A*, **110**, 4985.

52. Moilanen, D. E., Piletic, I. R., and Fayer, M. D. (2006) *J. Phys. Chem. A*, **110**, 9084.

53. Volkov, V. V., Palmer, D. J., and Righini, R. (2007) *J. Phys. Chem. B*, **111**, 1377.

54. Volkov, V. V., Palmer, D. J., and Righini, R. (2007) *Phys. Rev. Lett.*, **99**, 78302.

55. Rand, R. P., and Parsegian, V. A. (1989) *Biochim. Biophys. Acta*, **988**, 351.

56. Israelachvili, J., and Wennerstrom, H. (1996) *Nature*, **379**, 219.

57. Lipowsky, R., Grotehans, S. (1993) *Europhys. Lett.*, **23**, 599.

58. Pertsin, A., Platonov, D., and Grunze, M. (2007) *Langmuir*, **23**, 1388.

59. Chapman, D. (1975) *Q. Rev. Biophys.*, **8**, 185.

60. Mantsch, H. H., and McElhaney, R. N. (1991) *Chem. Phys. Lipids*, **57**, 213.

61. Blok, M. C., van Deenen, L. L. M., and de Gier, J. (1976) *Biochim. Biophys. Acta*, **433**, 1.

62. Clerc, S. G., and Thompson, T. E. (1995) *Biophys. J.*, **68**, 2333.

63. Hays, L. M., Crowe, J. H., Wolkers, W. F., and Rudenko, S. (2001) *Cryobiology*, **42**, 88.

64. Oliver, A. E., Tablin, F., Walker, N. J., and Crowe, J. H. (1999) *Biochim. Biophys. Acta*, **1416**, 349.

65. Crowe, J. H., Crowe, L. M., and Hoekstra, F. A. (1989) *J. Bioenerg. Biomembr.*, **21**, 77.

66. Leslie, S. B., Teter, S. A., Crowe, L. M., and Crowe, J. H. (1994) *Biochim. Biophys. Acta*, **1192**, 7.

67. Crowe, L. M., and Crowe, J. H. (1982) *Arch. Biochem. Biophys.*, **217**, 582.

68. Crowe, J. H., Tablin. F., Wolkers, W. F., Gousset, K., Tsvetkova, N. M., and Ricker, J. (2003) *Chem. Phys. Lipids*, **122**, 41.

69. Crowe, J. H., Carpenter, J. F., and Crowe, L. M. (1998) *Ann. Rev. Physiol.*, 6, 73.

70. Argüelles, J. C. (2000) *Arch. Microbiol.*, **174**, 217.

71. Mansure, J. J., Panek, A. D., Crowe, L. M., and Crowe, J. H. (1994) *Biochim. Biophys. Acta*, **1191**, 309.

72. Joo, W. H., Shin, Y. S., Lee, Y., Park, S. M., Jeong, Y. K., and Seo, J. Y. (2000) *Biotechnol. Lett.*, **22**, 1021.

73. Filip, R., Baena-Gonzalez, E., and Sheen, J. (2006) *Annu. Rev. Plant Biol.*, **57**, 675.

74. Paul, M. J., Primavesi, L. F., Jhurreea, D., and Zhang, Y. (2008) *Annu. Rev. Plant Biol.*, **59**, 417.

75. Crowe, J. H. (2007) In *Molecular Aspects of the Stress Response: Chaperones, Membranes and Networks* (ed. Csermely, P., and Vigh, L.), Springer Science+Business Media, New York.

76. Crowe, J. H., Crowe, L. M., Oliver, A. E., Tsvetkova, N. M., Wolkers, W. F., and Tablin, F. (2001) *Cryobiology*, **43**, 89.

77. Crowe, J. H., and Crowe, L. M. (1992) In *Liposome Technology*, 2nd ed. (ed. Gregoriadis, G), CRC Press, Boca Raton, FL.

78. Crowe, J. H., Hoekstra, F. A., Nguyen, K. H. N., and Crowe, L. M. (1996) *Biochim. Biophys. Acta*, **1280**, 187.

79. Crowe J. H., Crowe L. M., and Chapman D. (1984) *Science*, **223**, 701.

80. Cottone, G., Cicotti, G., and Cordone, L. (2002) *J. Cell. Phys.*, **117**, 9862.

81. Koster, K. L., Webb, M. S., Bryant, G.. and Lynch, D. V. (1994) *Biochim. Biophys. Acta*, **1193**, 143.

82. Wolfe, J., and Bryant, G. (1999) *Cryobiology*, **39**, 103.

83. Ricker, J. V., Tsvetkova, N. M., Wolkers, W. F., and Crowe, J. H. (2003) *Biophys. J.*, **84**, 3045.

84. Lee, C.W.B, Waugh, J. S., and Griffin, R. G. (1986) *Biochemistry*, **25**, 3737.

85. Tsvetkova, N. M., Phillips, B. L., Crowe, L. M., and Crowe, J. H. (1998) *Biophys. J.*, **75**, 2947.

86. Nakagaki, M., Nagase, H., and Ueda, H. (1992) *J. Mem. Sci.*, **73**, 173.

87. Luzardo, M. D., Amalfa, .F, and Nunez, A. M. (2000) *Biophys. J.*, **78**, 2452.

88. Chandrasekhar, I., and Gaber, B. P. (1988) *J. Biomol. Struct. Dyn.*, **5**, 1163.

89. Rudolph, B. R., Chandrasekhar, I., and Gaber, B. P. (1990) *Chem. Phys. Lipids*, **53**, 243.

90. Sum, A. K., Faller, F., and de Pablo, J. J. (2003) *Biophys. J.*, **85**, 2830.

91. Golovina, E. A., Golovin, A. V., Hoekstra, F. A., and Faller, F. (2009) *Biophys. J.*, **97**, 490.

92. Womersley C. (1990) In *Entomopathogenic Nematodes in Biological Control* (ed. Gaugler, R., and Kaya, H. K.), CRC Press, Boca Raton, FL.

93. Westh, P., and Ramløv, H. (1991) *J. Exp. Zool.*, **258**, 303.

94. Lapinski, J., and Tunnacliffe, A. (2003) *FEBS Lett.*, **553**, 387.

95. Caprioli, M., Katholm, A. K., and Melone, G. (2004) *Comp. Biochem. Physiol. A*, **139**, 527.

96. Tunnacliffe, A., and Lapinski, J. (2003) *Philos. Trans. R. Soc. London B*, **358**, 1755.

97. Crowe, J. H., Oliver, A. E., and Hoekstra, F. A. (1997) *Cryobiology*, **3**, 20.

98. Hill, D. R., Keenan, T. W., Helm, R. F., Potts, M., Crowe, L. M., and Crowe, J. H., and (1997) *J. Appl. Phycol.*, **9**, 237.

99. Buitink, J., Walters-Vertucci, C., and Hoekstra, F. A. (1996) *Plant Physiol.*, **111**, 235.

100. Hincha, D. K., Hellwege, E. M., and Meyer, A. G. (2000) *Eur. J. Biochem.*, **267**, 535.

101. Vereyken, I. J., Chupin, V., and Hoekstra, F. A. (2003) *Biophys. J.*, **84**, 3759.

102. Hincha, D. K., Zuther, E., and Heyer, A. G. (2003) *Biochim. Biophys. Acta*, **1612**, 172.

103. Hoekstra, F. A., and Golovina, E. A. (2002) *Comp. Biochem. Physiol.*, **131A**, 527.

104. Goodrich, R. P., Crowe, J. H., Crowe, L. M., and Baldeschwieler, J. (1991) *Biochemistry*, **30**, 2313.

105. Popova, A. V., and Hincha, D. K. (2005) *Glycobiology*, **15**, 1150.

106. Allison, S. D., Manning, M. C., Randolph, T. W., and Carpenter, J. F. (2000) *J. Pharm. Sci.*, **89**, 199.

107. Anchordoquy, T. J., Izutsu, K. I., Randolph, T. W., and Carpenter, J. F. (2001) *Arch. Biochem. Biophys.*, **390**, 35.

108. Guo, N., Puhlev, I., Brown, D. R., Mansbridge, J., and Levine, F. (2000) *Nat. Biotechnol.*, **18**, 168.

109. Eroglu, A., Russo, M. J., Biaganski, R., Fowler, A., Cheley, S., and Bayley, H. (2000) *Nat. Biotechnol.*, **18**, 163.

110. Beattie, G. M., Crowe, J. H., Lopez, A. D., Cirulli, V., Ricordi, C., and Hayek, A. (1997) *Diabetes*, **46**, 519.

111. De Castro, A. G., and Tunnacliffe, A. (2000) *FEBS Lett.*, **487**, 199.

112. Gordon, S. L., Oppenheimer, S. R., Mackay, A. M., Brunnabend, J., Puhlev, I., and Levine, F. (2001) *Cryobiology*, **43**, 182.

113. Chen, T., Acker, J. P., Eroglu, A., Cheley, S., Bayley, H., Fowler, A., and Toner, M. (2001) *Cryobiology*, **43**, 168.

114. Wolkers, W. F., Walker, N. J., Tablin, F., and Crowe, J. H. (2001) *Cryobiology*, **42**, 79.

115. Tablin, F., Oliver, A. E., Walker, N. J., and Crowe, J. H. (1996) *J. Cell. Physiol.*, **168**, 305.

116. Tablin, F., Wolkers, W. F., Walke, N. J., and Crowe, J. H. (2001) *Cryobiology*, **43**, 114.

117. Crowe, J. H., Tablin, F., and Tsvetkova, N. M. (1999) *Cryobiology*, **38**, 180.

118. Tsvetkova, N. M., Walker, N. J., and Crowe, J. H. (2001) *Mol. Mem. Biol.*, **17**, 209.

Part V

WATER AND CELLULAR STRUCTURE/PHYSIOLOGY

Chapter 12

INTERFACIAL WATER AND CELL ARCHITECTURE/FUNCTION

Frank Mayer
Bungenstrasse 29,
D-21682 Stade,
Germany
fmayer12@gmx.de

12.1 FEATURES OF THE ULTRASTRUCTURAL ORGANIZATION OF EUKARYOTIC AND PROKARYOTIC CELLS AT THE CELLULAR LEVEL

12.1.1 Eukaryotic Cells and Prokaryotic Cells: A Comparison

Typical eukaryotic cells are characterized by the fact that they house a number of organelles. Figure 12.1a depicts, in a diagrammatic view, such a cell. Within the cytoplasm, various organelles can be seen: membrane stacks, flat membrane bags, and vesicles of various sizes. The arrows in the drawing indicate the fact that most of the organelles—with the exception of mitochondria and chloroplasts—are constantly involved in membrane flow and membrane transformation. In Fig. 12.1b–f, various organelles of the eukaryotic cell are shown at higher magnification. It is obvious that all these organelles exhibit internal ultrastructural organizations with spaces between membranes and with volume sizes in the nanometer range.

In Fig. 12.2, several examples of prokaryotic cells are depicted. Examples of bacteria were chosen that exhibit ultrastructural features similar to those of chloroplasts and mitochondria of eukaryotic cells. Again, ultrastructural

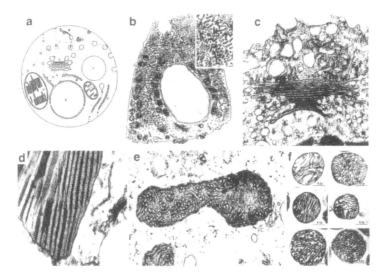

Figure 12.1 Structural organization of the eukaryotic cell. (a) A diagrammatic view of an idealized eukaryotic cell, exhibiting membrane-enclosed—narrow—compartments and their interactions by membrane flow and membrane transformation (arrows). Note: Chloroplasts and mitochondria are not involved in these dynamic processes. (b–f) Various organelles in a eukaryotic cell; (b) endoplasmic reticulum; (c) dictyosome; (d) chloroplast; (e,f) mitochondria. *Abbreviations*: B, C, vesicles; D, dictyosome; K, nucleus; M, mitochondrium; P, chloroplast; R, endoplasmic reticulum with attached ribosomes; V, vacuole.

organizations with spaces between membranes and with volume sizes in the nanometer range are present.

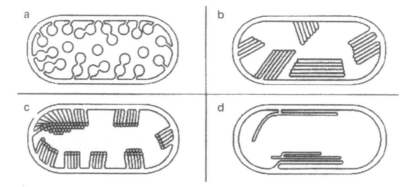

Figure 12.2 Intracytoplasmic membrane systems bearing the photosynthetic apparatus in phototrophic bacteria. (a) Vesicle type; (b) stacks; (c) bundled tubes; (d) thylakoids, partially stacked, partially irregularly arranged. Reproduced with permission from Ref. [27].

A structural similarity of these schematically drawn bacteria with chloroplasts and mitochondria is obvious. As shown in a multitude of analyses, this similarity is due to the reason that chloroplasts and mitochondria in recent eukaryots derived from precursors of today's prokaryotic cells that were taken up by precursors of today's eukaryotic cells. The principle of the structural features described earlier for eukaryotic cells appears to have been "invented" already at a time when eukaryotic cells did not yet exist.

12.1.2 Aspects of the Ultrastructural Organization of Prokaryotic Cells: Examples

Figure 12.3a depicts, in a schematic presentation, the ultrastuctural organization of the periphery of a typical gram-negative bacterial cell. Several features can be seen. The volume between the outer membrane (OM) and the cytoplasmic membrane (CM) (periplasmic space, containing a multitude of enzymes and other proteins [1]) is crossed by fibrillar elements (peptidoglycan [PG]), and structural proteins (lipoprotein [LP]) extend from the inner face of the outer membrane into the volume. This principle has the consequence that the periplasmic space is subdivided into very narrow subspaces in the nanometer range and that many surfaces are created within this space. A second feature can be deduced from this figure. Attached to the outer face of the outer membrane, fibrillar elements (lipopolysaccharide [LPS]) can be seen extending into the surrounding medium. It is known that they have functions in cell attachment to surfaces and in the specification of the cell type. However, when this structural feature is considered under the aspect of the presence or absence of interfacial water in the immediate vicinity of the cell surface, it may be assumed that the presence of so many fibrillar elements may considerably modify the conditions compared to the situation that would be given in this respect when only the surface of the outer membrane would be exposed.

In Fig. 12.3b–f, electron micrographs (Fig. 12.3b–e) and drawings on the basis of electron micrographs depict structural aspects of a bacterium (*Mycoplasma pneumoniae*) that is devoid of an outer membrane [2]. It is enclosed by a cytoplasmic membrane (arrow in Fig. 12.3b). After artificial removal of this membrane, stalks with knobs at their tip (Fig. 12.3c [top and center], electron micrographs, Fig. 12.3c [bottom], drawing according to Fig. 12.3c [center]) can be seen that extend from a so far unknown basement into the direction of the site where the cytoplasmic membrane was located prior to removal. A closer look (Fig. 12.3d,e) reveals that this basement is, in fact, composed of a regular network of meshes surrounding the entire cell body (Fig. 12.3f depicts a model of the basic structure), most probably made up of proteins organized as short fibrils. Though the identity of the components of the meshwork and the stalks is not definitively clear, their existence and their role as the basic support for holding the cytoplasmic membrane at its place are evident. Such a function may allow to define the network together with the stalks as a kind of bacterial cytoskeleton [3, 4].

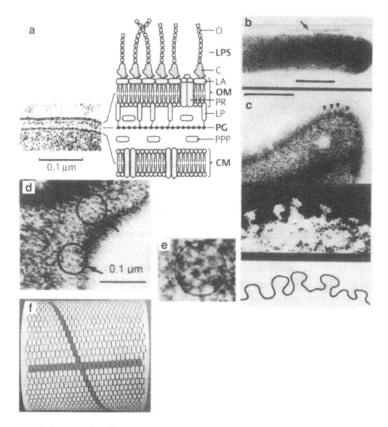

Figure 12.3 Bacterial cell envelope and cytoskeletal net in prokaryotes. (a) Electron micrograph of a cryosection through a cell of *Escherichia coli* (*left*); diagrammatic view of the macromolecular architecture of such a cell (*right*). The lines indicate the respective layers in the cryosection; (b–f) *Mycoplasma pneumoniae*, a wall-less bacterium; (b) the outermost layer of the cell is the cytoplasmic membrane (CM, arrow); (c) after artificial removal of the CM, stalks with terminal knobs can be seen extending toward the plane where the CM was located prior to its removal; (d,e) the basis of the stalks is attached to a proteinaceous network consisting of regular meshes; (f) diagrammatic view of this network, enclosing the cell body proper like a tube. *Abbreviations*: C, core region of the lipopolysaccharide; CM, cytoplasmic membrane; LA, lipid A; LP, lipoprotein; LPS, lipopolysaccharide; O, O-specific side chains of lipopolysaccharide; OM, outer membrane; PR, porin; PG, peptidoglycan (murein); PPP, periplasmic protein; T, transmembrane protein. Reproduced with permission: a from Ref. [27] and b–f from Refs. [2] and [31].

An additional aspect should also be kept in mind: the ultrastructural organization of the cell periphery of the bacterium *Mycoplasma pneumoniae* has an interesting consequence regarding the presence or absence of interfacial water and its influence on cell functions [5]. The volume of the cell in immediate vicinity of the inner face of the cytoplasmic membrane is substructured by the fibrillar elements of the cytoskeletal network into ultrastructures in the nanometer range, giving rise to a multitude of surfaces exposed to water, a prerequisite for the existence of interfacial water [6].

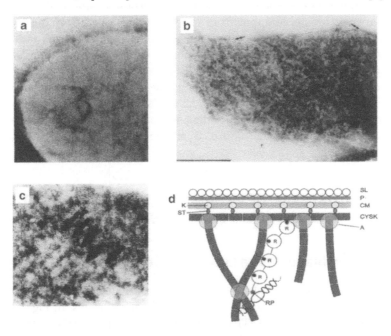

Figure 12.4 Bacterial cell envelope, cytoskeletal net, interaction of ribosomes with the cytoskeletal net. (a) Electron micrograph (negative staining) of the pole of an *Escherichia coli* cell; (b) cell body of an *Escherichia coli* cell after artificial removal of the entire cell envelope and the cytoplasmic membrane. Note that the cytoskeletal net is exposed (double arrows) and that the artificially flattened cell body, depicted in a face-on view, exhibits parallel, helically arranged striations; (c) a close inspection of these striations reveals that, along the striations, often rows of globular units with sizes of ribosomes are attached. This observation allows one to conclude that polysomes are attached to the components of the cytoskeletal net; (d) a diagrammatic view of the interaction of the CM of a typical eubacterium; depicted here is a gram-positive bacterium with a thick PG layer and a surface layer (SL) with the bacterial cytoskeletal net (CYSK); ST, stalks with terminal knobs (K), connecting the CM with the CYSK; A, postulated additional proteins stabilizing the contact of cytoskeletal fibrils. Note: The diagram also illustrates the interaction of ribosomes (R) with cytoskeletal elements; RP, RNA polymerase. Reproduced with permission: a–c from Ref. [31] and d from Ref. [4].

In Fig. 12.4a–c, electron micrographs of cell poles of the gram-negative bacterium *Escherichia coli* are shown. In Fig. 12.4a, a situation is depicted where the cell has lost, by experimental treatment, parts of its outer membrane, with the cytoplasmic membrane still present. In Fig. 12.4b, the cell is devoid of the cell envelope; also, the cytoplasmic membrane was removed. The salt used for electron microscopic contrasting (negative staining) had penetrated the remaining cell body. Fibrillar structures became visible, exhibiting an organization similar to a multistart helix surrounding the cell body. Again, such a structural organization allows to define the fibrillar structures as components of a bacterial cytoskeleton [4]. In Fig. 12.4c, parallel groups of ribosomes (polysomes) could be discovered, indicating that the polysomes were attached to fibrils of the multistart helix, that is, to the cytoskeleton. For eukaryotic cells, such an attachment is known [7]. For bacteria, its discovery was surprising. The observations depicted in Figs. 12.3 and 12.4a–c are combined in Fig. 12.4d. This schematic drawing exhibits the ultrastructural organization of the bacterial cell with many intracellular surfaces proteinaceous in nature, a prerequisite for the occurrence of interfacial water.

12.2 FEATURES OF THE ULTRASTRUCTURAL ORGANIZATION OF CELLS AT THE LEVEL OF ENZYME COMPLEXES

12.2.1 Membrane-Bound Enzyme Complexes

In Fig. 12.5a–c, examples of enzyme complexes are depicted that are known to function in close interaction with biological membranes. In Fig. 12.5a,b, adenosine triphosphatase (ATPase) complexes with general typical structural features of the F0F1 or the A0A1 type, present in prokaryotes as well as in the mitochondria of eukaryotic cells, are shown [8–10]. Without discussing detailed aspects of the functions of these enzymes, one structural feature is obvious: the active sites of the enzyme, located in the head part of the complex, have a distinct distance from the membrane surface. The respective enzyme subunits are connected, *via* stalks, to the membrane. Within the membrane, other functional components of the enzyme complex are located. In principle, a similar situation is encountered in the case of oxaloacetate decarboxylase from a bacterium (*Klebsiella pneumoniae*). By electron microscopic affinity labeling, it could be shown that its prosthetic biotin group is located at a specific distance from the membrane surface [11]. This distance is shorter compared with the respective distance of the active sites of the ATPases depicted in Fig. 12.5a,b. As in the case of these ATPases, the enzyme subunit carrying the prosthetic group is connected with a membrane-integrated subunit complex that plays a significant role in the overall enzyme function.

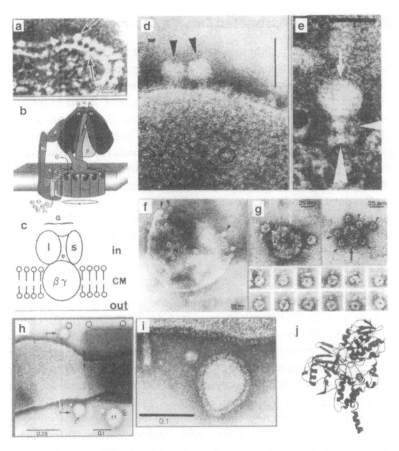

Figure 12.5 Structural organization and membrane attachment of enzyme complexes. Electron micrograph (a) and diagrammatic view (b) of ATPases of the FoF1and the AoA1 type (after Ref. [9], modified); (c) oxaloacetate decarboxylase from *Klebsiella pneumoniae*: size and shape of the enzyme and location of its prosthetic biotin group; (d,e) methano-reductosomes (arrowheads) attached to the cytoplasmic side of the (artificially reversed) cytoplasmic membrane of *Methanobacterium thermoautotrophicum*; (e) in addition to the "head," details of the stalk and the membrane integration of the basis of the stalk are visible; (f,g) F420-reducing hydrogenase complexes attached to the cytoplasmic side of the (artificially reversed) cytoplasmic membrane of *Methanobacteriumthermoautotrophicum*; (g) details of the stalks connecting the enzyme complex with the membrane are visible; (h,i) remnants of cells of *Thermoanaerobacterium thermosulfurogenes* EM1after removal of the cell envelope by specific growth conditions. The surface of the exposed cytoplasmic membrane and vesicles thereof is densely covered by α-amylase complexes attached to the membrane by stalks; (j) a ribbon diagram of human monoamine oxidase B (MAO B). The protein has a single transmembrane helix that anchors it to the outer membrane of the mitochondrium. Nevertheless, the protein is considered monotopic because the bulk of the 520 residues, including the active center, is outside the membrane. Reproduced with permission: c from Ref. [11], d,e from Refs. [12] and [13], g from Ref. [14], h,i from Ref. [15], and j from Ref. [16]. For color reference, see p. 381.

12.2.2 Membrane-Associated Enzyme Complexes

Remarkable structural similarities with the ATPase complexes described earlier are visible, at first view, in the enzyme complexes illustrated in Fig. 12.5d,e. The complex was named methano-reductosome [12, 13]. It contains, in its head part, numerous copies of the enzyme methyl-CoM methylreductase. Again, the head part has a distinct distance from the membrane surface (in this case, the cytoplasmic membrane of a methanogenic archaeon). A major difference compared with the ATPases described earlier is, as far as the function of the enzyme is concerned, that the enzyme methyl-CoM methylreductase proper is a typical "soluble" enzyme. It does not contain, within the enzyme complex, a component, involved in the process catalyzed by the enzyme, that is membrane integrated. Nevertheless, the functionally active head part is located at a specific distance from the surface of a membrane. The only known function of the stalk is that of keeping the head part at this specific distance.

Very similar is the situation in the case of the F420-reducing hydrogenase complex [14] depicted in Fig. 12.5f,g. Again, the enzyme complex is attached, by a stalk, at a distinct distance, to a membrane surface (here, again, the cytoplasmic membrane of a methanogenic archaeon). The enzyme proper is not known to contain a functionally important membrane-integrated part. Presumably, also in this enzyme the stalk has only the function of holding the enzyme complex with its catalytic centers at a certain distance from the membrane surface.

Figure 12.5h,i illustrates the attachment of the bacterial enzyme α-amylase to the outer surface of the cytoplasmic membrane. This attachment was surprising when it was discovered [15]. After all, this enzyme complex is a typical "soluble" enzyme, and it is involved in the degradation of a soluble substrate, amylase. Prior to the discovery of its attachment to a membrane, it was obtained, for technical purposes, from the supernatant of respective bacterial mass cultures, and the amount of enzyme activity harvested from the culture broth was not high, according to measurements by a specific test. This changed when it was realized that the soluble enzyme should not be harvested, but, instead, membrane fragments of degraded bacterial cells with the enzyme attached to the membrane fragments. From them, the enzyme could then be removed.

"Soluble" enzyme complexes that are associated to membranes are known not just in bacteria and archaea. Also in eukaryotic cells, similar cases can be found. An example is depicted in Fig. 12.5j. The figure shows the ribbon diagram of human monoamine oxidase [16]. The enzyme is monotopic. The bulk of the 520 residues, including the active center, is outside the membrane. It has characteristic features of a typical "soluble" enzyme. Nevertheless, as studies showed, the catalytic center of the enzyme is positioned at a distinct distance from the membrane surface. This is achieved by a stalklike extension that, according to current experimental evidence, is not involved in the function proper of the enzyme.

In conventional *in vitro* assays of specific activity applied for typical "soluble" enzymes, the results may not be comparable with the real specific activities of the enzymes *in vivo*. However, no proof for this assumption was obtained. The simple reason for this was that for typical "soluble" enzymes, it was assumed that they are also "soluble" inside the cell, and no need was seen to change this view. Hence, respective analyses were not undertaken. As shown later, this was very shortsighted.

12.2.3 Experiments Simulating, *in Vitro*, the *in Vivo* Spatial Situation Within Cells as Far as Membranes and Enzyme Complexes Are Concerned

From a multitude of experimental studies, it was deduced that various kinds of water structures do occur within the living cell [17, 18]. Water moieties in the immediate neighborhood of (membrane) surfaces—that is, in the nanometer range—were called interfacial water, and it was postulated that the specific—advantageous—properties of interfacial water influence cell functions.

As described earlier, several features of the structural organization of the cell, both at cellular and macromolecular levels, may indicate that cell functions might have been optimized during evolution with respect to water structures best suited for a most effective function [19]. Creating narrow spaces enclosed by membranes that have dimensions in the nanometer range might have been one approach. Placing the catalytic centers of enzymes very close (2–10 nm) to exposed membrane surfaces, by using stalklike structures for attachment, might have been a second advantageous feature. In such a situation, the actual distribution of the microdomains with various types of water structures around the enzyme protein would be determined both by the influence of the membrane surface and by the contribution of the enzyme proper [17]. In addition, one has to keep in mind that various areas of a protein surface, formed by specific amino acids in these areas, may induce different microdomains of water structure in the water shell surrounding the protein. A series of experiments was carried out, simulating situations in which enzymes were either positioned at different distances from membrane surfaces by the introduction of stalks of varying lengths (Fig. 12.6a) [20] or encapsulated in reversed micelles of various sizes, with dimensions in the nanometer range (Fig. 12.6c, left part) [21,22]. In both cases, a dependency of specific enzyme activity on the distance of the enzyme from the membrane surface could be measured (12.6b,c, right part).

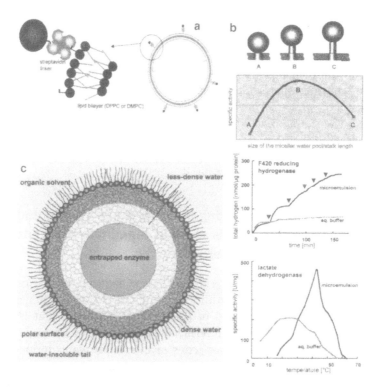

Figure 12.6 Experiments designed to simulate *in vivo* conditions of enzyme complexes. (a) Assembly of enzyme–liposome complexes. At the right side, a schematically drawn liposome with attached enzymes is depicted. Part of the diagram (circled) is drawn, at the left side, with more details: a Streptavidin linker connects an enzyme molecule with the surface of the liposome. The size of the linker determines the distance of the enzyme from the liposome. A biotin tag not occupied by a linker is also shown attached to the outside of the liposome. *Abbreviations*: DPPC, 1,2-palmitoyl-*sn*-glycero-3-phosphocholine; DMPC, 1,2-dimyristoyl-*sn*-glycero-3-phosphocholine. (b) Dependency of the measurable specific activity of an enzyme from the length of a linker as shown in Fig. 12.6a or from the inner surface of a reversed micelle (Fig. 12.6c) (original drawing by M. Hoppert). (c) (*left*) An enzyme molecule (size about 5–8 nm) placed inside a reversed micelle with a diameter of less than about 20 nm. The properties of the components making up the reversed micelle and the overall dimensions of the system were chosen in such a way that the system can be used for the simulation of the *in vivo* situation of an enzyme; (*right*) comparison of the specific enzyme activities of enzymes (F420-reducing hydrogenase, lactate hydrogenase), measured, in the conventional way, in aqueous buffer, or trapped inside a reversed micelle; (*top*) hydrogen production of the F420-hydrogenase at 60°C. The arrows indicate the points of time of methylviologen additions to both assays. The specific activity of the enzyme in the reversed micelle is very substantially higher compared to that in aqueous buffer; (*bottom*) thermal stability and specific activity of the enzyme are increased for enzymes in reversed micelles compared to in aqueous buffer. Reproduced with permission: a from Ref. [20] and c from Refs. [21] and [22]. For color reference, see p. 382.

12.3 CONCLUSION

This article was not intended to describe and discuss the occurrence of various water moieties within living cells. Mechanisms of the influence of surfaces of membranes and proteins on the dimensions of microdomains of various water moieties (low-density water, LDW; high-density water, HDW) in the immediate vicinity of enzyme molecules with their exposed surfaces and "clefts," or mechanisms of possible influences of various water moieties and solutes therein, on steps of metabolic pathways are described and discussed elsewhere in this book. Rather, specific features of the architecture of cells, both at the cellular and the macromolecular level, were described [23–28], with a view on the ways how living cells are adjusted to the specific properties of various types of water.

An additional aspect might be considered. The occurrence of various kinds of water structures in a living cell—some advantageous for specific cell functions, some less favorable—may have been an important factor during evolution [19]. This may be true not only for optimization of the catalyzing power of enzymes but also for regulation processes. After all, even slight modifications of amino acid sequences and, thus, the folding state of an enzyme complex, that is, of the "architecture" of individual enzymes, may alter the structural properties of water in the immediate neighborhood of the catalytic center [29, 30]. On the other hand, modifications of the properties of water microdomains adjacent to a protein can be assumed to alter the folding state and functional efficiency of an enzyme, for example, by changing the accessibility of "clefts" in an enzyme (i.e., the site where the active center of the enzyme is located) [17]. One could even speculate that transient or permanent alterations of the water structure around a protein, caused not by changes of the ultrastructural conditions at a given site in the cell but by translocation of the protein to a site with a different water structure, might alter the functional state or the folding state of that protein. An example could be the prion protein. This protein is known to be translocated within a cell (even across a membrane) under certain conditions. The observed irreversible alterations of its folding state, with their dramatic consequences, might be caused, as a consequence of the translocation, by a transition from a water moiety (containing specific solutes) that preserves the folding state of the protein, to a water moiety that does not protect the protein from changes in the folding state.

References

1. Beveridge, T. J. (1995) The periplasmic space and the periplasm in gram-positive and gram-negative bacteria, *ASM News*, **61**, 125–130.

2. Hegermann, J., Herrmann, R., and Mayer, F. (2002) Cytoskeletal elements in the bacterium *Mycoplasma pneumoniai*, *Naturwissenschaften*, **89**, 453–458.

3. Mayer, F., Vogt, B., and Poc, C. (1998) Immunoelectron microscopic studies indicate the existence of a cell shape preserving cytoskeleton in prokaryotes, *Naturwissenschaften*, **85**, 278–282.

4. Mayer, F. (2003) Cytoskeletons in prokaryotes, *Cell Biol. Int.*, **27**, 429–438.

5. Mayer, F., Wheatley D., and Hoppert M. (2006) Some properties of interfacial water: determinants for cell architecture and function? in *Water and the Cell* (ed., Pollack, G. H., Cameron, I. L., and Wheatley, D. N.), Springer, New York, pp. 253–271.

6. Wiggins, P. M. (2001) High and low density intracellular water, *Cell Mol. Biol.*, **47**, 735–744.

7. Hesketh, J. E. , and Pryme, I. F. (1991) Interaction between mRNA, ribosomes and the cytoskeleton, *Biochemistry*, **277**, 1–10.

8. Boyer P. (1997) The ATP synthase—a splendid molecular machine, *Annu. Rev. Biochem.*, **66**, 717–749.

9. Lingl, A., Huber, H., Stetter, K. O., Mayer, F., Kellermann, J., and Müller, V. (2003) Isolation of a complete A1A0ATPsynthase comprising nine subunits from the hyperthermophile *Methanococcus jannaschii*, *Extremophiles*, **7**, 249–257.

10. Reidlinger, J., Mayer, F., and Müller, V. (1994) The molecular structure of the Na+-translocating F1F0ATPase of *Acetobacterium woodii* as revealed by electron microscopy resembles that of the H+-translocating ATPases, *FEBS Lett.*, **356**, 17–20.

11. Däkena, P., Rohde, M., Dimroth, P., and Mayer, F. (1988) Oxaloacetate decarboxylase from *Klebsiella pneumoniae*: size and shape of the enzyme, and localization of its prosthetic biotin group by electron microscopy affinity labelling, *FEMS Microbiol. Lett.*, **55**, 35–40.

12. Mayer, F., Rohde, M., Salzmann, M., Jussofie, A., and Gottschalk, G. (1988) The methano-reductosome: a high molecular weight enzyme complex in the methanogenic bacterium strain Gö1 that contains components of the methylreductase system, *J. Bacteriol.*, **170**, 1438–1444.

13. Hoppert, M., and Mayer, F. (1990) Electron microscopy of native and artificial methylreductase high-molecular-weight complexes in strain Gö1 and Methanococcus voltae, *FEBS Lett.*, **267**, 33–37.

14. Braks, I. J., Hoppert, M., Roge, S., and Mayer, F. (1994) Structural aspects and immunolocalization of the F420-reducing and non-F420-reducing hydrogenases from *Methanobacterium thermoautotrophicum* Marburg, *J. Bacteriol.*, **176**, 7677–7687.

15. Antranikian, G., Herzberg, C., Mayer, F., and Gottschalk, G. (1987) Changes in the cell envelope structure of *Clostridium* sp. strain EM1 during massive production of α-amylase and pullulanase, *FEMS Microbiol Lett.*, **41**, 193–197.

16. Binda, C., Newton-Vinson, P., Hubalek, F., Edmonson, D. E., and Mattevi, A. (2002) Structure of human monoamine oxidase B, a drug target for the treatment of neurological disorders, *Nat. Struct. Biol.*, **9**, 22–26.

17. Wiggins, P. M., and MacClement, B. (1987) Two states of water found in hydrophobic clefts: their possible contribution to mechanisms of cation pumps and other enzymes, *Int. Rev. Cytol.*, **108**, 249–303.

18. Wiggins, P. M. (2002) Enzymes and two-state water, *J. Biol. Phys. Chem.*, **2**, 25–37.

19. Clegg, J. S., and Wheatley, D. N. (1991) Intracellular organization: evolutionary origins and possible consequences of metabolic rate control in vertebrates, *Am. Zool.*, **31**, 504–513.

20. Wichmann, C., Naumann, P. T., Spangenberg, O., Konrad, M., Mayer, F., and Hoppert, M. (2003) Liposomes for microcompartmentation of enzymes and their influence on catalytic activity, *Biochem. Biophys. Res. Comm.*, **310**, 1104–1110.

21. Hoppert, M., Braks, I. J., and Mayer, F. (1994) Stability and activity of hydrogenases of *Methanobacterium thermoautotrophicum* and *Alcaligenes eutrophus* in reversed micellar systems, *FEMS Microbiol. Lett.*, **118**, 249–254.

22. Hoppert, M., Mlejnek, K., Seiffert, B., and Mayer, F. (1997) Activities of microorganisms and enzymes in water-restricted environments: biological activities in aqueous compartments at μm-scale, in *Instruments, Methods, and Missions for the Investigation of Extraterrestrial Microorganisms*, SPIE Proceed Series, Vol. 3111, pp. 501–509.

23. Mayer, F. (1993) "Compartments" in the bacterial cell and their enzymes, *ASM News*, **59**, 346–350.

24. Mayer, F. (1993) Principles of functional and structural organization in the bacterial cell: "compartments" and their enzymes, *FEMS Microbiol. Rev.*, **104**, 327–346.

25. Mayer, F., and Hoppert M. (1996) Functional compartmentalization in bacteria and archaea: a hypothetical interface beween cytoplasmic membrane and cytoplasm, *Naturwissenschaften*, **83**, 36–39.

26. Hoppert, M., and Mayer, F. (1999a) Principles of macromolecular organization and cell function in bacteria and archaea, *Cell Biochem. Biophys.*, **31**, 247–284.

27. Mayer, F. (1999) Cellular and Subcellular Organization of Prokaryotes, in *Biology of the Prokaryotes* (ed. Lengeler, J. W., Drews, G., and Schlegel, H. G.), Thieme, Stuttgart, pp. 20–46.

28. Hoppert, M., and Mayer, F. (1999) Prokaryotes, *Am. Sci.*, **87**, 518–525.

29. Timasheff, S. N. (1993) The control of protein stability and association by weak interactions with water: how do solvents affect these processes? *Annu. Rev. Biophys. Biomol. Struct.*, **22**, 67–97.

30. Robinson, G., and Cho, C. (1999) Role of hydration water in protein unfolding, *Biophys. J.*, **77**, 3311–3318.

31. Mayer, F. (2006) Cytoskeletal elements in bacteria *Mycoplasma pneumoniae*, *Thermoanaerobacterium* sp., and *Escherichia coli* as revealed by electron microscopy, *J. Mol. Microbiol. Biotechnol.*, **11**, 228–243.

Chapter 13

WATER-BASED BIOLOGICAL HOMEOSTASIS IN MULTIPLE FLUID-FILLED COMPARTMENTS IN THE HIGHER ORGANISMS: REGULATION BY THE PARACELLULAR BARRIER–FORMING TIGHT JUNCTIONS OF EPITHELIAL CELL SHEETS

Yuji Yamazaki, Atsushi Tamura, and Sachiko Tsukita

Laboratory of Biological Science, Graduate School of Frontier Biosciences and Graduate School of Medicine, Osaka University, 2-2 Yamadaoka,
Suita, Osaka 565-0871, Japan
atsukita@biosci.med.osaka-u.ac.jp

The higher organisms have a hierarchical body plan in which the whole body system is composed of the organ system, organs, tissues, cells, and subcellular organelles, each of which is itself composed of various compartments (Fig. 13.1). This highly organized compartmentalization is a sophisticated means of housing biological systems according to their specific functions, and homeostasis is maintained in each compartment as a nonequilibrium open system.

The concept of homeostasis as the essence of life was proposed by Claude Bernard, a famous French physiologist, in the 1860s (Fig. 13.1). He described the *milieu interieur* as based on fluid, the main component of which is water.

> I think it was the first to urge the belief that animals have really two environments: a milieu extérieur in which the organism is situated, and a milieu intérieur in which the tissue elements live. The living organism does not really exist in the milieu extérieur (the atmosphere it breathes, salt or fresh water if that is the element) but in the liquid milieu intérieur formed

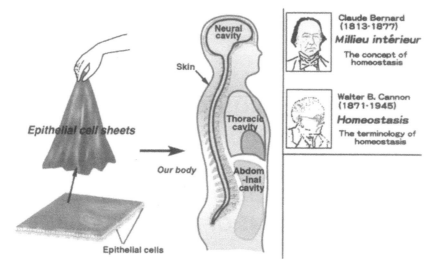

Figure 13.1 Illustration of the functions of epithelial cell sheets in higher organisms. Epithelial cells adhere to each other in a side-by-side fashion to form the sheetlike configuration shown. The epithelial cell sheet covers the surface of every compartment inside and outside the body of higher organisms to maintain the homeostasis of the separate fluid-filled compartments of multicellular organisms (*left*). The concept of homeostasis was defined by Claude Bernard, and the term, "homeostasis," by Walter Cannon (*right*). For color reference, see p. 383.

by the circulating organic liquid which surrounds and bathes all the tissue elements; this is the lymph or plasma, the liquid part of the blood which, in the higher animals, is diffused through the tissues and forms the ensemble of the intercellular liquids and is the basis of all local nutrition and the common factor of all elementary exchanges. A complex organism should be looked upon as an assemblage of simple organisms which are the anatomical elements that live in the liquid milieu intérieur. (Bernard, 1860s)

The term "homeostasis" was first used by Dr. Walter Bradford Cannon, an American physiologist, in the 1920s (Fig. 13.1), by combining the Latin terms "homeo" for "same" and "stasis," which means "standing" [1].

The coordinated physiological reactions which maintain most of the steady states in the body are so complex, and are so peculiar to living organism, that it has been suggested that a specific designation for these states be employed— homeostasis. Objection might be offered to the use of the term stasis, as implying something set and immobile, a stagnation. (Cannon, 1920s)

Thus, Cannon defined homeostasis as a unifying concept of human physiology [2].

The word "homeostasis" generally invokes the idea of the steady state in biological systems. Modern thermodynamics has revealed, however, that biological homeostasis is based on a nonequilibrium open system that has water as its major component. Therefore, the dynamic regulation of homeostasis is an important issue in biological systems. Recently, evidence that biological homeostasis is maintained in different hierarchical compartments has accumulated [3–6]. Here, we focus on the compartmentalization of the body plan, which depends on each compartment's being surrounded by an epithelial cell sheet to create spaces that are physically enclosed but thermodynamically open, to maintain water-based homeostasis. In particular, we focus on the role of epithelial TJs, which form the permselective paracellular barrier, in the creation of these enclosed spaces.

13.1 EPITHELIAL CELL SHEETS ORGANIZE THE INNER SPACE OF HIGHER ORGANISMS INTO SEPARATE COMPARTMENTS THAT INDIVIDUALLY REGULATE HOMEOSTASIS

The mission of epithelial cells is to form cell sheets, which both enclose and define biological compartments. To accomplish this mission, epithelial cells align side by side and adhere to each other through cell–cell adhesions, mediated by E-cadherin and related factors that are highly organized into beltlike cell–cell adherens junctions, termed the "zonula adherens" (zAJs) (Fig. 13.2) [7–10].

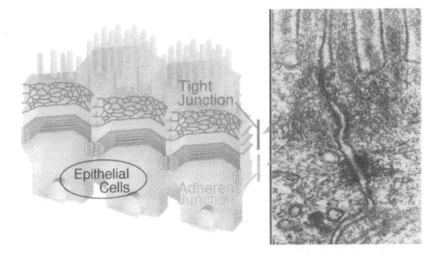

Figure 13.2 Schematic drawing of epithelial cell sheets and electron micrographs of adherens and tight junctions. Epithelial cells adhere to each other with specialized adhesion complexes, such as tight junctions, adherens junctions, and desmosomes. An electron microscopic image of tight junctions, adherens junctions in the intestinal epithelial cells is shown (*right*). For color reference, see p. 383.

To act as a barrier, the sheet must regulate permeation through the cells themselves (the transcellular barrier) and between the cells (the paracellular barrier). To allow needed elements to enter the system, selective permeability (permselectivity) is required. To establish permselectivity of the paracellular barrier, a highly specialized system has developed, apical to the zAJ, in which TJs form a beltlike barrier, termed the "zonula occludens"(zTJs) [11–14]. The beltlike arrangement of the zAJ and the zTJ is the fundamental basis of transepithelial permselectivity.

The zAJ–zTJ system is a typical example of the cell biological principle—where there is morphological organization, there is also a function. A classical electron microscopic observation revealed the structural specialization of the epithelial cell–cell adhesion apparatus of zAJs and zTJs, which hinted at their respective cell adhesion and paracellular barrier functions (Fig. 13.2) [7]. The most salient ultrastructural characteristic of zAJs is the consistent intercellular space of ~20 nm between the plasma membranes of adjacent cells, which is cytoplasmically associated with actin-related membrane scaffolds. The ultrastructural characteristics of zTJs are different and unusual. The outer leaflets of the plasma membranes of adjacent cells are brought together at the zTJs, with no space between them, but they are not fused. To explain this ultrastructural observation, an early proposal for the molecular organization of the zTJ was that it represented a specialized lipid-based structure [15]. This hypothesis was favored until occludin and claudins were identified as the essential adhesion molecules of the zTJ [5, 16, 17].

The claudins are a family of at least 24 integral membrane proteins that contain 4 transmembrane domains [5, 6]. By using cultured epithelial cell sheets to express specific claudins, many physiological analyses have shown that the specific characteristics of zTJs in terms of their permselectivity can be explained by the combination of claudins expressed by the epithelial cells [18–21]. *In vivo*, different tissues exhibit specific combinations of claudins at different ratios in their zTJs. Elucidating the algorithm for the combination of claudins involved at each zTJ is one of the most critical issues for understanding the function of each paracellular permselective barrier, but the overall picture of this algorithm is currently still unclear. The analyses of human diseases that are caused by mutations or deletions of the claudin genes, and *in vivo* analyses of claudin-knockout (KO) mice will be indispensable for understanding the physiological relevance of various claudin-based permselective barriers.

In this review, we focus on several examples in which specific claudins regulate the permselectivity of the paracellular barrier in different epithelial sheets *in vivo*. The five examples (cases) later illustrate several claudin functions: the paracellular barrier function in skin (case I), the paracellular barrier in the blood–brain barrier (BBB) (case II), paracellular ion-permselectivity through the barrier in the kidney (case III), paracellular permeability to water across the barrier in the kidney (case IV), and paracellular ion-permselectivity via the barrier in the small intestine, with linkage to morphogenesis (case V) (Fig. 13.3).

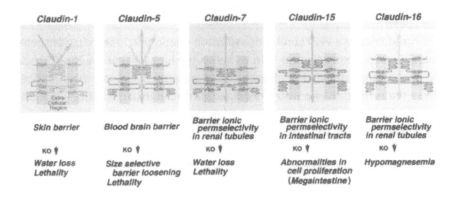

Figure 13.3 Schematic drawings of the function of particular claudins described in this review. Claudins are responsible for epithelial cell–cell adhesion and regulate the permselective barrier function of the paracellular space by providing selective permeability to water and other molecules according to their size and/or charge, as revealed by knockout mice analyses. For color reference, see p. 384.

13.1.1 Case I: Claudin-1 Prevents the Paracellular Leakage of Water Through the Skin, and Claudin-6 Allows It

It is well recognized that the epidermal barrier has an essential role in maintaining homeostasis in organisms. The epidermis covers the whole body, and one of its functions as a barrier is to prevent water from evaporating freely, thus preventing detrimental dehydration. It also protects the body from external dangers, such as invading microorganisms and harmful solutes. Histologically, the epidermis is a stratified epithelium in which the basal cells are attached to a basement membrane, the deepest structure of the epidermis, and epidermal cells at different stages of differentiation are overlaid by an epithelial sheet, called the stratum granulosum, and then by the outer layer of the skin, the stratum corneum, which forms the external barrier. However, the barrier function of the skin cannot be attributed solely to the stratum corneum. The paracellular barrier between the epidermal epithelial cells in the stratum granulosum plays a crucial role, which was largely unappreciated until the analysis of claudin-1-deficient mice was reported.

Claudin-1-KO mice exhibit a wrinkled skin phenotype, and they die within 24 hours of birth because of catastrophic paracellular water loss [22]. In humans, mutations in the claudin-1 gene result in neonatal sclerosing cholangitis, associated with ichthyosis caused by the disruption of the paracellular barrier of the skin [23]. In both cases, however, despite the absence of claudin-1, the TJ structure persists. Thus, claudin-1 is an essential functional component for the TJ strands to function properly as a barrier in the epidermis.

Paradoxically, a very similar dehydration phenotype is seen in transgenic mice overexpressing claudin-6. In this case, the high level of claudin-6 seems to disturb the preexisting claudin-based paracellular water barrier [24]. Thus, the formation of a water-resistant paracellular barrier is regulated by the combination of claudins expressed in the epidermal cells. These studies demonstrated that the epidermal barrier is not restricted to the cornified envelope of the stratum corneum, which relies on components such as lipids and cross-linked proteins, but is also provided by a transcellular and paracellular epidermal barrier in the stratum granulosum. Because of this finding, skin claudins are recognized as possible targets for therapeutic drugs.

13.1.2 Case II: Claudin-5 Maintains the Blood–Brain Barrier

The BBB prevents solutes or various harmful materials from crossing freely into the brain [25, 26]. Well-developed TJs, in addition to various transporters and the basement membrane, are crucial for establishing the BBB. The previous report demonstrated that claudin-5 plays a central role in the paracellular barrier function of the BBB as well as its size-selective permeability, a functional analysis that was possible because some barrier functions persist in claudin-5-KO mice [27]. In claudin-5-KO mice, a loosening of the size selectivity occurs at the BBB, allowing a fluorescent dye, biotin, and water to cross the BBB without causing edema of the brain. This finding suggests that claudins may regulate the size-selective permeability of the BBB, so that the claudin-5-KO mice die on postnatal day 1. Detailed studies on the molecular mechanisms of BBB formation using conditional claudin-5-KO mice will provide additional information on the functions of claudin-5 in the BBB in the future. These studies hint that modulating the TJ barrier to permit therapeutic agents to penetrate the BBB might be achievable by targeting claudin-5 at the BBB.

13.1.3 Case III: Claudin-16 and Claudin-19 Contribute to Paracellular Mg^{2+} *Permeation*

The kidney is indispensable for maintaining the balance of electrolytes in the body, which it does, in part, by reabsorbing ions and water. The reabsorption is attributed to the epithelial cell sheets of the renal tubules, and the routes of the reabsorption are transcellular, via channels and transporters, and paracellular, as permitted by the claudins at the zTJs. It was reported that loss-of-function mutations of claudin-16 and claudin-19 prevented the reabsorption of Mg^{2+} and led to hypomagnesemia and hypercalciuria, indicating that claudins are critical not only for maintaining the paracellular barrier but also in paracellular ion channel-like permselectivity [28–30]. However, recently, it was also reported that Mg^{2+} reabsorption in the paracellular pathway is dependent on the epithelial cells' ionic state and its effects on the diffusion voltage between the lumenal and interstitial sides of the renal tubules.

The diffusion voltage formed by the leakage of Na$^+$ across the paracellular barrier from the lumen to the interstitial space (in-out permeation) is thought to be the driving force for the reabsorption of Mg^{2+} (out-in permeation). If this is the case, the reabsorption of Mg^{2+} increases when the in-out permeation of Na$^+$ increases [31]. When claudin-16 or claudin-19 was exogenously expressed in cultured kidney epithelial (Madin–Darby canine kidney [MDCK]) cells, the paracellular permeabilities for cations increased, including the permeability for Na$^+$ and Mg^{2+} [32]. Thus, claudin-based paracellular permeability appears not simple. The detailed characterization of the permselectivity for cations of renal tubules awaits more delicate physiological analyses that take into consideration the various in-out states of ions in the epithelial cells.

13.1.4 Case IV: Claudin-7 Regulates the Paracellular Permeation of Water across the Epithelial Cell Sheets of *Renal Tubules*

Recent reports have shown that loss of claudin-7 function in the kidney causes severe dehydration. Claudin-7-KO mice die from severe salt wasting, chronic dehydration, and growth retardation [33]. Considering that the major expression of claudin-7 is in the distal convoluted tubules (DCTs) and collecting ducts (CDs) of the kidney, these phenotypes may be caused by the disruption of the paracellular barrier at these sites.

In a cell culture system, the expression of claudin-7 was reported to induce increased Cl$^-$ paracellular transport [18], which led to the hypothesis that defects in the reabsorption of Cl$^-$ at the DCT and CD of claudin-7-KO mice causes Na$^+$ wasting and thus continuous water loss. On the other hand, since the knockdown of claudin-7 in cultured epithelial (MDCK) cells increases cation conductance, claudin-7 may contribute to the barrier against cations in the kidney tubules [34]. Taking both of these findings at face value, even though the anion selectivity of claudin-7 was not examined critically, it is possible that both the claudin-7-based paracellular ionic permselectivities and the claudin-7-based barrier functions are determined by other claudins that are associated with claudin-7 in various ways. Future extensive physiological studies on renal tubule ion selectivities should resolve this issue.

13.1.5 Case V: Claudin-15 Contributes to Paracellular Permeation for Monocations

Recently, we reported that claudin-15-KO mice exhibit the megaintestine phenotype without tumorigenesis [35]. This phenotype demonstrates that the claudin-based permselective barrier can play a role not only in the maintenance of ionic homeostasis but also, at times, in morphogenesis associated with cell proliferation.

The small intestine is an important organ for maintaining the homeostasis of water and ions for the absorption of nutrients. The intestinal epithelial sheet that lines the villi and crypts separates the luminal space and the submucosa and has a nutrient-absorptive function. The main players in nutrient absorption by the epithelium are transporters and channels in the apical membranes of the epithelial cells and, in the case of the paracellular pathway, the claudins at the TJs. The microenvironment of the intestinal epithelial cells appears to be regulated largely by the claudins so as to maintain the action of the apical transporters and channels. In particular, the submucosa supplies the small intestinal lumen with Na^+, which is maintained at a concentration that is approximately the same as in plasma. To maintain the homeostasis of the luminal microenvironment, the intestinal epithelium is well known to be "leaky"; that is, the paracellular barrier is permeable to ions with relatively low transepithelial resistance [3, 4, 36].

Several kinds of claudins are expressed in the intestinal epithelium, and their combination determines the selectivity of the paracellular pathway. Among the claudins expressed in the small intestine are claudin-2 and claudin-15, which are thought to be leaky claudins, on the basis of the physiological analysis of cultured cells, in which their expression increases ionic conductance and permeability to solutes [20, 21, 30]. In the small intestine, these claudins are localized to the crypt and/or villi. A deficiency in claudin-15 significantly decreases the ionic conductance across the small intestine [35]. However, how the claudin-15-KO-induced abnormality in paracellular ion permeability, which is responsible for decreased transepithelial conductance, influences cell proliferation in the small intestine awaits future analysis.

From different directions, the roles of claudins in development have been reported in other organisms. The claudin-15 homolog in zebrafish contributes to gut luminal formation, and its knockdown causes the malformation of gut luminal tubules [38]. In *Drosophila*, mutants of a claudin homolog named *megatrachea* show a phenotype characterized by the enlargement of the tracheal tubules [39]. The morphogenesis of tubal structures is the basis for the organization of the body, and many organs develop from such structures. Thus, future reports will include investigations of claudin function in development.

13.2 CONCLUSIONS AND PERSPECTIVE

Water-based homeostasis is essential in multicellular biological systems, which are formed by the combination of various fluid-filled compartments with different characteristics. Claudins are structurally and functionally essential for the formation of the TJs responsible for paracellular barrier functioning and barrier-based selective permeability [6, 40]. The dynamic aspects of claudin function at the TJs are also becoming recognized [6, 41, 42]. Hence, the spatiotemporally dynamic regulation of biological homeostasis via claudins is changing the classical view of homeostasis as a static state. In principle, for an effective understanding of claudin function, both gain-of-function and loss-of-function analyses must converge to provide a consistent perspective. Gain-of-function analyses include the expression of each claudin in cultured cells and the generation of transgenic mice that express each claudin. Loss-of-function analyses include the knockdown of each claudin in cultured cells and the corresponding generation of KO mice. Here we focus on KO mouse analyses, which are beginning to elucidate the *in vivo* biological relevance of paracellular barriers/permeability when single or multiple claudins in different combinations are targeted. To fully understand the functions of claudins in different combinations and to identify new therapeutic targets for clinical applications, it will be critical to perform thorough gain-of-function and loss-of-function studies of these molecules.

References

1. Cannon, W. B. (1929) Organization for physiological homeostasis, *Physiol. Rev.*, **9**, 399–431.

2. Cooper, S. J. (2008) From Claude Bernard to Walter Cannon. Emergence of the concept of homeostasis, *Appetite*, **51**, 419–426.

3. Diamond, J. M. (1978) Channels in epithelial cell membranes and junctions, *Fed. Proc.*, **37**, 2639–2643.

4. Powell, D. W. (1981) Barrier function of epithelia, *Am. J. Physiol.*, **241**, 275–288.

5. Tsukita, Sh., Furuse, M., and Itoh, M. (2001) Multifunctional strands in tight junctions, *Nat. Rev. Mol. Cell Biol.*, **2**, 285–293.

6. Van Itallie, C. M., and Anderson, J. M. (2006) Claudins and epithelial paracellular transport, *Annu. Rev. Physiol.*, **68**, 403–429.

7. Farquhar, M. G., and Palade, G. E. (1963) Junctional complexes in various epithelia, *J. Cell Biol.*, **17**, 375–412.

8. Gumbiner, B. M. (2005) Regulation of cadherin-mediated adhesion in morphogenesis, *Nat. Rev. Mol. Cell Biol.*, **6**, 622–634.

9. Takeichi, M. (1991) Cadherin cell adhesion receptors as a morphogenetic regulator, *Science*, **251**, 1451–1455.

10. Nelson, W. J. (2003) Adaptation of core mechanisms to generate cell polarity, *Nature*, **422**, 766–774.

11. Abe, K., and Takeichi, M. (2008) EPLIN mediates linkage of the cadherin catenin complex to F-actin and stabilize the circumferential actin belt, *Proc. Natl. Acad. Sci. USA*, **105**, 13–19.

12. Perez-Moreno, M., Jamora, C., and Fuchs, E. (2003) Sticky business: orchestrating cellular signals at adherens junctions, *Cell*, **112**, 535–548.

13. Yamazaki, Y., Umeda, K., Wada, M., Nada, S., Okada, M., Tsukita, Sh., and Tsukita Sa. (2008) ZO-1- and ZO-2-dependent integration of myosin-2 to epithelial zonula adherens, *Mol. Biol. Cell*, **19**, 3801–3811.

14. Umeda, K., Ikenouchi, J., Katahira-Tayama, S., Furuse, K., Sasaki, H., Nakayama, M., Matsui, T., Tsukita, Sa., Furuse, M., and Tsukita, Sh. (2006) ZO-1 and ZO-2 independently determine where claudins are polymerized in tight-junction strand formation, *Cell*, **126**, 741–754.

15. Kachar, B., and Reese, T. S. (1982) Evidence for the lipidic nature of tight junction strands, *Nature*, **296**, 464–466.

16. Furuse, M., Fujita, K., Hiiragi, T. Fujimoto, K., and Tsukita, S. (1998) Claudin-1 and -2: novel integral membrane proteins localizing at tight junctions with no sequence similarity to occludinm, *J. Cell Biol.*, **141**, 1539–1550.

17. Gumbiner, B. M. (1993) Breaking through the tight junction barrier, *J. Cell Biol.*, **123**, 1631–1633.

18. Alexandre, M. D., Jeansonne, B. G., Renegar, R. H., Tatum, R., and Chen, Y. H. (2007) The first extracellular domain of claudin-7 effects paracellular Cl-permeability, *Biochem. Biophys. Res. Commun.*, **357**, 87–91.

19. Colegio, O. R., Van Itallie, C. M., McCrea, H. J., Rahner, C., and Anderson, J. M. (2002) Claudins create charge-selective channels in the paracellular pathway between epithelial cells, *Am. J. Physiol.*, **283**, C142–C147.

20. Van Itallie, C. M., Fanning, A. S., and Anderson, J. M. (2003) Reversal of charge selectivity in cation or anion-selective epithelial lines by expression of different claudins, *Am. J. Physiol.*, **285**, F1078–F1084.

21. Yu, A. S., Cheng, M. H., Angelow, S., Günzel, D., Kanzawa, S. A., Schneeberger, E. E., Fromm, M., and Coalson, R. D. (2009) Molecular basis for cation selectivity in claudin-2-based paracellular pores: identification of an electrostatic interaction site, *J. Gen. Physiol.*, **133**, 111–127.

22. Furuse, M., Hata, M., Furuse, K., Yoshida, Y., Haratake, A., Sugitani, Y., Noda, T., Kubo, A., and Tsukita, Sh. (2002) Claudin-based tight junctions are crucial for the mammalian epidermal barrier: a lesson from claudin-1-deficient mice, *J. Cell Biol.*, **156**, 1099–1111.

23. Hadj-Rabia, S., Baala, L., Vabres, P., Hamel-Teillac, D., Jacquemin, E., Fabre, M., Lyonnet, S., De Prost, Y., Munnich, A., Hadchouel, M., and Smahi, A. (2004) Claudin-1 gene mutations in neonatal sclerosing cholangitis associated with ichthyosis: a tight junction disease, *Gastroenterology*, **127**, 1386–1390.

24. Turksen, K., and Troy, T. C. (2002) Permeability barrier dysfunction in transgenic mice overexpressing claudin-6, *Development*, **129**, 1775–1784.

25. Pardridge, W. M. (1998) *Introduction to the Blood-Brain Barrier*, Cambridge University Press, Cambridge, p. 486.

26. Rubin, L. L., and Staddon, J. M. (1999) The cell biology of the blood-brain barrier, *Annu. Rev. Neurosci.*, **22**, 11–28.

27. Nitta, T., Hata, M., Gotoh, S., Seo, Y., Sasaki, H., Hashimoto, N., Furuse, M., and Tsukita, Sh. (2003) Size-selective loosening of the blood-brain barrier in claudin-5-deficient mice, *J. Cell Biol.*, **161**, 653–660.

28. Kang, J. H., Choi, H. J., Cho, H. Y., Lee, J. H., Ha, I. S., Cheong, H. I., and Choi, Y. (2005) Familial hypomagnesemia with hypercalciuria and nephrocalcinosis associated with CLDN16 mutations, *Pediatr. Nephrol.*, **20**, 1490–1493.

29. Konrad, M., Schaller, A., Seelow, D., Pandey, A. V. Waldegger, S., Lesslauer, A., Vitzthum, H., Suzuki, Y., Luk, J. M., and Becker, C. (2006) Mutations in the tight-junction gene claudin 19 (CLDN19) are associated with renal magnesium wasting, renal failure, and severe ocular involvement, *Am. J. Hum. Genet.*, **79**, 949–957.

30. Simon, D. B., Lu, Y., Choate, K. A., Velazquez, H., Al-Sabban, E., Praga, M., Casari, G., Bettinelli, A., Colussi, G., and Rodriguez-Soriano, J. (1999) Paracellin-1, a renal tight junction protein required for paracellular Mg2+ resorption, *Science*, **285**, 103–106.

31. Shan, Q., Himmerkus, N., Hou, J., Goodenough, D. A., and Bleich, M. (2009) Insights into driving forces and paracellular permeability from claudin-16 knockdown mouse, *Ann. N. Y. Acad. Sci.*, **1165**, 148–151.

32. Hou, J., Renigunta, A., Konrad, M., Gomes, A. S., Schneeberger, E. E., Paul, D. L., Waldegger, S., and Goodenough, D. A. (2008) Claudin-16 and Claudin-19 interact and form a cation-selective tight junction complex, *J. Clin. Invest.*, **118**, 619–628.

33. Tatum, R., Zhang, Y., Salleng, K., Lu, Z., Lin, J.J., Lu, Q., Jeansonne, B. G., Ding, L., and Chen, Y. H. (2010) Renal salt wasting and chronic dehydration in claudin-7-deficient mice, *Am. J. Physiol. Renal Physiol.*, **298**, F24–F32.

34. Hou, J., Gomes, A. S., Paul, D. L., and Goodenough, D. A. (2006) Study of claudin function by RNA interference, *J. Biol. Chem.*, **281**, 36117–36123.

35. Tamura, A., Kitano, Y., Hata, M., Katsuno, T., Moriwaki, K., Sasaki, H., Hayashi, H., Suzuki, Y., Noda, T., Furuse, M., Tsukita, Sh., and Tsukita, Sa. (2008) Megaintestine in claudin-15-deficient mice, *Gastroenterology*, **134**, 523–534.

36. Wright, E. M. (1966) Diffusion potentials across the small intestine, *Nature*, **212**, 189–190.

37. Furuse, M., Furuse, K., Sasaki, H., and Tsukita, Sh. (2001) Conversion of zonulae occludentes from tight to leaky strand type by introducing claudin-2 into Madin-Darby canine kidney I cells, *J. Cell Biol.*, **153**, 263–272.

38. Bagnat, M., Cheung, I. D., Mostov, K. E., and Stainer, D. Y. (2007) Genetic control of single lumen formation in the zebrafish gut, *Nat. Cell Biol.*, **9**, 954–960.

39. Behr, M., Riedel, D., and Schuh, R. (2003) The claudin-like megatrachea is essential in septate junctions for the epithelial barrier function in Drosophila, *Dev. Cell*, **5**, 611–620.

40. Tsukita, Sa., Yamazaki, Y., Katsuno, T., Tamura, A., and Tsukita, Sh. (2008) Tight junction-based epithelial microenvironment and cell proliferation, *Oncogene*, **27**, 6930–6938.

41. Heller, F., Florian, P., Bojarski, C., Richter, J., Christ, M., Hillenbrand, B., Mankertz, J., Glitter, A. H., Bürgel, N., Fromm, M., Zeitz, M., Fuss, I., Strober, W., and Schulzke, J. D. (2005) Interleukin-13 is the key effector Th2 cytokine in ulcerative colitis that affects epithelial tight junctions, apoptosis, and cell restitution, *Gastroenterology*, **129**, 550–564.

42. Shen, L., Weber, C. R. and Turner, J. R. (2008) The tight junction protein complex undergoes rapid and continuous molecular remodeling at steady state, *J. Cell Biol.*, **181**, 683–695.

Chapter 14

ROLE OF THE SEPTIN CYTOSKELETON FOR THE FUNCTIONAL ORGANIZATION OF THE PLASMA MEMBRANE

Makoto Kinoshita

Department of Molecular Biology, Division of Biological Sciences,
Nagoya University Graduate School of Science,

Furo, Chikusa, Nagoya 464-8602, Japan

kinoshita.makoto@c.mbox.nagoya-u.ac.jp

Biological systems exchange water and various other molecules with the environment across membranes. The direction and flux of water and solutes are determined by subcellular distribution of various channels and transporters. These membrane-bound proteins are freely diffusible in principle, but in cells their distribution is asymmetrical and nonhomogeneous for functional reasons. Membrane proteins are organized mainly by three cellular processes: (i) vesicle transport that targets the membrane proteins to specific membrane domains, (ii) containment of the proteins within each membrane domain by diffusion barriers, and (iii) tethering of the proteins at narrower subdomains by scaffolds. The cytoskeleton plays essential roles in each process; microtubules provide tracks for directed vesicle transport, while an actin meshwork beneath the plasma membrane provides a framework for the diffusion barriers and scaffolds. Getting an overview of these processes, this chapter focuses on the unique roles of the septin cytoskeleton, which helps organize the transport system in conjunction with the microtubule and actin cytoskeleton.

14.1 INTRODUCTION

Regulated transport of water, ions, and other biomolecules across membranes is a vital activity for all organisms [1]. In cells with high transport demand,

Water: The Forgotten Biological Molecule
Edited by Denis Le Bihan and Hidenao Fukuyama
Copyright © 2011 by Pan Stanford Publishing Pte. Ltd.
www.panstanford.com

distinct membrane domains are differentially organized into asymmetric compartments, as exemplified by an epithelial cell with apical/basolateral membrane domains or a neuron with axonal/somatodendritic membrane domains [2]. At a smaller scale, each membrane domain is not homogeneous but a mosaic of mesoscopic (nano~submicrometer scale) subdomains with a set of channels, transporters, and various accessory molecules [3]. Thus, targeting and recruitment of certain membrane-bound molecules to specific membrane domains and organization of functional membrane domains and subdomains by assembling additional molecules are fundamental issues of physiology.

The cytoskeleton provides the structural and mechanical framework for cell morphogenesis, motility, mitosis, and various other phenomena. It has been established that the major cytoskeletal components, microtubules and actin filaments, help organize the membrane transport system by providing (i) tracks for directional transport of vesicles that carry specific channels and transporters, (ii) diffusion barriers for the physical containment of channels and transporters in a given membrane domain, and (iii) scaffolds for tethering the membrane proteins and associated molecules to narrower subdomains. Recent studies have indicated that the unconventional cytoskeletal polymers composed of the septin family guanosine triphosphatases (GTPases) play unique roles for the three aspects of the functional organization of the plasma membrane as follows.

14.1.1 Targeted Vesicle Transport

All cells are more or less asymmetric or polarized in terms of the subcellular architecture. Above all, epithelial, endothelial, mesothelial, and neural cells develop a highly asymmetric subcellular distribution of organelles and molecules to fulfill their functions.

The membrane trafficking/sorting system plays a major role in the establishment of cell polarity [4–6], which is facilitated by directional vesicle transport via the microtubule network. Microtubules are rearranged in response to spatial and mechanical cues (e.g., cell–cell or cell–matrix contact sites, free surfaces, and/or mechanical stimuli) and reorient the endoplasmic reticulum (ER)/Golgi system where membrane proteins are produced and loaded onto vesicles [7]. Thus, the configuration of these organelles is crucial for the asymmetric sorting and targeting of membrane proteins.

In epithelial cells, most microtubules grow from the centrosomes near the free surface (apical domain) toward other surfaces contacting the basement membrane or adjacent cells (basolateral domains): the centrosomal (–) and growing (+) ends of most microtubules are in the apical and basolateral domains, respectively. Thus, (–) end-directed motor proteins play a major role in microtubule-mediated vesicle transport from trans-Golgi to the apical membrane domain [7].

Besides the spatial configuration of microtubules, a variety of covalent modifications and microtubule-associated proteins (MAPs) can regulate vesicle transport [8]. Recent studies have unexpectedly revealed that RNAi-mediated depletion of septins in epithelial cells can hyperstabilize microtubules [9] and interfere with vesicle transport (e.g., apical transport of the neurotrophin receptor and basolateral transport of Na/K-ATPase) [10]. Further analyses showed that (i) microtubules near the Golgi apparatus are polyglutamylated and associated with septins, (ii) septins and another MAP, MAP4, can interact and also compete for their binding to microtubules, and (iii) the association of MAP4 stabilizes microtubules [9, 10]. Hence septins on and off microtubules appear to keep off excess MAP4, which somehow facilitates vesicle transport. The causal relationship between septin binding and tubulin polyglutamylation and whether septins could directly facilitate vesicle transport in some other way are intriguing subjects for future studies.

14.1.2 Diffusion Barrier for Membrane Compartmentalization

14.1.2.1 Diffusion barriers at cell adhesion sites and beneath the plasma membrane

To maintain functional asymmetry in each cell, membrane proteins selectively transported to certain domains are prevented from diffusing across the domain boundaries. The nonselective diffusion barriers that contribute to compartmentalization of the plasma membrane are represented by the junctional complex [11] and the membrane skeleton [3].

The junctional complex, which separates the plasma membrane into apical and basolateral compartments, is found typically in an epithelial or an endothelial cell as a horizontal belt of cell adhesion molecules and associated molecules connecting to actin and intermediate filaments [2, 11].

The membrane skeleton is a ubiquitous, actin-based submembranous meshwork that separates mesoscopic membrane domains [3]. The diffusion barrier that separates the somatodendritic and axonal membrane compartments at the axon initial segment is a specialized density of actin-based membrane skeleton that abundantly contains specific scaffold proteins such as ankyrin-G [12–14].

14.1.2.2 Septin-based diffusion barriers

Submembranous structures composed mainly of septin filaments also serve as diffusion barriers. A septin-based palisade that encircles the cytoplasmic neck between dividing yeast cells is known as a *bona fide* diffusion barrier. The budding yeast mutants lacking either one of the components (Cdc3p, Cdc10p, Cdc11p, Cdc12p, or Shs1p) cannot form the septin ring, and the asymmetric partitioning of mother cell– or daughter cell–specific membrane proteins is abolished [15–18].

Another septin-based ring found in the flagellum of mammalian spermatozoon is the annulus, a submembranous ring that separates the proximal and distal parts of the flagellum [19–21]. Although the annulus limits the diffusion of an immunoglobulin-like membrane protein CD147/basigin/emmprin [22], the significance of this diffusion barrier on spermatogenesis remains unclear. (Another function of the annulus as the scaffold for an anion transporter is discussed in section 14.3.)

The third example is the dense submembranous clusters of septins found in neurons and glia, some of which are localized at the bases of dendritic spines [23, 24]. Several groups have visualized that neurotransmitter receptors and/or membrane-proximal signaling molecules in each spine are not freely diffusible into and from the dendritic shaft (and the adjacent spines), indicative of a diffusion barrier that compartmentalize each spine at the base or neck [14, 18, 25, 26]. An obvious hypothesis that a septin cluster at a spine base is the diffusion barrier is under critical testing with mice lacking a septin subunit. Considering the ubiquity of the septin cytoskeleton, other tissues and organisms should have many other types of septin-based diffusion barriers.

14.1.3 Scaffold for selective organization of proteins in small membrane domains

Transport vesicles targeted to a membrane domain fuse into the plasma membrane via the soluble *N*-ethylmaleimide-sensitive factor activating protein receptor (SNARE) complex and release the membrane protein cargoes into specific membrane subdomains. As the physical barriers are insufficient to enclose the membrane proteins within narrow membrane subdomains, biochemical interaction is employed to anchor the membrane proteins to low-mobility structures. The molecular anchors for adhesion molecules may also lie extracellularly, but membrane proteins are generally tethered by submembranous supramolecular complexes composed of diverse adapter molecules and a cytoskeleton, collectively termed scaffolds. To date, a variety of septin-containing scaffolds have been found by genetic studies as follows.

14.1.3.1 Yeast septin ring

The scaffold functions of yeast septin rings had been established prior to the discovery of the diffusion barrier function. The septin ring is a scaffold for dozens of proteins involved in cell cycle progression, bud formation, and other morphogenesis [27, 28]. The diverse and complex phenotypes of septin mutants indicate the importance of the spatial organization of the membrane and membrane-proximal molecules.

14.1.3.2 The annulus in the sperm flagellum

Mice lacking a septin subunit Sept4 are male-sterile, partly because their spermatozoa lack the annulus, where the flagella bend and break by their faint motion [19, 20]. The local fragility has indicated that the annulus is a septin-based ring to confer rigidity against flagellar motion.

The linkage between loss of the annulus and poor flagellar motility remains unclear. However, a clue came from a similar set of phenotypes of mice lacking Slc26a8, an annulus-localized transmembrane protein [29]. Although the hypothetical anion transporter function of Slc26a8 remains to be tested, it is reasonable to speculate that Sept4-null spermatozoa cannot tether and stabilize Slc26a8, which maintains the ionic homeostasis for the dynein-driven flagellar motion. The physical and functional interaction between Sept4 and Slc26a8 is a good illustration of the scaffold-dependence of the transporter system, providing insight into the mechanism underlying human asthenospermia (male infertility syndrome with immotile spermatozoa) [19, 30, 31].

14.1.3.3 Presynaptic scaffold

Sept4-null mice not only are male-sterile, but also their nigrostriatal dopaminergic (DA) neurotransmission is attenuated. Sept4-null DA neurons are normal in morphology, but they lack (10–40%) specific proteins for DA metabolism, especially transmitter release (the SNARE complex), and reuptake (the dopamine transporter) [32]. Previous studies have consistently indicated physical and functional interaction between septins and the SNARE complex, which is the membrane-fusion machinery composed of syntaxins, SNAP-25, and VAMP-2 [33–35]. These data collectively indicate that presynaptic septins are a scaffold for the SNARE complex. Intriguingly, however, immunoelectron microscopic analysis revealed that septins cluster away from the synaptic active zones where membrane fusion occurs at high frequency [Hagiwara, Kinoshita, *et al.*, unpublished]. Considering the fact that septin overexpression interferes with exocytosis [33, 35], septin scaffold-bound SNARE complexes may represent a reservoir, or a sequestered fraction. In this case, the septin scaffold may be a negative regulator, rather than an inert tether, of the exocytic machinery.

14.2 CONCLUSION

Since the original discovery of yeast septin mutants [36], studies on diverse organisms have been revealing that the septin system is closely associated with biological processes involving the plasma membrane: Live-imaging analyses in yeast have demonstrated the cell cycle–regulated stability of submembranous septin polymers in cells [37, 38], which may be suitable either as a membrane skeleton or as a cortical scaffold for the organization of specialized membrane domains [39]. Morphological analyses of the mouse brain have revealed complex clustering of septins beneath specific membrane domains in major neurons and astroglia [40], which are likely to serve as a diffusion barrier or scaffold for diverse membrane proteins. Studies on human neuropsychiatry disorders have revealed that qualitative and/or quantitative dysregulation of septins is associated with neurodegeneration (e.g., Parkinson's disease [32, 41], hereditary neuralgic amyotrophy [42]) and psychological disorders (e.g, schizophrenia and bipolar disorders) [43]. Future studies for better understanding of the submembranous septin system and its relationship with membrane proteins should deepen our insight into diverse physiological and pathological phenomena.

Acknowledgments

I appreciate Professor Denis Le Bihan and Professor Hidenao Fukuyama for providing the precious opportunity to contribute to this book commemorating the 150th anniversary of the relationship between France and Japan.

References

1. Borgnia, M., Nielsen, S., Engel, A., and Agre, P. (1999) Cellular and molecular biology of the aquaporin water channels, *Annu. Rev. Biochem.*, **68**, 425–458.
2. Nelson, W. J. (2003) Adaptation of core mechanisms to generate cell polarity, *Nature*, **422**, 766–774.

3. Kusumi, A., Nakada, C., Ritchie, K., Murase, K., Suzuki, K., Murakoshi, H., Kasai, R. S., Kondo, J., and Fujiwara, T. (2005) Paradigm shift of the plasma membrane concept from the two-dimensional continuum fluid to the partitioned fluid: high-speed single-molecule tracking of membrane molecules, *Annu. Rev. Biophys. Biomol. Struct.*, **34**, 351–378.

4. Rodriguez-Boulan, E., Kreitzer, G., and Müsch, A. (2005) Organization of vesicular trafficking in epithelia, *Nat. Rev. Mol. Cell. Biol.*, **6**, 233–247.

5. Fölsch, H. (2008) Regulation of membrane trafficking in polarized epithelial cells, *Curr. Opin. Cell Biol.*, **20**, 208–213.

6. Lafont, F., Simons, K., and Ikonen, E. (1995) Dissecting the molecular mechanisms of polarized membrane traffic: reconstitution of three transport steps in epithelial cells using streptolysin-O permeabilization, *Cold Spring Harb. Symp. Quant. Biol.*, **60**, 753–762.

7. Akhmanova, A., Stehbens, S. J., and Yap, A. S. (2009) Touch, grasp, deliver and control: functional cross-talk between microtubules and cell adhesions, *Traffic*, **10**, 268–274.

8. Westermann, S., and Weber, K. (2003) Post-translational modifications regulate microtubule function, *Nat. Rev. Mol. Cell Biol.*, **4**, 938–934.

9. Kremer, B. E., Haystead, T., and Macara, I. G. (2005) Mammalian septins regulate microtubule stability through interaction with the microtubule-binding protein MAP4, *Mol. Biol. Cell*, **16**, 4648–4659.

10. Spiliotis, E. T., Hunt, S., Hu, Q., Kinoshita, M., and Nelson, W. J. (2008) Morphogenesis of polarized epithelia requires Sept2-mediated coupling of vesicle transport to polyglutamylated microtubules, *J. Cell Biol.*, **180**, 295–303.

11. Tsukita, S., Furuse, M., and Itoh, M. (2001) Multifunctional strands in tight junctions, *Nat. Rev. Mol. Cell Biol.*, **2**, 285–293.

12. Nakada, C., Ritchie, K., Oba, Y., Nakamura, M., Hotta, Y., Iino, R., Kasai, R. S., Yamaguchi, K., Fujiwara, T., and Kusumi, A. (2003) Accumulation of anchored proteins forms membrane diffusion barriers during neuronal polarization, *Nat. Cell Biol.*, **5**, 626–632.

13. Conde, C., and Cáceres, A. (2009) Microtubule assembly, organization and dynamics in axons and dendrites, *Nat. Rev. Neurosci.*, **10**, 319–332.

14. Lasiecka, Z. M., Yap, C. C., Vakulenko, M., and Winckler, B. (2009) Compartmentalizing the neuronal plasma membrane from axon initial segments to synapses, *Int. Rev. Cell Mol. Biol.*, **272**, 303–389.

15. Barral, Y., Mermall, V., Mooseker, M. S., and Snyder, M. (2000) Compartmentalization of the cell cortex by septins is required for maintenance of cell polarity in yeast, *Mol. Cell.*, **5**, 841–851.

16. Takizawa, P. A., DeRisi, J. L., Wilhelm, J. E., and Vale, R. D. (2000) Plasma membrane compartmentalization in yeast by messenger RNA transport and a septin diffusion barrier, *Science*, **290**, 341–344.

17. Shcheprova, Z., Baldi, S., Frei, S. B., Gonnet, G., and Barral, Y. (2008) A mechanism for asymmetric segregation of age during yeast budding, *Nature*, **454**, 728–734.

18. Caudron, F., and Barral, Y. (2009) Septins and the lateral compartmentalization of eukaryotic membranes, *Dev. Cell*, **16**, 493–506.

19. Ihara, M., Kinoshita, A., Yamada, S., Tanaka, H., Tanigaki, A., Kitano, A., Goto, M., Okubo, K., Nishiyama, H., Ogawa, O., Takahashi, C., Itohara, S., Nishimune, Y., Noda, M., and Kinoshita, M. (2005) Cortical organization by the septin cytoskeleton is essential for structural and mechanical integrity of mammalian spermatozoa, *Dev. Cell*, **8**, 343–352.

20. Kissel, H., Georgescu, M. M., Larisch, S., Manova, K., Hunnicutt, G. R., and Steller, H. (2005) The Sept4 septin locus is required for sperm terminal differentiation in mice. The Sept4 septin locus is required for sperm terminal differentiation in mice, *Dev. Cell*, 2005, **8**, 353–364.

21. Guan, J., Kinoshita, M., and Yuan, L. (2009) Spatiotemporal association of DNAJB13 with the annulus during mouse sperm flagellum development, *BMC Dev. Biol.*, **9**, 23, 1–9.

22. Cesario, M. M., and Bartles, J. R. (1994) Compartmentalization, processing and redistribution of the plasma membrane protein CE9 on rodent spermatozoa. Relationship of the annulus to domain boundaries in the plasma membrane of the tail, *J. Cell Sci.*, **107**, 561–570.

23. Xie, Y., Vessey, J. P., Konecna, A., Dahm, R., Macchi, P., and Kiebler, M. A. (2007) The dendritic spine-associated GTP-binding protein Septin7 is a regulator of dendrite branching and dendritic spine morphology, *Curr. Biol.*, **17**, 1746–1751.

24. Tada, T., Simonetta, A., Batterton, M., Kinoshita, M., Edbauer, D., and Sheng, M. (2007) Role of Septin cytoskeleton in spine morphogenesis and dendrite development in neurons, *Curr. Biol.*, **17**, 1752–1758.

25. Svoboda, K., Tank, D. W., and Denk, W. (1996) Direct measurement of coupling between dendritic spines and shafts, *Science*, **272**, 716–719.

26. Ashby, M. C., Maier, S. R., Nishimune, A. and Henley, J. M. (2006) Lateral diffusion drives constitutive exchange of AMPA receptors at dendritic spines and is regulated by spine morphology, *J. Neurosci.*, **26**, 7046–7055.

27. Gladfelter, A. S., Pringle, J. R., and Lew, D. J. (2001) The septin cortex at the yeast mother-bud neck, *Curr. Opin. Microbiol.*, **4**, 681–689.

28. McMurray, M. A., and Thorner, J. (2009) Septins: molecular partitioning and the generation of cellular asymmetry, *Cell Div.*, **4**, 18, 1–14.

29. Touré, A., Lhuillier, P., Gossen, J. A., Kuil, C. W., Lhote, D., Jégou, B., Escalier, D., and Gacon, G. (2007) The testis anion transporter 1 (Slc26a8) is required for sperm terminal differentiation and male fertility in the mouse, *Hum. Mol .Genet.*, **16**, 1783–1793.

30. Sugino, Y., Ichioka, K., Soda, T., Ihara, M., Kinoshita, M., Ogawa, O., and Nishiyama, H. (2008) Septins as diagnostic markers for a subset of human asthenozoospermia, *J. Urol.*, **180**, 2706–2709.

31. Lhuillier, P., Rode, B., Escalier, D., Lorès, P., Dirami, T., Bienvenu, T., Gacon, G., Dulioust, E., and Touré, A. (2009) Absence of annulus in human asthenozoospermia: case report, *Hum. Reprod.*, **24**, 1296–1303.

32. Ihara, M., Yamasaki, N., Hagiwara, A., Tomimoto, H., Kitano, A., Tanigaki, A., Hikawa, E., Noda, M., Takanashi, M., Mori, H., Hattori, N., Miyakawa, T., and Kinoshita, M. (2007) Sept4, a component of presynaptic scaffold and Lewy bodies, is required for the suppression of alpha-synuclein neurotoxicity. *Neuron*, **53**, 519–533.

33. Beites, C. L., Xie, H., Bowser, R., Trimble, W. S. (1999) The septin CDCrel-1 binds syntaxin and inhibits exocytosis, *Nat. Neurosci.*, **2**, 434–439.

34. Dent, J., Kato, K., Peng, X. R., Martinez, C., Cattaneo, M., Poujol, C., Nurden, P., Nurden, A., Trimble, W. S., and Ware, J. (2002) A prototypic platelet septin and its participation in secretion, *Proc. Natl. Acad. Sci. USA*, **99**, 3064–3069.

35. Beites, C. L., Campbell, K. A., and Trimble, W. S. (2005) The septin Sept5/CDCrel-1 competes with alpha-SNAP for binding to the SNARE complex, *Biochem. J.*, **385**, 347–353.

36. Hartwell, L. H. (1967) Macromolecule synthesis in temperature-sensitive mutants of yeast, *J. Bacteriol.*, **93**, 1662–1670.

37. Dobbelaere, J., Gentry, M. S., Hallberg, R. L., and Barral, Y. (2003) Phosphorylation-dependent regulation of septin dynamics during the cell cycle, *Dev. Cell*, **4**, 345–357.

38. Caviston, J. P., Longtine, M., Pringle, J. R., and Bi, E. (2003) The role of Cdc42p GTPase-activating proteins in assembly of the septin ring in yeast, *Mol. Biol. Cell*, **14**, 4051–4066.

39. Kinoshita, M. (2006) Diversity of septin scaffolds, *Curr. Opin. Cell Biol.*, **18**, 54–60.

40. Kinoshita, A., Noda, M., and Kinoshita, M. (2000) Differential localization of septins in the mouse brain. *J. Comp. Neurol.*, **428**, 223–239.

41. Ihara, M., Tomimoto, H., Kitayama, H., Morioka, Y., Akiguchi, I., Shibasaki, H., Noda, M., and Kinoshita, M. (2003) Association of the cytoskeletal GTP-binding protein Sept4/H5 with cytoplasmic inclusions found in Parkinson's disease and other synucleinopathies, *J. Biol. Chem.*, **278**, 24095–24102.

42. Kuhlenbäumer, G., Hannibal, M. C., Nelis, E., Schirmacher, A., Verpoorten, N., Meuleman, J., Watts, G. D., De Vriendt, E., Young, P., Stögbauer, F., Halfter, H., Irobi, J., Goossens, D., Del-Favero, J., Betz, B. G., Hor, H., Kurlemann, G., Bird, T. D., Airaksinen, E., Mononen, T., Serradell, A. P., Prats, J. M., Van Broeckhoven, C., De Jonghe, P., Timmerman, V., Ringelstein, E. B., and Chance, P. F. (2005) Mutations in SEPT9 cause hereditary neuralgic amyotrophy, *Nat. Genet.*, **37**, 1044–1046.

43. Pennington, K., Beasley, C. L., Dicker, P., Fagan, A., English, J., Pariante, C. M., Wait, R., Dunn, M. J., and Cotter, D. R. (2008) Prominent synaptic and metabolic abnormalities revealed by proteomic analysis of the dorsolateral prefrontal cortex in schizophrenia and bipolar disorder, *Mol. Psychiatry*, **13**, 1102–1117.

Chapter 15

WATER MOVEMENTS ASSOCIATED WITH NEURONAL ACTIVITY REVEALED BY OPTICAL METHODS: FOCUS ON DIGITAL HOLOGRAPHIC MICROSCOPY

P. Marquet,[a,c] **P. Jourdain,**[a] **C. Depeursinge,**[b] **and P. J. Magistretti**[a,c]

[a] *Brain Mind Institute,* [b] *Institute of Microengineering, Ecole Polytechnique Fédérale de Lausanne, CH-1015 Lausanne, Switzerland*
[c] *Centre de Neurosciences Psychiatriques, Centre Hospitalier Universitaire Vaudois, Département de Psychiatrie, Site de Cery, CH-1008 Prilly/Lausanne, Switzerland*
pierre.marquet@chuv.ch and pierre.magistretti@epfl.ch

15.1 INTRODUCTION

Water represents up to 80% of the body mass; it is distributed evenly across the intra-and extracellular compartments. Since the membranes of animal cells are highly permeable to water [1], its movement across membranes is in large part dictated by osmotic pressure gradients. Changes in extracellular fluid (ECF) osmolality cause water to flow across cell membranes to equilibrate the osmolality of the cytoplasm with that of the ECF.

However, the membranes of animal cells cannot tolerate substantial hydrostatic pressure gradients implying the existence of subtle mechanisms to ensure an appropriate balance of ions across the membrane. In particular, a variety of cell functions, including enzyme activity, gene expression mechanisms, second messenger cascades, and hormone release are all extremely dependent on a homeostatic intracellular ionic environment [1–3]. Consequently, the ability to control cell volume is pivotal for cell function.

Accordingly, volume regulatory mechanisms involving multiple ion transport processes, such as channels, cotransporters, and exchangers, operate in order to maintain a constant ionic balance and hydrostatic pressure gradients [1, 4].

Nevertheless, even at constant extracellular osmolarity, volume constancy of any mammalian cell is permanently challenged [1, 5] by the normal cellular activity of the cells including, for example, transport of osmotically active substances, cell metabolism, and movements of different ionic species across the cell membrane, which all induce osmotic water movements.

15.2 BRAIN ACTIVITY AND WATER TRANSPORT

Within the brain, water is distributed between blood, cerebrospinal fluid, and interstitial and intracellular compartments, and moves across these compartments following the differences in osmotic and hydrostatic pressures. At constant hydrostatic pressure gradients, water movement depends largely on the osmotic gradients, which are created by the osmolyte concentration at the extracellular and intracellular compartments.

Principal osmolytes in brain cells are the electrolytes present in high levels in the cytosol and ECFs, for example, Na^+, K^+, and Cl^-, and small organic molecules of heterogeneous structure, including amino acids and derivatives (taurine, glutamate, glycine, GABA, and N-acetylaspartate), polyalcohols (myoinositol and sorbitol), polysaccharides such as glycogen, and amines (glycerophosphoryl choline, betaine, creatine/P-creatine, and phosphoethanolamine) [1].

At the microscopic level, water transport is involved in cell volume regulation and in controlling the dimensions of the extracellular space (ECS) [6, 7]. Neuronal activity involves ion movements resulting from the opening of conductances as well as neurotransmitter release including glutamate, glycine, and GABA. This implies that the transport of water across the plasma membrane, the regulation of the ECS as well as cell volume will depend on the level of neural activity.

Minor changes in the composition of ions in the brain's extracellular or intracellular fluids can significantly affect the function of neurons, which rely on the existence of precise ionic gradients across their plasma membranes to trigger the changes in membrane potential that underlie action potential generation and propagation, as well as postsynaptic potentials. Hence, water transport in the brain is tightly regulated as it is critical to maintain neuronal excitability and to prevent damage derived from brain swelling or shrinkage. Practically, osmolyte fluxes accompanying water movement permit the reestablishment of an osmotic balance and volume recovery in anisosmoticelicited cell volume changes. However, in different pathological conditions leading to homeostasis perturbations, these osmolyte-mediated water movements are also the causal factor per se of brain cell swelling or shrinkage, which may have particularly dramatic consequences. Thus, hyponatremia, epilepsy, ischemia, and excitoxicity are pathologies involving different triggering mechanisms, yet sharing common events including energy failure, depolarisation, high extracellular K^+ concentration, extracellular glutamate rise, lactacidosis, and ionic overload, which induce water movements resulting in cell swelling and brain oedema [8].

Water crosses cell membranes through several routes. One is by mere diffusion through the lipid bilayer, a relatively slow process. The second is by way of specialized water channels (the aquaporins, AQP). The third is via proteins usually associated with other functions such as uniports and cotransporters. Research on AQP is a rapidly developing field: the first aquaporin was functionally defined in 1992 [9], and the first evidence for the presence of AQP4 in brain was reported in 1994 [10, 11]. AQP4 is now known to be the predominant aquaporin in brain. The precise mechanism of water transport through cotransporters is not entirely understood. It is generally accepted that cotransporters act as passive water channels, but an active mode of transport has also been suggested. A transmembrane osmotic gradient induces passive water transport through cotransporters, as would be expected from a conventional water channel. But during cotransport, another component of water transport emerges in parallel and independently of the passive water transport. By this mechanism, water can move uphill against the osmotic gradient since the energy contained in the substrate gradients is transferred to the transport of water [12].

15.2.1 Brain Water Movements Measured with Intrinsic Optical Signals

Within this framework, the analysis of intrinsic optical imaging has provided new insights about the relationship between water movements in brain tissue and neuronal activation. Indeed, intrinsic optical signals (IOS) of brain tissues are correlated with neuronal activity [13–15], allowing, for example, the visualization of the functional organization of the primate visual cortex [16].

Specifically, photons that enter tissue may undergo several types of interactions with the tissue, including (i) *absorption*, which may lead to radiationless loss of energy into the medium, or induce fluorescence, and (ii) *scattering*. Both processes represent the main interactions underlying the generation of IOS. Numerous analyses of IOS, performed in many situations in mammalian brain *in vivo* as well as *in vitro*, have contributed to relate absorption and scattering to specific physiological events.

In vivo IOS are primarily sensitive to the characteristic absorption patterns in the visible and in near-infrared light range of oxygenated and desoxygenated haemoglobin. Therefore, on the basis of light absorption measurements, changes in the relative concentration of these molecules as well as blood flow changes have been successfully measured during functional brain activation with relatively high spatial and temporal resolution [17–21].

On the other hand, activity-related light-scattering changes have been measured in isolated axons [13, 22–25], neuronal cell cultures [23, 26], brain slices [27–32], isolated neurohypophysis *in vitro* [33], as well as *in vivo* [20, 34–36]. In fact, even noninvasive measurements in human adults have been reported [37–39]. In contrast to absorption, the light-scattering signal remains more difficult to interpret in terms of the underlying biological processes modifying the tissue structure. However, as light scattering occurs at borders of media with different refractive indices, it seems plausible to postulate that events occurring in neuronal membranes and volume changes of cellular compartments or of organelles associated with brain activity can influence light scattering.

Accurate analysis of the light-scattering signal has revealed two types of signals associated with brain cell activity: a fast and slow signal. The fast signal [13, 22–26, 33, 36, 40, 41] is independent of wavelength and has response times in the order of the millisecond. In isolated or cultured neurones, this fast light-scattering signal depends linearly on the transmembrane potential; accordingly axon potentials can be observed in this manner. The origin of this fast scattering signal remains poorly understood and is likely to result from several biophysical processes. Thus, reorientation of membrane proteins, channels, and phospholipids with voltage or mechanical changes can result in a transient change in the rotation of polarized light by neural tissue, measured as a change in birefringence [13, 23]. Transient changes in the membrane refractive index correlate with membrane potential and have also been associated with fast signal measurements [26]. Water influx in response to ionic currents through gated channels during depolarization causes cellular swelling that also produces changes in light scattering [13, 40–42]. These different mechanisms may represent the physiological processes underlying the generation of the noninvasively measured fast optical signals also called event-related optical signal in human subjects.

In contrast, the slow scattering signal with response times in the order of a few seconds has been described in hippocampal brain slices as well as in intact animals [20, 28–32, 43–47]. Specifically, the activity-dependent increased light transmittance across brain slices [7, 28, 43–45, 48] has allowed to provide dynamic measurements of glial swelling associated with excitatory synaptic input and action potential discharge. It is likely that one of the principal cause of cell swelling is the increase in K^+ caused by repetitive neuronal activity [49] and the principal cell type to swell is the astrocyte [50, 51]. In contrast, shrinkage of the cells following exposure to hypertonic solutions has been associated with a reduction of transmitted light.

It is well known that astrocytes respond to decreases in external osmolarity by rapid swelling, followed by a corrective process leading to cell volume recovery, usually referred to as regulatory volume decrease (RVD) [52, 53]. RVD involves the efflux of metabolites and activation of swelling-activated Cl^+ and K^+ currents [54, 55]. Noticeably, IOS correlate with the extent and time course of the change of the ECS size. They have a high signal-to-noise ratio and allow measurements of IOS changes in the order of a few percent [56]. Activity-induced changes of ECS volume in excitable tissues are a well-known phenomenon. Following neuronal activity, and in parallel to an extracellular potassium elevation, ECS shrinkage can be observed in several preparations *in vivo* [57] or *in vitro* [45, 50, 58]. The potassium-induced swelling of glial cells [59, 60] seems to provide a significant contribution to the ECS shrinkage [45]. However, it cannot be excluded that, in addition, neuronal cell volume changes take place.

Neural tissues possess several mechanisms to control the ensuing extracellular accumulation of potassium: a Na/K-exchanger is present in glial cells and neurons including a glial NaK_2Cl-cotransporter mentioned above and resulting in a net-uptake of KCl [61]. In addition, the so-called *spatial buffer mechanism* may contribute to potassium homeostasis [57, 62–67]. The glial cells are indeed connected by an extensive network of gap junctions [68, 69], forming highly conductive membrane channels that are permeable to K^+ in particular. Thus, this astrocytic syncytium works as a draining system, which takes potassium up in regions of high potassium and liberates it in low potassium regions.

According to Nagelhus *et al.* 2004 [6], a possible coupling between K^+ buffering and water transport in molecular terms can be explained as follows (Fig. 15.1). The neuronal activity-mediated release of K^+ causes a depolarization of the adjacent astrocyte membrane. This favors an uptake of sodium and bicarbonate through the electrogenic Na^+/HCO_3 cotransporter (NBC), which, in turn, causes the intracellular osmolarity to increase and drives water into the glial cells through AQP4, thus contributing to a reduction of the ECS volume. Other mechanisms can be also involved in such a reduction, including the astrocytic uptake of glutamate via excitatory amino acid transporters (EAAT), associated with a considerable water flux, as well as the KCl uptake via the glial Na-K-2Cl cotransporter [32, 45]. Removal of water from the ECS causes a local increase in the concentration of all solutes. This will facilitate glutamate uptake into astrocytes and of lactate into neurons [70], and may also boost spatial buffering of K^+. Finally, in the classical model of K^+ buffering, the driving force for the redistribution of K^+ is equivalent to the difference in K^+ between the ECS at the site of neuronal activity and the remote ECS to which the excess K^+ is directed.

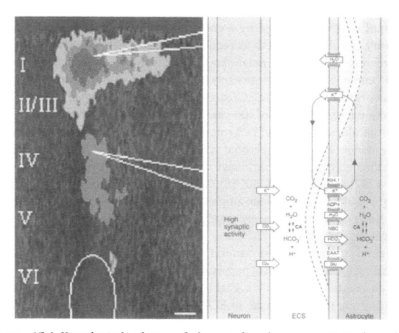

Figure 15.1 Hypothetical scheme of the coupling between activity-dependent changes of extracellular (ECS) volume, and water and ion transport across glial membranes. *Left*: Changes in intrinsic optical signals (IOS) following afferent stimulation of rat neocortical slice. Slices were electrically stimulated through an electrode in layer VI (indicated) with stimulus trains lasting 2 s. IOS were captured 6 s after stimulation was started. The bright signal (red) in layer IV (which is target of the afferent stimulation) corresponds to a reduction of the extracellular space. Darkening of the IOS ("black wave", blue)) corresponds to a widening of the extracellular space and is seen in the superficial cortical layers. Electrodes for measurement of potassium activity in layers I and IV are indicated. *Right* (scale bar 200 μm): Activity dependent volume changes and ion transport. Active neurons (orange) release K^+ and glutamate and produce increased amounts of CO_2. As indicated in left panel, activity is associated with a shrinkage of the extracellular space locally and with a widening of the extracellular space distal to the site of stimulation (i.e., in layer I). We assume that these volume changes reflect water influx and efflux, respectively, primarily across astrocyte plasma membranes (dashed lines indicate position of astrocyte membrane after neuronal activation; cf. left panel). Reproduced from Ref. [6] by permission. For color reference, see p. 385.

A working spatial buffer mechanism causes characteristic alterations of the ECS. The ECS should shrink at regions of potassium uptake and should widen at regions where potassium is liberated (Fig. 15.1). The analysis of the spatial extent and time course of the IOS—stimulation of afferent fibers to layer IV of the neocortex produced a local shrinkage in layer IV and an extracellular volume increase in the layer I–II—has allowed to visualize this spatial buffer mechanism, which appeared as a radial propagating water wave in a cortical slice preparation [45, 71]. Since K$^+$ channels do not admit water to any significant extent, the accompanying water flux must be mediated through a distinct channel [72]. In addition, the time course of the volume changes is indicative of rapid water movements, consistent with an involvement of AQP4 (the only aquaporin expressed in significant quantity by cortical glial cells [73]). The available data [56] suggest that AQP4 facilitates glial water uptake at sites of neuronal activation and water efflux at distant sites. Coherently, the radial water fluxes are facilitated by vasopressin and vasopressin receptor V1a agonists, known to regulate aquaporin expression and translocation in the collecting ducts of the kidney (Fig. 15.2).

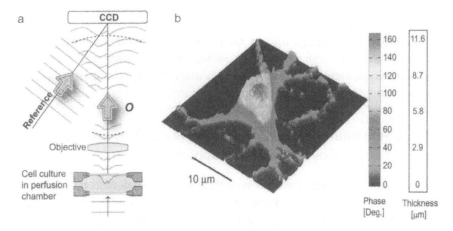

Figure 15.2 Digital holographic microscopy (DHM) of living mouse cortical neurons in culture. (a) Schematic representation of cultured cells mounted in a closed perfusion chamber and (b) trans-illuminated 3D perspective image in false colors of a living neuron in culture. Each pixel represents a quantitative measurement of the phase retardation or cellular optical path length (OPL) induces by the cell (with a sensitivity corresponding to a few tens of nanometers). By using the measured mean value of the neuronal cell body refractive index, resulting from the decoupling procedure, scales (at the right), which relate OPL (Deg) to morphology in the z axis (μm), can be constructed. For color reference, see p. 386.

Practically, the bright signal correlated to the reduction of the ECS, is likely to result from the decrease of the intracellular refractive index of astrocytes due to the dilution of their content by the influx of water. Indeed, the tissue scattering that contributes to reduce the transmitted light results mainly from the mismatch between the intracellular and the extracellular refractive indices. Coherently, the darkening of the IOS is due to an increased refractive index mismatch resulting from an efflux of water from the astrocytes. Concretely, for a more or less physiological range of activity, IOS reveals changes of the ECS in the order of a few percent. In addition, within a few seconds after the onset of neuronal activity, the K^+ spatial buffering process is accompanied by movements of water spreading over distances of several hundred of micrometers.

Whereas IOS has provided new insights into water transport between glial cells and the ECS during neuronal activation, the effective contribution of the neuronal transmembrane water movements to the changes of the ECS during brain activation remains poorly understood. Tasaki *et al.* 2002 [74] have however reported, by measuring slow scattering signals in hippocampal slices, a synaptically evoked neuronal swelling involving not only soma but also focal areas along dendrites and axons mediated by Cl influx through GABA-A receptors.

In summary, although optical methods have allowed the measurement of fast scattering signals directly associated with neuronal membrane depolarisation, their origin remains poorly understood, particularly regarding the possible involvement of the transmembrane water movements.

15.2.2 Fast Neuronal Transmembrane Water Movements Revealed by Digital Holographic Microscopy

Recently, we have developed a quantitative phase microscopy technique derived from digital holographic microscopy (DHM) [75], allowing to visualize, with a few millisecond temporal resolution and with a nanometric axial sensitivity, living cells in culture. Practically, information about cell structure and dynamics is encoded in the wavefront phase retardation that the observed specimen induces on the transmitted wavefront. This phase information, resulting from an interference between the wavefront transmitted by the object and a reference wave, is accurately coded in a hologram recorded on a CCD camera. An original and robust numerical process, based on the electromagnetic wave propagation theory, allows to reconstruct holograms, that is, to provide an exact replica of the wave diffracted by the object and consequently to calculate in particular the phase retardation induced by the specimen.

Due to its interferometric nature, DHM provides quantitative phase images with a nanometric sensitivity along the optical axis of the microscope, allowing to reveal extremely detailed information about the observed specimen. Consequently, the quantitative phase signal acts as an endogenous contrast agent allowing to optically probe transparent specimens at the nano-scale. DHM provides quantitative information and does not suffer of any optical artifacts [76] that could interfere with the interpretation of the phase signal in relation to the underlying biological processes. This is unlike the classical phase contrast (PhC) technique, initially proposed by Zernicke [77] as well as the Normarski's differential interference contrast (DIC) available for high-resolution light microscopy [78] and widely used in biology to observe transparent specimens.

The quantitative DHM phase signal Φ is rich in information about the observed specimen, including cell morphology, internal structure, as well as cellular micro-movements. The phase signal is related to the biophysical cell parameters by the following expression:

$$\Phi \approx (\bar{n}_c - n_m)d, \tag{15.1}$$

where d is the cellular thickness, $\bar{n}_c$ is the intracellular mean refractive index, and n_m the refractive index of the surrounding medium.

In order to exploit these different cell parameters, we have developed a decoupling procedure [79, 80] allowing to separately measure from the phase signal the information about cell morphology (volume, shape, nano-movements) and about the intracellular refractive index, a parameter which is highly sensitive to transmembrane water movements which contribute to dilute or concentrate the intracellular content particularly in terms of protein concentration [81]. Accordingly, experiments combining electrophysiology and DHM, have allowed us to accurately study the neuronal transmembrane movements of water associated with neuronal electrical activity.

Thus, local application of the excitatory neurotransmitter glutamate (500 μM, 200 ms) on primary cultures of mouse cortical neurons, induced both a strong inward current as well as a decrease of phase signal, whose amplitudes are proportional to the concentration of glutamate and to the duration of the application (Fig. 15.3). This inward current is consistent with glutamate-mediated activation of ionotropic receptors, namely, N-methyl-D-aspartate (NMDA), 2-amino-3-(3-hydroxy-5-methylisoxazol-4-yl), proprionate (AMPA), and kainate receptors, which induce the opening of their associated ion channel, allowing influxes of Ca^{2+} and Na^+ down to their electrochemical gradient. The receptor-specific nature of these phase and current signals has been demonstrated by their suppression resulting from the co-application of MK801 and CNQX, two specific antagonists of NMDA and AMPA/kainate ionotropic receptors, respectively. In addition, the decoupling procedure has permitted to confirm that glutamate pulses provoke a neuronal swelling as well as a significant decrease of the intracellular refractive index in agreement with an entry of water accompanying the ion influx that dilutes the intracellular content.

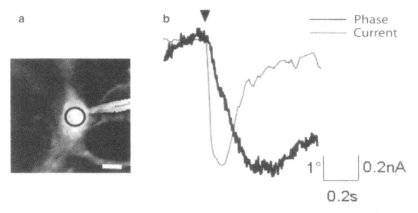

Figure 15.3 Phase shift associated with glutamate-mediated neuronal activity. (a) Quantitative phase image of a patched mouse cortical neuron in primary culture recorded by DHM. The full circles in the middle of the cell correspond to the region of interest where the phase signal is recorded (scale bar: 10μm). (b) Local application of glutamate (500 μM, 200 ms; arrow head) on the neuron triggered both a strong transient decrease of the phase signal associated to an inward current. Phase is expressed in degrees.

The phase being primarily dependent on the refractive index [79], it follows that water entry is associated with a phase decrease despite a cell swelling. In contrast, an exit of water leads to an increase of the phase signal. The phase signal can therefore directly monitor water movements associated with transmembrane ionic fluxes mediated by agonist-activated receptor-gated channels.

In addition, for glutamate applications lasting a few hundred milliseconds, the corresponding phase response can be decomposed into two components, a rapid one accompanying glutamate mediated current (I_{GLUT})—the phase decrease in Fig. 15.3—and a slow one corresponding generally to a phase recovery while $I_{GLUT} = 0$. The phase recovery, which is much slower than the other one is likely to correspond to a nonelectrogenic neuronal volume regulation involving several mechanisms. Specifically, preliminary results obtained by DHM revealed the involvement of the cation/chloride cotransporters KCC2 in the regulatory volume processes (data not shown).

The time-course presented in Fig. 15.3 shows that the water movements are not significantly delayed—at the tenth of a second scale at least—relative to the recorded current and can be interpreted as a temporal integration of the water fluxes accompanying the transmenbrane current. Furthermore, the measurements of I_{GLUT} as well as the corresponding intracellular refractive index and neuronal volume changes allow to estimate the parameter "ε_{GLUT}" [mL/C] representing the volume of the water movement associated with the net charge transported through the cell membrane. Practically, values of ε_{GLUT} lie within the range of 60–120 $\mu m^3/nC$, which is equivalent to 340–620 water molecules transported per ion having crossed the membrane. This order of magnitude is in good agreement with a cotransport of water, whose typical coupling ratios are of around 150–500 water molecules per charge translocated by the effector protein [12].

The typical intracelluar refractive index change induced by a glutamate pulse (500 μM, 0.2 s) is around 0.002–0.003, corresponding to a substantial variation (7–10%) of the scattering potential proportional to the parameter $\left(\bar{n}_c^2 - n_m^2 \right)$ from which the scattering coefficient can be evaluated. The associated neuronal swelling is around 100 femtoliter (fL) for a typical neuronal cell body of 1500 fL corresponding to a 6–7% cell volume variation.

In addition, on the basis of the phase expression given by Eq. (1), the hypothesis that the n_c value depends linearly on the concentration of the different intracellular components—an assumption well-established for living cells—and the knowledge of the ε_{GLUT} value, a mathematical relationship has been derived allowing to quantitatively calculate, from the optical DHM phase signal, the transmembrane current associated with the movement of water. According to this relationship, each pixel of the quantitative phase image can be seen as a virtual electrode allowing to locally and noninvasively record the transmembrane current.

15.3 CONCLUSION

IOS as well as the DHM phase signals have allowed to reveal two kinds of water movements associated with neuronal activity. The first one, involving mainly astrocytes, represents a "slow" movement of water—within the range of seconds—when compared to the time course of electrical phenomena associated with neuronal activity. In practice, this "slow" water movement accompanies the K^+ spatial buffer mechanism and is characterized by a propagation of water over distances of several hundred of micrometers within the neocortex, preferentially along the radial direction. In contrast, "rapid" water movements as revealed by DHM—at the scale of hundredth of a second—associated with glutamate-mediated neuronal activity lead to a significant neuronal swelling. Furthermore these "rapid" water movements, inducing intracellular refractive index changes, and consequently a modification of the light scattering process, are very likely to be one of the underlying mechanisms involved in the generation of the even-related optical signals measured in human subjects [39]. Besides, the dynamics of these two different movements of water require the presence of a cell membrane highly permeable to water, stressing the key role played by water channels such as AQP, as well as the water movements involving uniports and cotransporters. This result is in good agreement with the observation obtained with an ultrafast nuclear magnetic resonance imaging technique indicating that cell membranes in the brain, although likely to hinder the process of water diffusion, seem highly permeable to water [82].

These two "fast" and "slow" water movements relative to the neuronal activity need to be further characterized in order to assess their individual contribution to the water diffusion process in the brain. At this stage, it is indeed difficult to directly link these water movements to the two different pools of water diffusion—representing a slow and fast diffusion phase—revealed by the water diffusion–sensitized MRI signal [83].

A better understanding of water movement mechanisms in relation with membrane permeability properties and neuronal activity represent important dynamic parameters underlying the biexponential function which adequately fits the measured water diffusion decay in the brain and will complement at the cellular level the information provided by the water diffusion–sensitized MRI signal.

References

1. Lang, F., Busch, G. L., Ritter, M., Völkl, H., Waldegger, S., Gulbins, E., and Häussinger, D. (1998) Functional significance of cell volume regulatory mechanisms, *Physiol. Rev.*, **78**, 247–306.

2. Waldegger, S., and Lang, F. (1998) Cell volume and gene expression, *J. Membr. Biol.*, **162**, 95–100.

3. Lang, F. *et al.* (2000) Cell volume in the regulation of cell proliferation and apoptotic cell death, *Cell. Physiol. Biochem.*, **10**, 417–428, doi:cpb10417 [pii].

4. Hoffmann, E. K., Lambert, I. H., and Pedersen, S. F. (2009) Physiology of cell volume regulation in vertebrates, *Physiol. Rev.*, **89**, 193–277, doi:89/1/193 [pii] 10.1152/physrev.00037.2007.

5. Strange, K. (2004) Cellular volume homeostasis, *Adv. Physiol. Edu.*, **28**, 155–159, doi:28/4/155 [pii] 10.1152/advan.00034.2004.

6. Nagelhus, E. A., Mathiisen, T. M., and Ottersen, O. P. (2004)Aquaporin-4 in the central nervous system: cellular and subcellular distribution and coexpression with KIR4.1, *Neuroscience*, **129**, 905–913, doi:10.1016/j.neuroscience.2004.08.053.

7. Syková, E., Vargová, L., Kubinová, S., Jendelová, P., and Chvàtal, A. (2003) The relationship between changes in intrinsic optical signals and cell swelling in rat spinal cord slices, *NeuroImage*, **18**, 214–230.

8. Pasantes-Morales, H., and Cruz-Rangel, S. (2009) Brain volume regulation: osmolytes and aquaporin perspectives, *Neuroscience*, doi:S0306-4522(09)01985-X [pii] 10.1016/j.neuroscience.2009.11.074.

9. Preston, G. M., Carroll, T. P., Guggino, W. B., and Agre, P. (1992) Appearance of water channels in Xenopus oocytes expressing red cell CHIP28 protein, *Science*, **256**, 385–387.

10. Jung, J. S. *et al.* (1994) Molecular characterization of an aquaporin cDNA from brain: candidate osmoreceptor and regulator of water balance, *Proc. Natl. Acad. Sci. USA*, **91**, 13052–13056.

11. Hasegawa, H., Ma, T., Skach, W., Matthay, M. A., and Verkman, A. S. (1994) Molecular cloning of a mercurial-insensitive water channel expressed in selected water-transporting tissues, *J. Biol. Chem.*, **269**, 5497–5500.

12. MacAulay, N., Hamann, S., and Zeuthen, T. (2004) Water transport in the brain: role of cotransporters, *Neuroscience*, **129**, 1031–1044, doi:10.1016/j.neuroscience.2004.06.045.

13. Cohen, L. B. (1973) Changes in neuron structure during action potential propagation and synaptic transmission, *Physiol. Rev.*, **53**, 373–418.

14. Grinvald, A., Lieke, E., Frostig, R. D., Gilbert, C. D., and Wiesel, T. N. (1986) Functional architecture of cortex revealed by optical imaging of intrinsic signals, *Nature*, **324**, 361–364, doi:10.1038/324361a0.

15. Lieke, E. *et al.* (1989) Optical imaging of cortical activity: real-time imaging using extrinsic dye-signals and high resolution imaging based on slow intrinsic-signals, *Annu. Rev. Physiol.*, **5**, 543–559.

16. Ts'o, D. Y., Frostig, R. D., Lieke, E. E., and Grinvald, A. (1990) Functional organization of primate visual cortex revealed by high resolution optical imaging, *Science*, **249**, 417–420.

17. Chance, B., Zhuang, Z., UnAh, C., Alter, C., and Lipton, L. (1993) Cognition-activated low-frequency modulation of light absorption in human brain, *Proc. Natl. Acad. Sci. USA*, **90**, 3770–3774.

18. Hoshi, Y., and Tamura, M. (1993) Dynamic multichannel near-infrared optical imaging of human brain activity, *J. Appl. Physiol.*, **75**, 1842–1846.

19. Villringer, A., Planck, J., Hock, C., Schleinkofer, L., and Dirnagl, U. (1993) Near infrared spectroscopy (NIRS): a new tool to study hemodynamic changes during activation of brain function in human adults, *Neurosci. Lett.*, **154**, 101–104.

20. Malonek, D., and Grinvald, A. (1996) Interactions between electrical activity and cortical microcirculation revealed by imaging spectroscopy: implications for functional brain mapping, *Science*, **272**, 551–554.

21. Dunn, A. K. *et al.* (2003) Simultaneous imaging of total cerebral hemoglobin concentration, oxygenation, and blood flow during functional activation, *Opt. Lett.*, **28**, 28–30.

22. Hill, D. K., and Keynes, R. D. (1949) Opacity changes in stimulated nerve, *J. Physiol.*, **108**, 278–281.

23. Tasaki, I., Watanabe, A., Sandlin, R., and Carnay, L. (1968) Changes in fluorescence, turbidity, and birefringence associated with nerve excitation, *Proc. Natl. Acad. Sci. USA*, **61**, 883–888.

24. Carter, K. M., George, J. S., and Rector, D. M. (2004) Simultaneous birefringence and scattered light measurements reveal anatomical features in isolated crustacean nerve, *J Neurosci. Methods*, **135**, 9–16, doi:10.1016/j.jneumeth.2003.11.010.

25. Yao, X. C., Foust, A., Rector, D. M., Barrowes, B., and George, J. S. (2005) Cross-polarized reflected light measurement of fast optical responses associated with neural activation, *Biophys. J.*, **88**, 4170–4177, doi:10.1529/biophysj.104.052506.

26. Stepnoski, R. A, *et al.* (1991) Noninvasive detection of changes in membrane potential in cultured neurons by light scattering, *Proc. Natl. Acad. Sci. USA*, **88**, 9382–9386.

27. Lipton, P. (1973) Effects of membrane depolarization on light scattering by cerebral cortical slices, *J. Physiol.*, **231**, 365–383.

28. MacVicar, B. A., and Hochman, D. (1991) Imaging of synaptically evoked intrinsic optical signals in hippocampal slices, *J. Neurosci.*, **11**, 1458–1469.

29. Andrew, R. D., Adams, J. R., and Polischuk, T. M. (1996) Imaging NMDA- and kainate-induced intrinsic optical signals from the hippocampal slice, *J. Neurophysiol.*, **76**, 2707–2717.

30. Muller, M., and Somjen, G. G. (1999) Intrinsic optical signals in rat hippocampal slices during hypoxia-induced spreading depression-like depolarization, *J. Neurophysiol.*, **82**, 1818–1831.

31. Trent, A., Basarsky, D., and MacVicar, B. (1999) Glutamate release through volume-activated channels during spreading depression, *J. Neurosci.*, **19**, 6439–6445.

32. MacVicar, B. A., Feighan, D., Brown, A., and Ransom, B. (2002) Intrinsic optical signals in the rat optic nerve: role for K(+) uptake via NKCC1 and swelling of astrocytes, *Glia*, **37**, 114–123, doi:10.1002/glia.10023.

33. Salzberg, B. M., and Obaid, A. L. (1988) Optical studies of the secretory event at vertebrate nerve terminals, *J. Exp. Biol.*, **139**, 195–231.

34. Federico, P., Borg, S. G., Salkauskus, A. G., and MacVicar, B. A. (1994) Mapping patterns of neuronal activity and seizure propagation by imaging intrinsic optical signals in the isolated whole brain of the guinea-pig, *Neuroscience*, **58**, 461–480, doi:0306-4522(94)90073-6.

35. Villringer, A., and Chance, B. (1997) Non-invasive optical spectroscopy and imaging of human brain function, *Trends Neurosci.*, **20**, 435–442, doi:S0166-2236(97)01132-6.

36. Rector, D. M., Poe, G. R., Kristensen, M. P., and Harper, R. M. (1997) Light scattering changes follow evoked potentials from hippocampal Schaeffer collateral stimulation, *J. Neurophysiol.*, **78**, 1707–1713.

37. Gratton, G., Corballis, P. M., Cho, E., Fabiani, M., and Hood, D. C. (1995) Shades of gray matter: noninvasive optical images of human brain responses during visual stimulation, *Psychophysiology*, **32**, 505–509.

38. Gratton, G. *et al.* (1997) Fast and localized event-related optical signals (EROS) in the human occipital cortex: comparisons with the visual evoked potential and fMRI, *NeuroImage*, **6**, 168–180, doi:10.1006/nimg.1997.0298.

39. Gratton, G., and Fabiani, M. (2001) The event-related optical signal: a new tool for studying brain function, *Int. J. Psychophysiol.*, **42**, 109–121, doi:10.1016/S0167-8760(01)00161-1.

40. Tasaki, I., and Byrne, P. M. (1992) Rapid structural changes in nerve fibers evoked by electric current pulses, *Biochem. Biophys. Res. Commun.*, **188**, 559–564.

41. Rector, D., Rogers, R., Schwaber, J., Harper, R., and George, J. (2001) Scattered-light imaging *in vivo* tracks fast and slow processes of neurophysiological activation, *NeuroImage*, **14**, 977–979.

42. Yao, X. C., Rector, D. M., and George, J. S. (2003) Optical lever recording of displacements from activated lobster nerve bundles and Nitella internodes, *Appl. Opt.*, **42**, 2972–2978.

43. McManus, M., Fischbarg, J., Sun, A., Hebert, S., and Strange, K. (1993) Laser light-scattering system for studying cell volume regulation and membrane transport processes, *Am. J. Physiol.*, **265**, C562–C570.

44. Andrew, R. D., and MacVicar, B. A. (1994) Imaging cell volume changes and neuronal excitation in the hippocampal slice, *Neuroscience*, **62**, 371–383, doi:10.1016/0306-4522(94)90372-7.

45. Holthoff, K., and Witte, O. W. (1996) Intrinsic optical signals in rat neocortical slices measured with near-infrared dark-field microscopy reveal changes in extracellular space, *J. Neurosci.*, **16**, 2740–2749.

46. Isokawa, M. (2000) Altered pattern of light transmittance and resistance to AMPA-induced swelling in the dentate gyrus of the epileptic hippocampus, *Hippocampus*, **10**, 663–672, doi:10.1002/1098-1063(2000)10:6<663::AID-HIPO1004>3.0.CO;2-S.

47. Lee, J., Tommerdahl, M., Favorov, O., and Bl, W. (2005) Optically recorded response of the superficial dorsal horn: dissociation from neuronal activity, sensitivity to formalin-evoked skin nociceptor activation, *J. Neurophysiol.*, **94**, 852–864.

48. Andrew, R. D., Jarvis, C. R., and Obeidat, A. S. (1999) Potential sources of intrinsic optical signals imaged in live brain slices, *Methods*, **18**, 185–196, 179, doi:10.1006/meth.1999.0771.

49. Ransom, C. B., Ransom, B. R., and Sontheimer, H. (2000) Activity-dependent extracellular K+ accumulation in rat optic nerve: the role of glial and axonal Na$^+$ pumps, *J. Physiol.*, **522**(Pt 3), 427–442.

50. Ransom, B. R., Yamate, C. L., and Connors, B. W. (1985) Activity-dependent shrinkage of extracellular space in rat optic nerve: a developmental study, *J. Neurosci.*, **5**, 532–535.

51. Ransom, B. R., and Orkand, R. K. (1996) Glial-neuronal interactions in non-synaptic areas of the brain: studies in the optic nerve, *Trends Neurosci.*, **19**, 352–358, doi:10.1016/0166-2236(96)10045-X.

52. Pasantes-Morales, H., Alavez, S., Sanchez Olea, R., and Moran, J. (1993) Contribution of organic and inorganic osmolytes to volume regulation in rat brain cells in culture, *Neurochem. Res.*, **18**, 445–452.

53. Strange, K., Emma, F., and Jackson, P. S. (1996) Cellular and molecular physiology of volume-sensitive anion channels, *Am. J. Physiol.*, **270**, C711–C730.

54. Kimelberg, H., Rutledge, E., Goderie, S., and Charniga, C. (1995) Astrocytic swelling due to hypotonic or high K+ medium causes inhibition of glutamate and aspartate uptake and increases their release, *J. Cereb. Blood Flow Metab.*, **15**, 409–416.

55. Darby, M., Kuzmiski, J. B., Panenka, W., Feighan, D., and MacVicar, B. A. (2003) ATP released from astrocytes during swelling activates chloride channels, *J. Neurophysiol.*, **89**, 1870–1877, doi:10.1152/jn.00510.2002.

56. Witte, O. W., Niermann, H., and Holthoff, K. (2001) Cell swelling and ion redistribution assessed with intrinsic optical signals, *An. Acad. Bras. Cienc.*, **73**, 337–350, doi:S0001-37652001000300005.

57. Dietzel, I., Heinemann, U., Hofmeier, G., and Lux, H. D. (1980) Transient changes in the size of the extracellular space in the sensorimotor cortex of cats in relation to stimulus-induced changes in potassium concentration, *Exp. Brain. Res.*, **40**, 432–439.

58. Svoboda, J., and Sykova, E. (1991) Extracellular space volume changes in the rat spinal cord produced by nerve stimulation and peripheral injury, *Brain Res.*, **560**, 216–224, doi:10.1016/0006-8993(91)91235-S.

59. Walz, W., and Hinks, E. C. (1985) Carrier-mediated KCl accumulation accompanied by water movements is involved in the control of physiological K1 levels by astrocytes, *Brain Res.*, **343**, 44–51.

60. Walz, W. (1987) Swelling and potassium uptake in cultured astrocytes, *Can. J. Physiol. Pharmacol.*, **65**, 1051–1057.

61. Walz, W., and Hinks, E. C. (1986) A transmembrane sodium cycle in astrocytes, *Brain Res.*, **368**, 226–232, doi:10.1016/0006-8993(86)90565-2.

62. Orkand, R. K., Nicholls, J. G., and Kuffler, S. W. (1966) Effect of nerve impulses on the membrane potential of glial cells in the central nervous system of amphibia, *J. Neurophysiol.*, **29**, 788–806.

63. Dietzel, I., Heinemann, U., Hofmeier, G., and Lux, H. D. (1982) Stimulus-induced changes in extracellular Na^+ and Cl^- concentration in relation to changes in the size of the extracellular space, *Exp. Brain. Res.*, **46**, 73–84.

64. Dietzel, I., and Heinemann, U. (1986) Dynamic variations of the brain cell microenvironment in relation to neuronal hyperactivity, *Ann. NY Acad. Sci.*, **481**, 72–86.

65. Gardner-Medwin, A., and Nicholson, C. (1983) Changes of extracellular potassium activity induced by electric current through brain tissue in the rat, *J. Physiol. Lond.*, **335**, 375–392.

66. Gardner-Medwin, A. (1986) A new framework for assessment of potassium buffering mechanisms, *Ann. N.Y. Acad. Sci.*, **481**, 287–302.

67. Dietzel, I., Heinemann, U., and Lux, H. D. (1989) Relations between slow extracellular potential changes, glial potassium buffering, and electrolyte and cellular volume changes during neuronal hyperactivity in cat brain, *Glia*, **2**, 25–44.

68. Cotrina, M. L. *et al.* (1998) Connexins regulate calcium signaling by controlling ATP release, *Proc. Natl. Acad. Sci. USA*, **95**, 15735–15740.

69. Giaume, C., Koulakoff, A., Roux, L., Holeman, D., and Rouach, N. (2010) Astroglial networks: a step further in neuroglial and gliovascular interactions, *Nat. Rev. Neurosci.*, **11**, 87–99.

70. Magistretti, P. J. (2006) Neuron-glia metabolic coupling and plasticity, *J. Exp. Biol.*, **209**, 2304–2311, doi:10.1242/jeb.02208.

71. Niermann, H., Amiry-Moghaddam, M., Holthoff, K., Witte, O. W., and Ottersen, O. P. (2001) A novel role of vasopressin in the brain: modulation of activity-dependent water flux in the neocortex, *J. Neurosci.*, **21**, 3045–3051.

72. Nagelhus, E. A. *et al.* (1999) Immunogold evidence suggests that coupling of K+ siphoning and water transport in rat retinal Muller cells is mediated by a coenrichment of Kir4.1 and AQP4 in specific membrane domains, *Glia*, **26**, 47–54, doi:10.1002/(SICI)1098-1136(199903)26:1<47::AID-GLIA5>3.0.CO;2-5.

73. Badaut, J., Lasbennes, F., Magistretti, P. J., and Regli, L. (2002) Aquaporins in brain: distribution, physiology, and pathophysiology, *J. Cereb. Blood Flow Metab.*, **22**, 367–378, doi:10.1097/00004647-200204000-00001.

74. Tasaki, I. (2002) Spread of discrete structural changes in synthetic polyanionic gel: a model of propagation of a nerve impulse, *J. Theor. Biol.*, **21**, 497–505.

75. Depeursinge, C., Cuche, E., Magistretti, P. J., and Marquet, P. DHM: Digital holographic microscopy, a novel modality in microscopy, *Nat. Photonics* (submitted).

76. Pluta, M. (1989 [1988]) *Advanced Light Microscopy, Vol. 2: Specialized Methods*, Elsevier, New York, pp. 146–197.

77. Zernike, F. (1942) Phase-contrast, a new method for microscopic observation of transparent objects, *Physica*, **9**, 686–698.

78. Nomarski, G. (1955) Differential micro interferometer with polarized waves, *J. Phys. Radium*, **16**, 9–13.

79. Rappaz, B. *et al.* (2005) Measurement of the integral refractive index and dynamic cell morphometry of living cells with digital holographic microscopy, *Opt. Express.*, **13**, 9361–9373, doi:10.1364/OPEX.13.009361.

80. Rappaz, B., Charriere, F., Depeursinge, C., Magistretti, P. J., and Marquet, P. (2008) Simultaneous cell morphometry and refractive index measurement with dual-wavelength digital holographic microscopy and dye-enhanced dispersion of perfusion medium, *Opt. Lett.*, **33**, 744–746, doi:10.1364/OL.33.000744.

81. Barer, R. (1953) Determination of dry mass, thickness, solid and water concentration in living cells, *Nature*, **172**, 1097–1098.

82. Le Bihan, D., Turner, R., and Douek, P. (1993) Is water diffusion restricted in human brain white matter? An echo-planar NMR imaging study, *Neuroreport*, **4**, 887–890.

83. Le Bihan, D. (2007) The "wet mind": water and functional neuroimaging, *Phys. Med. Biol.*, **52**, R57–R90.

Part VI

WATER, HEALTH, AND LIFE

Chapter 16

WATER AND THE MIND: ROLE OF WATER, CELL MEMBRANES, AND DIFFUSION IN BRAIN FUNCTION IMAGING

Denis Le Bihan

NeuroSpin, CEA Saclay, France
Human Brain Research Center, Kyoto University Graduate School of Medicine, Kyoto, Japan
denis.lebihan@cea.fr

Over the last 30 years functional neuroimaging has emerged as an important approach to study the brain and the mind. This has been possible because of significant advances mainly in two imaging modalities, namely, positron emission tomography (PET) and magnetic resonance imaging (MRI). Although those two modalities are based on radically different physical approaches (detection of β^+ radioactivity for the first one and of nuclear magnetization for the latest), both make brain activation imaging possible through measurements involving water molecules. This should come as no surprise, given that water constitutes nearly 80% of the brain weight and 90% of its molecules. So far, PET functional imaging and functional MRI (fMRI) have relied on the same principle that neuronal activation and blood flow are coupled through metabolism [1]: Blood flow increases locally in activated brain regions. In the case of PET one uses H_2O^{15} radioactive water, which is produced by using a cyclotron and injected into the subject's vasculature. In activated brain regions the increase in blood flow leads to a local increase in the tissue radioactive water content detected and localized by the PET camera [2–4]. With MRI the hydrogen nuclei of brain-endogenous water molecules are magnetized by a strong external magnetic field. In activated regions the increase in blood flow results in an increase of blood oxygenation, which slightly modifies, in and around blood vessels, the magnetization relaxation properties of the water molecules detected by the MRI scanner (so-called the "BOLD"—Blood Oxygen Level Dependent—effect [5, 6]). In

Water: The Forgotten Biological Molecule
Edited by Denis Le Bihan and Hidenao Fukuyama
Copyright © 2011 by Pan Stanford Publishing Pte. Ltd.
www.panstanford.com

both approaches water is, thus, merely an indirect means to observe the changes in cerebral blood flow that accompany brain activation (Fig. 16.1).

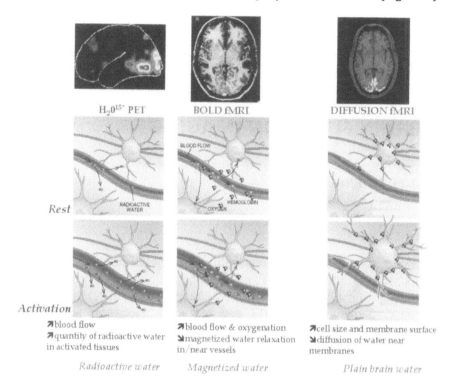

Figure 16.1 General principles of PET and MRI functional neuroimaging methods. *Left*: Brain activation mapping with PET. Radioactive water (H_2O^{15}) is used as a tracer to detect neuronal activation–induced increases in blood flow. *Middle*: Brain activation mapping with BOLD fMRI. Water magnetization in and around small vessels is modulated by the flow of red blood cells containing paramagnetic deoxyhemoglobin. (Adapted from Ref. [4]). *Right*: Brain activation mapping with diffusion fMRI. The reduction in water diffusion, which occurs during activation, is thought to originate from a membrane-bound water layer, which expands during activation-induced cell swelling. While with PET and BOLD fMRI water is only an indirect means to detect changes in blood flow through the imaging scanner, the changes in water properties seen with diffusion fMRI seem to be an intrinsic part of the activation process. For color reference, see p. 387.

Although PET and BOLD fMRI have been extremely successful for the functional neuroimaging community [7], they present well-known limitations. The extent, dynamics, and underlying mechanisms of this neurovascular coupling are not yet fully understood [8], and two main streams of hypotheses have been proposed, the so-called neurogenic and metabolic hypotheses [9, 10]. Besides neurons, astrocytes and glutamate seem to play a very important role [11, 12], and the BOLD signal depends on several parameters, so its link with neuronal activation remains indirect [13–15]. Furthermore, neuronal responses might still occur while astrocyte and hemodynamic responses are abolished using drugs, such as isofluorane [16]. Similarly, although the coupling between neuronal activation and hemodynamics, as observed with BOLD fMRI, has been shown [17], it may also fail in some pathological conditions [18], although brain function remains normal.

Also, it has been pointed out that the spatial functional resolution of vascular-based functional neuroimaging is necessarily limited because vessels responsible for the increase of blood flow and blood volume feed or drain somewhat large territories, which include clusters of neurons with potentially different functions [19]. Similarly the physiological delay necessary for the mechanisms triggering the vascular response to work intrinsically limits the temporal resolution of BOLD fMRI, although some vascular-related signals, such as the cerebral blood volume [20] or the total local hemoglobin concentration measured by diffuse optical imaging [21], could precede typical BOLD time courses.

On the other hand, a fundamentally new paradigm has emerged to look at brain activity through the observation with MRI of the *diffusion* behavior of the water molecules [22]. It has been shown that the diffusion of water slightly slows down in the activated brain cortical areas. This slowdown, which occurs several seconds before the hemodynamic response detected by BOLD fMRI [23, 24], has been described in terms of a phase transition of the water molecules from a somewhat fast to a slower diffusion pool in the cortex undergoing activation and tentatively ascribed to the membrane expansion of cortical cells, which undergo swelling during brain activation (Fig. 16.1). This hypothetical mechanism, which remains to be confirmed, would mark a significant departure from the former blood flow–based PET and MRI approaches and would potentially offer improved spatial and temporal resolution, due to its more intimate link to neuronal activation. However, the step might even extend further: in contrast with the former approaches based on changes in *artificially* induced changes in the atoms constituting the water molecules, namely, the radioactivity of the oxygen atom and the magnetization of the hydrogen atoms, required for external PET or MRI detection, the new, diffusion-based approach merely uses MRI as a means to reveal changes in *intrinsic* physical properties of the water molecules. These changes in the diffusion behavior of water during activation might be indeed an *active* component of this process, as water homeostasis and water movement have without any doubt a central role in brain physiology.

The importance of the physical properties of water for biological tissues has been reviewed in the previous chapters. The aim of this chapter is to evaluate their relevance to brain imaging with diffusion MRI on the one hand and to cellular events underlying brain function on the other hand. To this end, the literature on brain activation biophysical mechanisms is then assessed to shed light on their intimacy with the physical properties of water.

16.1 WATER DIFFUSION MRI: OUTSTANDING ISSUES

16.1.1 Principles of Diffusion MRI

Noninvasive imaging methods usable in living animals and humans currently have a macroscopic (millimeter) resolution. Hence, access to dynamic tissue microstructure must be provided through physical processes encompassing several spatial scales. Molecular diffusion is an exquisite example of such a multiscale integrated process by which fluctuations in molecular random motion at microscopic scale can be inferred from observations at a much larger scale using statistical physical models, although the individual molecular structure and pathway are completely ignored. This powerful multiscale approach allowed Einstein indirectly to demonstrate the existence of atoms through the identification of diffusion with Brownian motion in the framework of the molecular theory of heat [25]. Over the last 20 years it has been shown that MRI could provide *macroscopic* and quantitative maps of water molecular diffusion [26, 27], especially in the brain [28], to make indirect inferences on the *microstructure* of biological tissues.

As the diffusion coefficient of water in brain tissue at body temperature is about 1×10^{-3} mm^2/s, Einstein's diffusion equation ($<z^2> = 2DT_d$, where $<z^2>$ is the mean free quadratic displacement in one direction, D the diffusion coefficient, and T_d the diffusion time) [25] indicates that about two-thirds of diffusion-driven molecular displacements are within a range not exceeding 10 micrometers during diffusion times currently used with MRI (around 50 ms), well beyond typical image resolution. Indeed, water molecules move in the brain while interacting with many tissue components, such as cell membranes, fibers, or macromolecules, and the indirect observation of these displacements embedded into the diffusion coefficient provides valuable information on the microscopic obstacles encounters by diffusing molecules and, in turn, on the structure and geometric organization of cells in tissues, such as cell size or cell orientation in space (see [29] for a review).

However, the overall signal observed in a "diffusion" MRI image volume element (voxel), at a *millimetric* resolution, results from the integration, on a statistical basis, of all the *microscopic* displacement distributions of the water molecules present in this voxel, and some modeling is necessary to make inferences between those two scales. As a departure from earlier diffusion studies in biological studies, when efforts were made to depict the physical elementary diffusion process [30–32], it was suggested for simplicity to portray the complex diffusion patterns that occur in a biological tissue within a voxel by using the free diffusion physical model (where the distribution of molecular displacements obeys a Gaussian law) but replacing the physical diffusion coefficient, D, with a global parameter, the *apparent diffusion coefficient* (ADC) [28]. In practice the MRI signal is made sensitive to the diffusion-driven water molecular displacements through pulsed variations in space of the magnetic field, the so-called magnetic field gradients (Fig. 16.2). In the presence of these gradients, any molecular displacement occurring during a given time interval (diffusion time) produces a phase shift of the associated MRI signal. Due to the very large number of water molecules present in each image voxel, the phase shifts are distributed, closely reflecting the diffusion-driven (Gaussian) displacement distribution, resulting in a loss of coherence and, hence, an attenuation of the MRI signal amplitude.

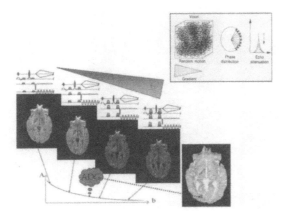

Figure 16.2 Diffusion-weighted MRI principles. *Right corner insert*: The overall, random displacement of water molecules due to diffusion in the presence of a space-varying magnetic field (gradient) results in a dispersion of the phases of the spinning magnetization carried by water hydrogen nuclei. This phase dispersion results, in turn, in an overall attenuation of the MRI ("echo") signal, depending on the diffusion coefficient and the intensity of the magnetic field gradient (b value). *Bottom*: Set of diffusion-weighted images are obtained using different b values, which are obtained by changing the intensity of the diffusion gradient pulses in the MRI sequence. In diffusion-weighted images the overall signal intensity decreases with the b value. Tissues with high diffusion (such as ventricles) get darker more rapidly when the b value is increased and become black. Tissues with low diffusion remain with a higher signal. As diffusion-weighted images also contain relaxation (T_1 and T_2) contrast, one may want to calculate pure diffusion (or ADC) images. To do so the variation of the signal intensity, $A(x, y, z)$, of each voxel with the b value is fitted using Eq. 16.1 to estimate the ADC for each voxel. In the resulting image the contrast is inverted: bright corresponds to fast diffusion and dark to low diffusion. For color reference, see p. 388.

This attenuation, *A*, can then be simply formulated as

$$A = S/S_0 = \exp(-b\,\text{ADC}),\qquad\qquad(16.1)$$

where *b* is the degree of diffusion sensitization (as defined by the amplitude and time course of the magnetic field gradient pulses used to encode molecular diffusion displacements [33]), *S* is the signal at a particular *b*-value, and S_0 is the signal at *b* = 0. However, one should bear in mind that, strictly speaking, the MRI signal is actually sensitive to the diffusion path of the hydrogen nuclei carried by water molecules, not the diffusion coefficient per se.

The ADC concept has been widely adopted in the literature. Potential applications of water diffusion MRI, which were suggested very early on [28] are many [29], but the most successful clinical application since the early 1990s has been acute brain ischemia, as the ADC sharply drops in the infarcted regions, minutes after the onset of the ischemic event [34] (Fig. 16.3).

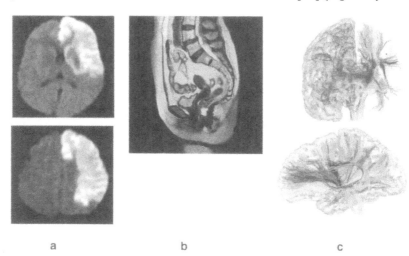

a   b   c

Figure 16.3 Major current applications of diffusion MRI. (a) Diffusion MRI in brain acute stroke. The region with a bright signal corresponds to brain regions where water diffusion is reduced, as a result of acute cerebral ischemia and associated cytotoxic edema. (Image courtesy of Dr. Openheim [Radiology Department, Hôpital Saint-Anne, France]). (b) Diffusion MRI in cancer. Colored areas correspond to regions where the water diffusion coefficient is decreased. Such regions have been shown to match areas where malignant cells are present (primary lesions or metastases). (Image courtesy of Dr. Koyama, Radiology Department, Kyoto University, Graduate School of Medicine, Kyoto, Japan). (c) Diffusion and brain white matter fiber tracking. Water diffusion in brain white matter is anisotropic. As a result it is possible to determine for each voxel of the image the direction in space of the fibers. Using postprocessing algorithms the voxels can be connected to produce color-coded images of the putative underlying white matter tracts. (Images courtesy of Y. Cointepas and M. Perrin, SHFJ/CEA, Orsay, France). For color reference, see p. 389.

With its unmatched sensitivity diffusion MRI provides some patients with the opportunity to receive suitable treatment at a stage when brain tissue might still be salvageable [35]. However, the exact mechanism responsible for this ADC drop has remained unclear, although cell swelling through cytotoxic edema seems to play a major role [36]. Another potentially important clinical application is the detection of cancer and metastases. The water ADC is significantly decreased in malignant tissues [37]. Here also the origin of this diffusion anomaly is not clear, but it is somewhat linked to the cell proliferation, and body diffusion MRI is currently under evaluation as a potential alternative approach to fluoro-deoxyglucose (FDG)-PET [38, 39] to detect malignant lesions (Fig. 16.3). FDG-PET relies on the hypothesis that hypermetabolic cells, such as cancer cells, uptake radioactive FDG. As FDG cannot be metabolized by cells it accumulates inside cells and can be detected using the PET camera. Besides the use of radioactivity and the limited spatial resolution of PET, a pitfall is that the method is not highly specific, as any hypermetabolic tissue (such as inflammation) will also exhibit an FDG uptake.

On the other hand, as diffusion is a three-dimensional process, molecular mobility in tissues may not be the same in all directions. Diffusion anisotropy was observed at the end of the 1980s in brain white matter [40]. Diffusion anisotropy in white matter originates from its specific organization in bundles of more or less myelinated axonal fibers running in parallel: diffusion in the direction of the fibers is faster than in the perpendicular direction, although the exact contribution of the white matter elements involved (intra- and extraaxonal compartments, membranes, myelin, etc.) is still elusive. It appeared quickly, nevertheless, that this feature could be exploited to map out the orientation in space of the white matter tracks in the brain [41]. With the introduction of the more rigorous formalism of the *diffusion tensor* [42], diffusion anisotropy effects could be fully extracted, characterized, and exploited, providing even more exquisite images of white matter tracks (Fig. 16.2) and brain connectivity (see [29, 43] for reviews). Diffusion MRI is currently used to map the wiring diagram of the entire living human brain ("Human Brain Connectome"). By systematically collecting brain imaging data from hundreds of subjects, one will gain insight into how brain connections underlie brain function. This will open up new lines of inquiry for human neuroscience and for treating brain disorders. Mapping the circuits and linking these circuits to the full spectrum of brain function in health and disease is an old challenge but one that can finally be addressed rigorously with diffusion MRI in combination with other powerful, emerging technologies. By knowing more about the normal or abnormal connections within the brain, we would know more about brain dysfunction in aging, mental health disorders, addiction, and neurological disease. For example, there is evidence that the growth of abnormal brain connections during early life contributes to autism and schizophrenia [44, 45]. Changes in connectivity also appear to occur when neurons degenerate, as a consequence either of normal aging or of diseases such as Alzheimer's [46].

The most recent development of diffusion MRI is its application to the field of brain *functional* imaging [23]. Using heavily diffusion-sensitized MRI, a transient increase of the diffusion-weighted signal has been observed in the occipital cortices of human subjects [23] and cats submitted to visual stimulation [47]. This diffusion response is characterized by a steep onset and a temporal precedence relative to the hemodynamic response detected by BOLD MRI suggesting a nonvascular origin, although this hypothesis has been challenged [48–50]. This signal increase reflects a decrease of the ADC of water. The exact physiological mechanisms underlying this transient water diffusion response, which have entirely to be clarified, have been thought to be linked to dynamic tissue changes associated with neuronal activation, a much more direct approach than BOLD fMRI (Fig. 16.1). There is already a large body of knowledge out in the literature, suggesting that these diffusion changes might well result from the properties of water in biological tissues and on their association with the biophysical events underlying neuronal activation.

16.1.2 Water Diffusion in Biological Tissues

The ADC in the brain is 2 to 10 times smaller than free-water diffusion in an aqueous solution (which is 2.272 mm^2/s at 25°C), as measured by nuclear magnetic resonance [51–54] or neutron-scattering methods [55]. High viscosity, macromolecular crowding, and restriction effects have been proposed to explain the water diffusion reduction in the intracellular space [53] and tortuosity effects for water diffusion in the extracellular space [56, 57]. Restricted diffusion effects, for instance, may be evaluated by changing the diffusion time [32, 58]: the displacement of the molecules becomes limited when they reach the boundaries of close spaces, and the diffusion coefficient artificially goes down with longer diffusion times. Variations of the ADC with the diffusion times have been reported in the brain [59, 60]. However, no clear restriction behavior has been observed *in vivo* for water in the brain, as the diffusion distance seems to increase well beyond cell dimensions with long diffusion times [61, 62] or even in cell suspensions [63]. Furthermore studies have established long ago that the overall low diffusivity of water in cells could not be fully explained by compartmentation (restriction) effects from cell membranes nor by scattering or obstruction (tortuosity) effects from cellular macromolecules [64–66]. This strongly suggests that the cellular components responsible for the reduced diffusion coefficient in biological tissues are much smaller than the diffusion length currently used with MRI. In summary, the cell membranes in the brain are likely the main obstacles to "hinder" the water diffusion process but do not completely restrict diffusion within cells; they are highly permeable to water, either passively or through transporters, such as the specific aquaporin channels that have been found in the brain [67].

Indeed, many studies have experimentally established that the water diffusion–sensitized MRI signal attenuation in brain tissue (and other tissues as well) as a function of the sensitization (*b*-value) could not be well described by a single exponential decay, as would have been expected (Eq. 16.1) in an unrestricted, homogenous medium (free Brownian diffusion) [29]. Furthermore, diffusion data gathered using the "q-space" approach, a technique that provides estimates of the distribution of the diffusion-driven molecular displacements, clearly demonstrates that the water diffusion process cannot be modeled by a single Gaussian distribution [68] (Fig. 16.).

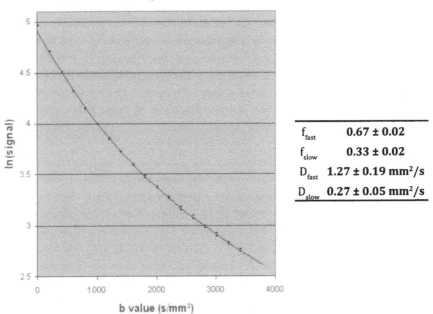

Water diffusion parameters in the human visual cortex

f_{fast}	**0.67 ± 0.02**
f_{slow}	**0.33 ± 0.02**
D_{fast}	**1.27 ± 0.19 mm²/s**
D_{slow}	**0.27 ± 0.05 mm²/s**

Figure 16.4 Water diffusion in the brain is not a free random walk process. Plot of the signal attenuation as a function of the degree of diffusion sensitization (*b* value) in the human visual cortex. The plot is clearly not linear and best described using a biexponential model (Eq. 16.2) with a fast and a slow diffusion water pool.

In most cases data has been very well fitted, however, with a biexponential function corresponding to two water diffusion pools or phases in slow exchange (i.e., the residence in each phase time is much longer than the diffusion time used for the measurements) with a fast and a slow diffusion coefficient [60, 69]:

$$S = S_0 f_{slow} \exp(-b\, D_{slow}) + S_0 f_{fast} \exp(-b\, D_{fast}). \qquad (16.2)$$

Here, f and D are the volume fraction and the diffusion coefficient associated with the slow diffusion phase (SDP) and the fast diffusion phase (FDP), respectively, with $f_{slow} + f_{fast} = 1$ (in this simple model differences in T2 relaxation are not taken into account). When the exchange regime becomes "intermediate" (the residence in each phase time and the diffusion time are similar), Eq. 16.2 becomes far more complex, as one has to replace the values for $f_{slow,fast}$ and $D_{slow,fast}$ in Eq. 16.2 by corrected values, taking into account the residence time, τ, of the molecules in the fast and slow compartments relative to the measurement time [70]. The "diffusion coefficients" and "volume fractions" in Eq. 16.2 then takes a much more complicated form, as they now depend each on the residence times, the diffusion coefficients, and volume fractions of both pools, but the overall signal decay retains an overall biexponential shape.

Studies performed by Niendorf *et al.* [69] in the rat brain *in vivo* (with b factors up to 10,000 s mm^{-2}) using this model yielded $D_{fast} = (8.24 \pm 0.30) \times 10^{-4}$ mm^2/s and $D_{slow} = (1.68 \pm 0.10) \times 10^{-4}$ mm^2/s with $f_{fast} = 0.80 \pm 0.02$ and $f_{slow} = 0.17 \pm 0.02$. Similar measurements have been made in the human brain using b factors up to 6,000 s mm^{-2}. The estimates for those diffusion coefficients and the respective volume fractions of those pools (Fig. 16.4) have been strikingly consistent across literature [23, 71–73], providing at least some phenomenological validation of the biexponential model.

It has been often considered that the extracellular compartment might correspond to the FDP, as water would be expected to diffuse more rapidly there than in the intracellular, more viscous compartment. However, the volume fractions of the two water phases obtained using the biexponential model ($f_{slow,fast}$) do not agree with those known for the intra- and extracellular water fractions ($F_{intra} \geq 0.80$ and $F_{extra} \leq 0.20$, respectively [74], even by taking into account differences in T2 relaxation contributions between those compartments), so the nature of those phases has yet remained unclear [75]. Furthermore, some careful studies have shown that such a biexponential diffusion behavior could also be seen solely within the intracellular compartment, pointing out that both the SDP and the FDP probably coexist within the intracellular compartment. Such studies were sometimes conducted with ions or molecules much larger than water, such as *N*-acetylaspartate [60] or fluoro-deoxy-glucose, or in particular biological samples, such as giant oocytes [76, 77], so extrapolation to water diffusion in neuronal tissues requires some caution. Theoretical models have shown that restriction caused by cylindrical membranes can also give rise to a pseudo-biexponential diffusion behavior in nerves [78]. Other models have been introduced, for instance, based on a combination of extraaxonal water undergoing hindered diffusion and intraaxonal water undergoing restricted diffusion [79]. Although such models could account for a pseudo-biexponential diffusion behavior and diffusion anisotropy in white matter, it remains to be seen how it could be applied to the brain cortex, given that true restricted diffusion effects have not been really observed for water in the brain (see above).

In summary, there is growing indication that a *direct* relationship between the intracellular and extracellular volumes and the biexponential parameters of the diffusion attenuation could probably not be established, and several groups have underlined the important role of dynamic parameters, such as membrane permeability and water exchange [70, 80, 81], and geometrical features, such as cell size distribution or axons/dendrite directional distribution [80, 82–84]. Noticeably, however, these distinct models lead to a diffusion signal decay that is nevertheless well approximated by a biexponential fit [80, 82, 85].

16.1.3 Variations of Water Diffusion with Cell Size

On the other hand, a second ensemble of experimental findings suggest that the *changes* of the volume fractions of the intra- and extracellular spaces that result from cell swelling and shrinking in different physiological, pathological, or experimental conditions always lead to *variations* of the ADC. The drop of ADC, which is observed during acute brain ischemia, has been clearly correlated with the cell swelling associated with cytotoxic edema [86, 87]. Variations in the tortuosity coefficient, λ, within the extracellular space, linked to the increased diffusion path lengths caused by obstructing cells, have been considered as a potential source of diffusion reduction within the extracellular space (the diffusion reduction, D/D_0, where D_0 is the free diffusion coefficient, would scale as $\sim 1/\lambda^2$ [88]. It is not questionable that the tortuosity of the extracellular space modulates the diffusion process for some molecules or ions, such as tetramethylammonium (TMA⁺), which is much larger than water (and, hence, more sensitive to hindrance effects) and must be directly introduced into the extracellular space [57, 74]. Extracellular space tortuosity for molecules such as metabolites or neurotransmitters has an important role in brain physiology and neural function. However, the importance of this mechanism is not so clear for water. Water diffusion studies (mainly conducted with MRI), whether *in vivo* or in tissue preparations, and variations of the extracellular space tortuosity have always been induced by changes in the cellular volume. Hence, the observations of ADC changes are linked to both the changes in cellular volume and the resulting extracellular space, and there is no way to untangle those two effects. So one cannot establish for water that the changed tortuosity is at the origin of the ADC change, but merely that it is correlated with it. The change in cellular volume might just as well be responsible per se (see below).

Further works have also established that the variations in *size* of the intra- and extracellular compartments correlate well with the observed changes in the fraction of the slow and fast diffusion pools of the biexponential model [69, 87, 89–92]. For instance, the ADC decrease that results from ouabain-induced cell swelling in perfused rat hippocampal slices has been shown to result in an increase of the SDP fraction, but the SDP and FDP diffusion coefficients do not change [93] (Fig. 16.5). These results suggest that the global water ADC decrease does not result from the increase in extracellular tortuosity induced by the shrinking of the extracellular space caused by cell swelling, but rather from a shift of balance between the fast and slow diffusion water pools. An increase in tortuosity would lead to a decrease of the fast diffusion coefficient (assuming the FDP corresponds to the extracellular space, which seems doubtful, as explained earlier). The idea that obstruction by cells does not seem to be the principal source of diffusion reduction for water is also supported by the fact that the highest observable ratio, D/D_0, for water in the brain would be around $1.3/3 \approx 0.43$ ($D_0 \approx 3.10^{-3}$ mm^2/s at brain temperature), corresponding to $\lambda \leq 1.5$, in the range of the values found experimentally for small molecules by several groups [94]. This is smaller than the theoretical tortuosity index (1.63–1.72) found by Thornes and Niholson [89] for infinitely small molecules solely based on geometric considerations. Clearly, other mechanisms than obstruction by cell geometry should be considered.

Furthermore, earlier work on animal models has also shown that a decrease in water diffusivity could be visualized using MRI during intense neuronal activation, such as during status epilepticus induced by bicuculline [95] or cortical electroshocks [96] (Fig. 16.5). This diffusivity drop propagates along the cortex at a speed of about 1–3 mm/min, consistent with spreading depression [97–101]. Here also the diffusion drop [98–101] has been correlated to cell swelling [102–104].

More recently, the biexponential model has also been used to explain the diffusion changes observed in the activated brain visual cortex [23]. As in the study by Buckley *et al.* [93] it has also been found that the SDP fraction was increasing, at the expense of the FDP, but that the SDP and FDP diffusion coefficients remained unchanged, which means that some water molecules

a

	Value in aCSF mean (SD)		Value in 1 mM ouabain mean (SD)		P value[a]
S_0 (arb. units)	1600	(71)	1697	(71)	<0.0005
f_2 (no units)	0.533	(0.064)	0.438	(0.050)	<0.0005
D_2 ($\times 10^{-3}$ mm²/s)	1.015	(0.155)	1.027	(0.178)	0.68
D_1 ($\times 10^{-3}$ mm²/s)	1.015	(0.155)	1.027	(0.178)	0.68
Single ADC[b] ($\times 10^{-3}$ mm²/s)	0.450	(0.098)	0.366	(0.056)	<0.006

[a] Comparing parameter values obtained in aCSF and 1 mM ouabain using a paired *t* test.
[b] Single exponential ADC estimated from the five lowest *b*-value images.

b

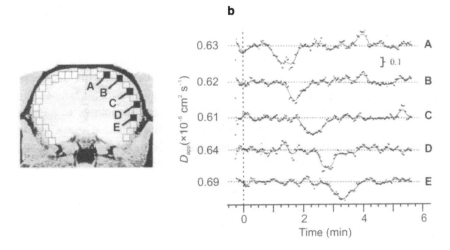

Figure 16.5 ADC decrease in extraphysiological brain challenges. (a) Analysis of the data of MRI diffusion-sensitized images (*b* = 1,980 s/mm²) of a rat brain hippocampus slice in artificial CSF and in 1 mM ouabain solution. Using the biexponential (biphasic) model indicates that the water diffusion coefficients of the slow (D_1) and the fast (D_2) components do not change during the osmotic challenge. The decrease in ADC is solely the result of a decrease in the fast diffusing fraction (f_2) or, in other words, an increase of the slow diffusion fraction. Adapted from Ref. [93]). (b) Time course of the diffusion coefficient in five regions (A through E shown on the anesthetized rat brain MRI image) separated by 7 mm in space during a spreading depression wave induced by a KCl application. The decrease in the diffusion coefficient is about 35% and lasts about 1 minute. The decreased diffusion wave propagates along the cortex at a speed of 3.5 mm/min. Adapted from Ref. [99]. *Abbreviations*: CSF, cerebrospinal fluid; KCl, potassium chloride.

undergo a change from a fast to a slow diffusing phase. At this point, it becomes obvious that the origin of the biphasic behavior of water diffusion in brain tissue must be reconsidered with a fresh look on the known status of water in cells and the experimental variations of the water diffusion coefficient with cell size. As this behavior seems to be general to most tissue types, a valid model should accommodate both brain cortex and white matter, as well as body tissues, and account for physiological and pathological observations.

16.2 MEMBRANES AND WATER IN BIOLOGICAL TISSUES

16.2.1 Current Cell-Membrane Model

In the nineteenth century it was recognized that the cell contents form a *gelatinous substance*. The concept of the membrane was introduced slightly later to account for the absence of mixing between the cytoplasm and the surrounding solution (the membrane, which is about 7.5–10 nm in thickness, could not be seen, of course, at that time). The membrane soon became *semipermeable*, allowing water to pass but not solutes. Later, the discovery that some ions, notably K^+, could also pass through the membrane led to the concept of *membrane channels*, which allow specific solutes to pass in some conditions [105]. This simple model apparently explained how K^+ could accumulate within cells to partially compensate the negative charges of the proteins (so-called Donnan equilibrium), while Na^+ would remain largely excluded from the cytoplasm (Table 16.1).

Table 16.1. Concentration of major ions in intra- and extracellular compartments (from http:/www.lsbu.ac.uk/water/)

Ion	Ionic radius (Å)	Surface charge density	Molar ionic volume (cm^3)[a]	Intracellular (mM)	Extracellular (mM)
Ca^{2+}	1.00	2.11	−28.9	0.1	2.5
Na^+	1.02	1.00	−6.7	10	150
K^+	1.38	0.56	+3.5	159	4

[a]Molar aqueous ionic volume, cm^3 mol^{-1}, 298.15 K; negative values indicate contraction in volume (i.e., addition of the ions reduces the volume of the water).

With this model, cell volume variations would result from changes in the osmotic balance between the intra- and extracellular compartments. Since then, the membrane has been physically observed and identified as a phospholipid bilayer with protein insertions, and a considerable number of types of channels have been identified and isolated, including water channels (aquaporins) [106, 107], which have been also found in the brain [67]. Because so many solutes were found to pass through the membrane channels, the concept of "pumps" was introduced [108]. Those pumps take charge of cell housekeeping, maintaining the right concentration gradients between the cytoplasm and the surrounding medium, and a very large number of pumps have been proposed to accommodate many substances, ions, sugars, amino acids, etc. In summary the cell membrane model has become extraordinarily complex, with the presence of a large quantity of channels and pumps at its surface, and water and ionic transmembrane shifts are required to maintain the ion homeostasis during neuronal activity.

Yet, some authors have questioned this model, or more exactly its functional role [109, 110]. The first of their arguments relates to the cell energy supply. It has been estimated that the Na^+ pump, in the current scheme, would consume by itself a third to a half of the cell energy supply to maintain the extra-/intracellular Na^+ gradient [111]. As there are even many other membrane channels and pumps (including those on the cell organelle surface, particularly high-membrane-density mitochondria), it is unclear how the cell produces the necessary amount of energy to maintain all of its concentration gradients solely from pumps [112, 113]. Indeed, when the cell is deprived of energy, the normal intracellular concentrations of Na^+ and K^+ are maintained for hours [114], suggesting that the basic solute partitioning is assumed by other, less-energy-demanding mechanisms. The channel pump systems would be involved to carry more specialized tasks or transient perturbations, recoveries, or modulations of the general balance, perhaps even in specific regions of the cell. Other groups have also demonstrated that the Na^+/K^+ and other gradients could be maintained for hours and the cell function normally even after the cell membrane was largely disrupted or even removed by different technical means [115], as long as the cell proteins remain in the cytoplasm [116, 117]. Actually, diffusion of macromolecules in the crowded cytoplasm is extremely slow [118–120], and macromolecules are expected to stay within the cytoplasm hours after the cell membrane has been removed.

In view of these observations it has been suggested that, whereas the membrane could be seen as essential to avoid the loss of proteins, adenosine triphosphate (ATP), and other important molecules over a long time range, it could perhaps not be the main or only reason for the baseline solute concentration gradients to exist between the cytoplasm and the surrounding medium [112, 113, 121]. The membrane could have other important roles, for instance, keeping the overall cell architecture, shape, and integrity in cooperation with the cytoskeleton scaffolding (see below) or managing the interactions between the cells' internal functional compounds and those in the extracellular space (cell signaling). In this view, the physical properties of the cytoplasm itself are considered to support its own content. Clearly the properties of the protein–ion–water matrix gels that form the cytoplasm should have a key role that makes the cytoplasm radically different from a banal aqueous solution with free diffusing solutes.

16.2.2 Cellular Water and Polar Interfaces

Water should not be overlooked as a mere structureless, space-filling background medium where biological events occur. Bilateral interactions between water molecules and cell compounds are very strong, as explained in the previous chapters. As a result the overall physical properties of the cell medium (cytoplasm) drastically differ from those of pure bulk liquid water in a glass. For instance, temperatures well below zero are needed for cell water to freeze [122, 123], and cell water viscosity is very high [119, 124, 125].

The fact that the cytoplasm content remains largely intact, although the cell membrane has been disrupted [115], suggests that some strong attractive forces must exist within the cytoplasm, which maintain its cohesion and prevent its water and content from leaking out. The roots of these forces can be found in the interactions between the negatively charged surfaces of cytoplasm proteins and the dipolar water molecules.

In liquid water each water molecule can bind to four first neighbors from its four charged sites (two donating and two accepting hydrogen bonds) forming a tetrahedral arrangement (see Chapter 1). This geometry is critical, although this tetrahedral network is not perfect in liquid water: while the four-coordination motif with "linear" hydrogen bonds is the dominant configuration, there are significant local defects where the coordination is either greater or less than four with "bifurcated" hydrogen bonds [126], which introduces local environment variability and prevents any long-range order, as expected in a liquid. An important consequence relevant to biology of this network "defect" model, as will be seen later, is that in liquid water, in contrast to other liquids, an *increase of structural order* of the liquid leads to a reduced density and *decreased diffusion mobility*, as the number of network defects declines [127, 128].

Indeed, liquid water may form other types of three-dimensional arrays in the presence of interfaces with charged materials. In cells, proteins have an especially profound effect on water due the presence of their charge, which results in protein–water adsorption [129, 130]. It has been suggested that the first layer of dipolar water molecules oriented by protein charges may, in turn, orient other water molecules in successive layers in a cooperative manner

[109, 131]. This multilayer polarization model [121, 132, 133] has remained extremely controversial in the physiology community (mainly because it put into question the role of the cell membrane but also because it seemed inconsistent with some experimental and theoretical results accepted by the majority of the scientific community; see below). Nevertheless, the presence of structured water in cells has been claimed by some groups [134–136].

Depending on the mechanisms of water protein–surface interactions and the interaction distance (the influence seems to decay exponentially as a function of distance), different vocabularies have been used, such as "bound," "hydration," "vicinal," or "interfacial" water [134, 137]. It is generally considered that van der Waals forces account for only one or two layers of water molecules at the protein surface (0.4g/g water/protein covering 13% or more of the protein surface), but hydration water could represent much more inside cells [138], up to about 2 to 4 g/g (dry weight) [137], which represents 70% to 80% of the cell mass or volume. Hence, water structuring and related effects on water mobility could perhaps extend several water molecule layers away from the protein surface, perhaps as much as 50 Å [139]. Another explanation would be that the amounts of the first layers of interfacial water on internal proteins are larger than expected, up to 1.6 g/g in a collagen model [140]. Water diffusion in the 5 Å thick hydration shell has been found to be around 0.5×10^{-3} mm^2/s at 25°C and shown to result from a stronger hydrogen-bonding structure than in bulk water [141].

Given the high degree of macromolecular packing in cytoplasm (surface-to-surface gaps between macromolecules are in the range of a few nanometers) and the exchange between water pools, one may consider than at least a substantial fraction of its water content is structured, although perhaps to various degrees. A study in plant cells showed 30% of water molecules greatly reduced in mobility, 70% to a lesser extent, and essentially no bulk water [142]. This water adsorption by the hydrophilic protein matrix gives the cytoplasm the structure of a gel that retains its water and its volume. Structured water also contributes to slow intracellular diffusion in addition to the presence of obstacles [143], as the hydrogen-bonding defects are reduced and less efficient for proton mobility. Diffusion rates of solutes, such as high-energy phosphates [144, 145] or proteins [146], are also much reduced.

As for surfaces and membranes, recent and elegant physics studies have, indeed, confirmed that water polarization exists near charged surfaces and builds up considerable forces [147–150]. Through cooperative effects the protein-trapping range has been reported to extend over distances up to 200 nm beyond physical surfaces, which accounts for up to hundreds of water molecule layers [151–154]. The spatial extent of this water-structuring process has not been established in biological tissues, but the reinforcement effects of the density and distribution of charges along a plane surface, particularly when the periodic pattern of positive and negative charges coincides with the dimensions of the structured water elementary blocks (about 16Å) [155], could help propagate the structuring effect on water molecules beyond several layers. Recent studies have pointed out the importance of the hydration process on the structure and function of biological membranes (head group and acyl chain motion [142]), but in turn, membranes deeply influence water behavior.

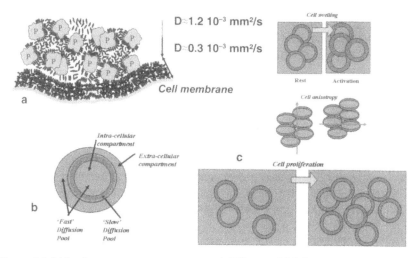

Figure 16.6 Membranes, water structure, and diffusion. (a) Schematic representation of the structuring effect of charged proteins (P) and membranes on water molecules. Bulk-water molecules are exchanging rapidly with the water molecules in the protein hydration shells. Other water molecules are trapped in a membrane-bound layer. Charges of the protein membranes and the underlying cytoskeleton strongly influence the water network in this layer, resulting in an increased order, a lower density, and a slower diffusion coefficient. (b) Conceptual biphasic water diffusion model. The slow diffusion pool is made of a water layer trapped by the electrostatic forces of the protein membranes and associated cytoskeleton, as indicated in (a). The remaining of the water molecules, whether in the intra- or in the extracellular compartment, constitutes the fast diffusion pool (which remains, however, slower that free water). (c) Diagram showing some predictions of the biphasic water diffusion model. Water diffusion would be reduced through an increase of the slow diffusion pool fraction associated with membrane expansion, as during cell swelling (brain activation, *top*) or cell proliferation (cancer, *bottom*). It should be noted that MRI diffusion measurements are always performed along one particular direction. In most cases, tissues are isotropic. However, in tissues with anisotropic cells (*middle*), such as brain white matter, the number of membrane surface intersections with the measurement direction will vary according the respective angle of the measurement direction and the long axis of the cells. The slow diffusion phase is, thus, expected to be the largest when those directions are perpendicular. For color reference, see p. 390.

Measurements in phospholipid membrane models have revealed a strong interaction of the lipids and the membrane proteins with its first hydration layer resulting in a reduced water diffusion parallel to the surface membrane, about five times smaller than in free water (0.44×10^{-3} mm^2/s) [156]. The water diffusion coefficient varies according to the water content (0.12 to 0.4×10^{-3} mm^2/s for water concentrations increasing from 4.9 to 18.6 mol water/lipid), which means that the diffusion coefficient is further reduced near the membrane surface [157]. Diffusion parallel to the membrane is unrestricted, as it does not depend on the diffusion time. On the other hand, the water diffusion coefficient in this membrane-bound layer is highly anisotropic, with $D_{perpendicular}$ as low as 10^{-6} mm^2/s in somewhat impermeable membranes but higher when bilayer defects or channels are present.

16.2.3 Implication for Water Diffusion MRI: A Conceptual Model

The non-Gaussian diffusion behavior in brain tissue could well result from the strong interactions that exist between water, proteins, phospholipids, etc., within the cytoplasm and at the interface with membranes. For all the reasons detailed earlier, one might speculate that the (~70%) "fast" diffusion pool would correspond to the tissue bulk water in fast exchange with the water hydration shell around proteins and macromolecules (whether in the intra- or the extracellular space [Fig. 16.6], although the contribution from the latter is probably much smaller), hence its reduced value ($D_{fast} \approx 1.2 \ 10^{-3}$ mm^2/s) compared to free water ($D_{bulk} \approx 3.0 \times 10^{-3}$ mm^2/s at 37°C). Considering both protein obstruction and hydration effects [66] one gets for D_{fast}:

$$D_{fast} = D_{bulk} \{1/[1-(1 - C_{hydr}/C_{bulk})\varphi]\}(1 - \beta\varphi)/(1 + \beta\varphi/2) \qquad (16.3)$$

where $\beta = (C_{bulk} D_{bulk} - C_{hydr} D_{hydr})/ (C_{bulk} D_{bulk} + C_{hydr} D_{hydr} /2)$, and D_{hydr} is the water diffusion coefficient in the $\approx$ 5Å hydration layer ($D_{hydr} \approx 0.3$–0.5×10^{-3} mm^2/s), C_{bulk} the concentration of the bulk water ($C_{bulk} = 1$ g/cm^3), C_{hydr} the concentration of hydration water, and φ the fractional volume occupied by proteins (which can be determined from the average cell protein mass, specific volume, ≈ 0.75 cm^3/g, and shape). For spherical proteins one gets the expected value for D_{fast} with a water hydration around 1.6 g/g of protein, which is consistent with the literature [140].

As for the extracellular space, experimental evidence suggests that it could be modeled as fluid pores of 38–64 nm [88]. While such pores are clearly hindering the diffusion of TMA and dextrans (a few nanometers in diameter), they represent huge spaces for water molecules (3Å in diameter).

It is, therefore, not unconceivable that the reduced value for water diffusion in the extracellular space mainly results from interactions with the extracellular matrix rather than from geometric factors (pore size, topology) caused by obstructing cells (although this tortuosity component probably also partially contributes, in particular during ischemia, as the pore size goes down to around 10 nm). This observation suggests why the extracellular water diffusion coefficient contributing to the FDP might not be so different than the intracellular FDP water diffusion coefficient, at least not enough to be separable with current diffusion MRI settings, because both share the same basic mechanisms of diffusion reduction, namely, bulk water in fast exchange with the water hydration shell around macromolecules, although such macromolecules would dramatically differ in nature between the intra- and extracellular spaces. Tracers that are not bound to macromolecules and diffuse freely also have similar diffusion coefficients inside and outside cells [158, 159], but geometrical effects certainly contribute, as such tracers cannot cross cell membranes (restricted diffusion).

The (~30%) slow diffusion water pool originates from packets of highly structured water molecules that are trapped within a membrane-bound water network and the three-dimensional cell microtrabecular network [143] (Fig. 16.6). The spatial distribution of charges at the membrane surface would result in an increase of the structural order of water, which leads to reduced density and decreased diffusion mobility, as outlined earlier [126, 127]. One should keep in mind that the cell membrane is not just a 10 nm bilayer with phospholipids and proteins. This membrane structuring effect could well be reinforced by the relatively thick and rigid matrix that runs contiguously and extends a few tens of nanometers on both sides of the membrane, the glycocalyx (made of tangled strands of glycoproteins) on the outside and the cytoskeleton (a dense polymer–gel matrix of cross-linked actin filaments and microtubules) on the inside. Hence, water-structuring effects could occur undisturbed on relatively long ranges, because interstices within those networks are likely protein and ion free [137, 139]. Besides, as the membrane is not totally permeable, a fraction of water molecules "bounce back" when hitting the membranes, which contributes to increasing their residence time within the layer. Both the SDP fraction (~30%) and its diffusion coefficient, D_{slow} (~0.3 × 10^{-3} mm^2/s), agree well with literature values for this membrane interfacial structured water [139, 142, 156].

In summary, the FDP and the SDP would correspond to two differently structured water pools rather than specific water compartments. In the proposed model, both the SDP and the FDP originate partly in the intracellular space and partly in the extracellular space. The presence of both an SDP and an FDP pool has, indeed, been found in an oocyte model [77]. Those two water pools are in slow or intermediate exchange and separable with current diffusion MRI settings (*b* values, diffusion times), resulting in a biexponential diffusion decay behavior. A more extensive model would require to take into account the residence time of water molecules within the SDP and the FDP (intermediate exchange rate regime) using the Karger equations (Eq. 16.3 to allow for the SDP and FDP fractions to change slightly with the diffusion time).

Given the important surface/volume ratio of most cells, the cell membrane–bound water certainly constitutes an important fraction of the SDP. In these conditions it might not be so surprising that any fluctuation in cell size, whether swelling or shrinking, would induce a large variation of the total membrane-bound water volume (the total cell membrane surface scales with the square of its radius), making diffusion-sensitized MRI, and especially its derived SDP fraction, very sensitive to cell size variations, as supported from the literature [69, 87, 89–92].

Using a very simple tissue model and assuming the membrane-bound water pool corresponds to 50% of the SDP, the estimated water layer thickness consistent with diffusion MRI measurements in the brain approaches 50 nm, that is, about 25 nm on each side of the membrane [160]. Small variations in cell size or shape would greatly affect the SDP fraction, as measured with diffusion MRI at the voxel level. Changes in the SDP fraction must occur at the expense of the overall (intra- and extracellular) FDP fraction so that "tortuosity" in the extracellular space must change. This mechanism explains the apparent link between the water ADC and the extracellular space tortuosity; however, obstruction by cells would not be the prime effect. This model would also, at least qualitatively, explains why the ADC is reduced in cancer or metastases. Because of the cell proliferation the density of membranes increases, as well as the membrane surface, and the related SDP volume in each voxel increases almost linearly with the number of cells per voxel, resulting in a decreased ADC linked to cell proliferation. This mechanism would suggest that diffusion MRI should be more specific to cancer states than FDG-PET, which relies on the unspecific increase of metabolism in cancer cells.

Another interesting prediction of this model is that the *volume* of the SDP would be *anisotropic* in oriented tissues, such as brain white matter, which is rather counterintuitive: the SDP fraction should be larger when diffusion measurements are made in a direction that maximizes membrane surface intersections, for example, when diffusion is measured perpendicularly to white matter fibers. Statistically, at the voxel level, water molecules that diffuse across cells will stay longer, on average, during the diffusion time, in the membrane-bound layer than when diffusion is measured in the direction of the fibers. Hence, the SDP fraction is expected to increase and the FDP fraction to decrease (and D_{slow} to decrease further, in an intermediate exchange rate regime), also because the diffusion coefficient in the membrane-bound layer is itself highly anisotropic [156]. Conversely, the FDP fraction would increase (and the SDP fraction decrease) and D_{fast} increase for measurements in the direction of the fibers. This anisotropic effect in SDP and FDP volumes has, indeed, been observed in the human brain [161], as the SDP fraction in white matter varies between 10% and 50% from a measurement direction parallel to the fibers to a direction perpendicular to them.

Finally, going back to the finding that water diffusion decreases during cortical activation, it has also been shown that this slowdown in diffusion solely reflects an increase of the SDP volume at the expense of the FDP [23]. Such results have been confirmed in a model made of slices of rat brain hippocampus [162, 163]. Neuronal activation by kainate results an overall decrease of the ADC caused by an increase of the FDP fraction. Effects are suppressed in the presence of kaneate antagonists and exacerbated when the slices are prepared with N-methyl-D-aspartate (NMDA) receptor antagonists. In view of the proposed conceptual model this SDP volume inflation implies an extension of the membrane-bound water layer, hence membrane unfolding and cell swelling. Swelling likely occurs in limited parts of the cells (e.g., dendrites or spines), similarly, but to a much lesser extent, to the patterns of varicosities that have been observed during cytotoxic edema [164].

At this stage, if it seems plausible to find mechanisms that explain how water diffusion could be reduced in tissues affected by cell swelling, it remains to be seen whether, why, and how cells might swell during physiological activation. This is the topic of the next section.

16.3 WATER AND NEURONAL ACTIVATION

The classical model of neuronal activation, built on the seminal observations of Hodgkin and Huxley in the early 1950s, rests mainly on the action potential that arises from discrete membrane-channel currents. In the action potential model, activation triggers a flux of sodium down its concentration gradient through the membrane, which inverts the cell membrane potential. Potassium then flows out the cell, restoring its negative potential, while the sodium/potassium pumps restore the initial solute balance.

Yet, although sodium and potassium seem to play crucial roles, their absence does not prevent the action potential to occur [165–167]. Hence, while in classical neurophysiology great importance was placed on transient electrical changes associated with the excitation processes, there is now compelling evidence that activation in nervous tissues is accompanied by other important physical phenomena and that the Hodgkin and Huxley model perhaps does not account for everything [168].

16.3.1 Evidence for Cell Swelling upon Activation

Structural changes in excited tissues have been observed, first from optical birefringence measurements [169–171] and later more directly using piezoelectric transducers [172, 173]. Intrinsic optical imaging has revealed that in the brain, cell swelling is one of the physiological responses associated with neuronal activation [174–176]. In neural tissues these volume variations have been observed during both intense [177–180] and normal [181, 182] neuronal activation. Conversely, changes in blood osmolarity modify brain cortical excitability in animal models [175, 183–185] and in humans [186]. In the cat cortex transient changes in ionic transmembrane fluxes, especially K^+, are accompanied by the movement of water and cellular swelling partly due to osmotic imbalance [187], while Cl^- influx through γ-aminobutyric acid (GABA)-A receptors contributes to a synaptically evoked cell swelling in the hippocampus [182]. Such swelling not only involves neuronal soma but also focal areas along dendrites and axons [182, 188], as well as glial cells [189–193]. Rat cortical neurons can recover from osmotic swelling only in the presence of an NMDA receptor antagonist [194].

Hence, cortical cell swelling and its active regulation appear to have fundamental importance to neuronal function. Noticeably, these mechanical changes start simultaneously with the electric response with the peak of the mechanical response coinciding accurately with the action potential peak [167, 195, 196]. The response is asymmetric, as the swelling presents a sharp increase, while the return to baseline is smooth and monotonic [167] (Fig. 16.7).

Interestingly, this shape is very similar to that of the intrinsic response function derived from the diffusion MRI signal in the human visual cortex after deconvolution from the activation time course pattern (Fig. 16.8) [198], suggesting that both methods probably address the same physiological processes.

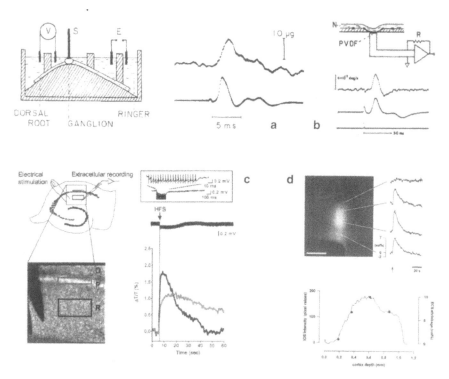

Figure 16.7 Cellular events accompanying activation. (a) Diagram of the setup for recording mechanical responses of a dorsal root ganglion (S: stylus; E: stimulating electrode; V: input for recording the action potential). The plots show the mechanical response of the ganglion (*top*) and the associated action potential (*bottom*). Reprinted with permission from Ref. [227]). (b) Diagram of the setup for recording thermal responses of garsfish olfactory nerve (PVDF: thin pyroelectric film of polyvinylidene fluoride, R: feedback resistor of operational amplificatory). The plots show the action potential (*top*) and the thermal response (*bottom*) (horizontal bar: 30 ms, vertical bar: 4 m°/s). Reprinted with permission from Ref. [196]. (c) Transient increases in near-infrared (IR) light transmittance in the rat hippocampal CA1 region. *Left*: Experimental setup and sample image (O: stratum oriens; P: stratum pyramidale; R: stratum radiatum). The black (dendritic region) and gray (somatic region) indicate areas for measurements. *Right*: Traces from a single trial (*top*: extracellular field potential; *bottom*: IR transmittance changes). HFS indicates the time of high-frequency stimulation. Reprinted with permission from Ref. [182]. (d) *Top*: IOS in rat brain slice 4 s after begin of stimulation (pulses of 200 μs in a train of 50 Hz for 2 s) showing differential involvements within cortical layers (scale bar: 300 μm). The time course of the IOS and corresponding extracellular space shrinking is shown in different cortical layers. *Bottom*: Intensity plot of the IOS and corresponding ECS shrinkage along the cortical depth. Reprinted with permission from Ref. [180]. *Abbreviation*: IOS, intrinsic optical signal.

16.3.2 Heat Release and Phase Transition

Another established process, first observed by Abott [199], is heat production concomitant to the action potential followed by partial (45–85%) reabsorption of the heat [200, 201]. Recent measurements made in the garfish olfactory nerves using synthetic pyroelectric polymers as heat sensors show that the thermal response also starts and peaks with the action potential, resulting in a temperature rise by $23\mu°C$ [196, 202–205] (Fig. 16.6). Such heat release has also been observed during spreading depression [202]. The combined observations of abrupt volume increase (swelling) and heat release suggest that a transient structural change, in the form of a physical first-order phase transition, occurs in the tissue during the generation of the action potential. However, rapid temperature changes, which may be much larger in magnitude than those implied by this proposed phase transition, also arise from the biochemical reactions associated with neuronal activity, neuroenergetics, and the kinetic design of excitatory synapses [206]. The concept that the action potential is an electrochemical manifestation of structural changes at the membrane surface rather that solely membrane electric changes is, indeed, not new [165, 207]. In short, neurons might be considered tiny piezoelectric sensors, changes in their membrane polarization resulting in changes in their shape. But this effect should be reversible with changes in cell shape (caused by neighboring cells, for instance) causing depolarization. One can envision that information could be transferred as a rather high rate through this direct electromechanical coupling, presenting an alternative to the classical synaptic (electrochemical) transmission (see below).

16.3.3 Mechanism of Cortical Cell Swelling: Water and the Cytoskeleton

Ionic transmembrane shifts must be accompanied by water. However, osmotic changes linked only to fluxes of Na^+ and K^+ during the action potential cannot account quantitatively for the observed volume changes [208]. In the small nonmyelinated fibers of the garfish olfactory nerves (0.2–0.3 μm), swelling is in the order of 2×10^{-2} μm^3. Such a large excitation-induced swelling, which has been observed in various nerve tissues, including mammalian ganglion cells [209], cannot easily be ascribed to a simple translocation of water from the extra- to the intracellular compartment but can be better explained by an overall decrease in the density of the excited tissue [208].

Further studies have identified that such nonelectric effects, namely, cell swelling and temperature increase in the excited tissue, take place mainly in a thin gel layer attached to the cell membrane and containing a high density of macromolecules, Ca^{2+}, and structured water [171, 210, 211]. This dense polymer–gel matrix of cross-linked actin filaments and microtubules runs contiguously to the cell membrane [212, 213] with a high negative surface charge responsible for the baseline negative membrane potential [211]. It has been shown that agents that disrupt microtubules or solubilize gels, and thus eliminate this cytoskeleton, suppress the action potential [214, 215]. Cytoskeletal integrity is necessary for the action potential to occur [212].

Within this layer, structured water and Ca^{2+} seem to play a crucial role [216]. The presence of calcium is vital, and its external concentration must be kept as a high level for the action potential to appear [165–167]. Calcium is a divalent cation and under baseline conditions has a sufficiently high concentration in the anionic gel cytoskeleton to maintain Ca^{2+} bridges that are formed between negative sites of the protein strands. The cytoskeleton polymer–gel matrix is kept condensed, as there is a cooperation effect between neighboring anionic sites to form bridges [167]. During excitation the matrix loosens and expands, which is responsible for the swelling and a massive movement of water and an immediate rearrangement (reordering) of water molecules around the exposed anionic charges of the unfolded proteins of the cytoskeleton and the cell membrane. Water dipoles build one upon another, wedging strands farther apart, which expose even more anionic sites, and so on. The cytoskeleton and attached membrane expands due to the enhancement of repulsive electrostatic forces between protein strands near the membrane, resulting in a swollen, lower-density structure with reduced diffusion (see above). Hence, water may be an active component of the swelling effect that occurs during cell activation. The associated phase transition that necessarily accompanies such reordering of the water within the cell membrane–bound gel layer could be partially responsible for the liberation of heat, which has been reported earlier. This process terminates, as covalent bonds within the cytoskeleton prevent further expansion, and the accumulation of cationic ions starts to neutralize anionic sites, competing against water structuring and finally triggering a reverse phase transition of the membrane-layer water at the peak of the action potential: Na^+ is excluded from the cell, restoring the cell membrane potential, while Ca^{2+} recondenses the cytoskeleton.

Furthermore, these physical changes in the otherwise electrically charged cyotoskeleton could participate in the voltage transition observed in the action potential [165, 217]. At the end of the action potential the "melting" of the structured water partially reabsorbs the released heat. Only the very large number of water molecules involved in the process could explain the level of observed changes, in terms of either swelling or heat release. Overall this cycling process with a spontaneous recovery is largely "free" in terms of energy demand, although some energy would still be irreversibly lost in the process. This scheme is well supported by the observation of analogous transient swelling events in synthetic negatively charged polymer gels containing both Na^+ and Ca^{2+} ions [218, 219].

There is a further point of interest related to the heat release associated with the activation-induced phase transition. The substantial increased order in the water structure within the activated region, given the considerable number of water molecules involved, is associated with a decrease in entropy. To offset this and satisfy basic thermodynamic laws the activated region must give rise to a release of heat (loss of enthalpy) and a rise in the temperature of the region, assuming the system is isolated (no change in free energy) during the very fast onset of activation (Fig. 16.7). The restoration of the initial water status and the return of the cell volume to baseline would partially result from a reuptake of the heat and a production of internal energy through cell metabolism and membrane pump activation, which could occur at a slower pace (Fig. 16.7). It is widely supposed that one evolved function of the increase in blood flow accompanying local neuronal activity (which forms the bases of H_2O PET and BOLD fMRI) is to maintain temperature homeostasis [220]. Certainly this restricts temperature changes during activation to well below 0.1°C. Overheating might adversely affect action potentials [221], although a mild fever of 1°C clearly has a limited impact on brain function. One group had already suggested that activation-induced changes in temperature could be at one origin of the vascular events detected by BOLD fMRI [222], but the large temperature changes implied by this model have not been observed [223, 224].

16.3.4 Implications for Functional Neuroimaging

Cell swelling and the associated water phase transitions, thus, seem to play a prominent role in the physiology of cell activation, and these phenomena have been put forward as a possible mechanism to explain the observed changes in water diffusion during brain activation [23]. According to the biphasic diffusion model, (small) signal changes in the human visual cortex on the order of $\cong 1.5\%$ would result from an increase of 0.65% of the SDP [23] (Fig. 16.8).

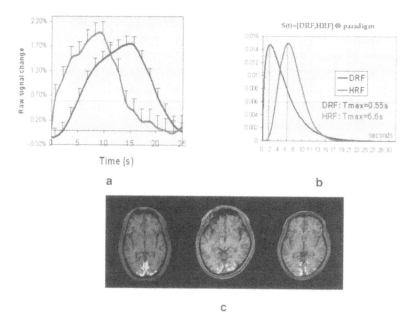

Figure 16.8 Diffusion fMRI. (a) Time course of the BOLD and diffusion fMRI responses in visual cortex (raw signal changes) for a 10 s stimulus and an image resolution of 2 s. The amplitudes of the responses are comparable, but the diffusion response is clearly ahead of the BOLD response by several seconds, both at onset and at offset. (b) Intrinsic response functions derived from the raw signal time courses of (a) after deconvolution of the signal from the stimulation time course. These intrinsic response functions, DRF for water diffusion and HRF for the hemodynamic response, represent the theoretical responses that should be observed from impulse stimuli. The DRF picks much earlier than the HRF (T_{max}) and presents a more asymmetric shape with a sharp increase and a slightly slower return to baseline, a pattern also observed in IOS responses (Fig. 16.7). (c) Activation maps obtained with diffusion fMRI during visual stimulation in three subjects, showing that the localization of the voxels where water diffusion is reduced are well localized within the brain cortex (Courtesy of D. Le Bihan, S. Urayama, T. Aso, T. Hanakawa, and H. Fukuyama, Human Brain Research Center, Kyoto University). *Abbreviations*: DRF, diffusion response function; HRF, hemodynamic response function. For color reference, see p. 391.

This effect would only require about 15% of the elements in the voxel to swell by 4% in radius. Hence, only a small fraction of cells need to swell in order to be detected by diffusion MRI, which would be an extremely sensitive functional neuroimaging tool, provided, of course, that the basic signal to noise ratio is not too low, which is, unfortunately, often the case with diffusion MRI, hence the aim for very high field MRI. Such variations in cell size linked to neuronal depolarization have also been observed in brain slices where confounding blood effects are removed, using proton-density-weighted MRI [225] and diffusion MRI [163].

16.4 CONCLUSION

Water has been, so far, the molecule of choice for functional neuroimaging, whatever the modality, PET or MRI. This is clearly no accident, given its ubiquitous distribution in living organisms and in the brain. Life cannot exist without liquid water, but it remains amazing that, although a great many number of theories have been published about the unusual physical properties of liquid water, sometimes with great controversy, we still do not understand the special relationship this tiny and apparently simple molecule has with our lives. The special features carried by the characteristics of water molecules are necessary for the operation of cellular biological mechanisms within cells to happen, but conversely, these mechanisms and the content of the cells have a profound impact on the water molecular structure itself, and this reciprocal relationship has often been overlooked in the past. Among those peculiar features is the molecular diffusion process of water. The mechanisms underlying water diffusion in itself at the molecular level are still a subject of research. However, here also life has given its imprint, as even this process, random by definition, seems to have been modified or even exploited in living organisms. Because of the presence of numerous structuring elements, water diffusion over time ranges of milliseconds is different in cells than in bulk water and reflects with great intimacy the subtle changes in water and cell structure that accompany various physiological and pathological status, hence the considerable success of diffusion MRI, which can monitor such changes with accuracy in space and time.

So water may not just be a passive player in cell physiology. This is particularly true in the brain cortex, where intricate biophysical mechanisms between membrane events and water movements and structure could take place during brain activation. Water diffusion has been found as a very early physiological surrogate marker of neuronal activation, which can be monitored to produce maps of cortical activation in the human brain with MRI. Many issues, though, have to be investigated to understand the actual mechanisms underlying the observed water diffusion changes. Although it appears that the swelling of a slow diffusion water phase could be linked to the expansion of a membrane-bound water layer upon activation-induced cell swelling, the physical nature of this slow diffusion water phase and the relationship between its volume variations and cell size changes, as well as the biophysical mechanisms underlying the observed swelling must be clarified. However, the possibility to observe transient cell swelling events in the brain of intact animals and humans would provide new opportunities to study the elementary processes underlying cortical activation and the existence of "nonsynaptic" mechanisms, as recently pointed out in animal models [184, 185] and in humans [186]. Such mechanisms could partly be responsible for the synchronized activation of neurons through changes in the size of the extracellular space [184] and result in faster responses than synaptic mechanisms [183]. What is the contribution to brain tissue function of those rapid "mechanical" changes that have been noticed in tissue microstructure (such as the twitching of the dendrite spines [226])? Could we envision that, because laws of nature are most often reversible, neurons are also acting as piezoelectric sensors that get depolarized when "feeling" the movements of neighboring cells, a fast alternative to synaptic transmission for information fluxes? Clearly, many questions remain without answers yet, but the aim of this chapter would be fully reached if it contributes to trigger some readers' interest in the puzzling interaction that could exist between the functioning of our brain and water, the "molecule of the mind."

References

1. Roy, C. W., and Sherrington, C. S. (1890) On the regulation of the blood supply of the brain, *J. Physiol.*, **11**, 85–108.

2. Posner, M. I., Peterson, S. E., Fox, P. T., and Raichle, M. E. (1988) Localization of cognitive operations in the human brain, *Science*, **240**, 1627–1631.

3. Fox, P. T., Mintun, M. A., Raichle, M. E., Miezin, F. M., Allman, J. M., and Van Essen, D. C. (1986) Mapping human visual cortex with positron emission tomography, *Nature*, **323**, 806–809.

4. Raichle, M. E. (1994) Visualizing the mind, *Sci. Am.*, **270**, 58–64.

5. Ogawa, S., Menon, R. S., Tank, D. W., Kim, S. G., Merkle, H., Ellermann, J. M., and Ugurbil, K. (1993) Functional brain mapping by blood oxygenation level-dependent contrast magnetic resonance imaging. A comparison of signal characteristics with a biophysical model, *Biophys. J.*, **64**, 803–808.

6. Kwong, K. K., Belliveau, J. W., and Chesler, D. A. (1992) Dynamic magnetic resonance imaging of human brain activity during primary sensory stimulation, *Proc. Natl. Acad. Sci. USA*, **89**, 5675–5679.

7. Raichle, M. E., and Mintun, M. A. (2006) Brain work and brain imaging, *Annu. Rev. Neurosci.*, **29**, 449–476.

8. Logothetis, N. K., and Wandell, B. A. (2004) Interpreting the BOLD signal, *Annu. Rev. Physiol.*, **66**, 735–769.

9. Gordon, G. R., Choi, H. B., Rungta, R. L., Ellis-Davies, G. C., and Macvicar, B. A. (2008) Brain metabolism dictates the polarity of astrocyte control over arterioles, *Nature*, **456**, 745–749.

10. Sirotin, Y. B., and Das, A. (2009) Anticipatory haemodynamic signals in sensory cortex not predicted by local neuronal activity, *Nature*, **457**, 475–479.

11. Iadecola C, Nedergaard M (2007) Glial regulation of the cerebral microvasculature. *Nat. Neurosci.*, **10**, 1369-1376.

12. Magistretti, P. J., and Pellerin, L. (1999) Cellular mechanisms of brain energy metabolism and their relevance to functional brain imaging, *Philos. Trans. R. Soc. Lond. B. Biol. Sci.*, **354**, 1155–1163.

13. Buxton, R. B., Uludag, K., Dubowitz, D.J., and Liu, T.T. (2004) Modeling the hemodynamic response to brain activation, *NeuroImage*, **23**, S220–S233.

14. Van Zijl, P. C. M., Eleff, S. M., Ulatowski, J. A., Oja J. M. E., Ulug, A. M., Traystman, R.J., and Kauppinen, R. A. (1998) Quantitative assessment of blood flow, blood volume and blood oxygenation effects in functional magnetic resonance imaging, *Nat. Med.*, **4**,159–167.

15. Malonek, D., Dirnagl, U., Lindauer, U., Yamada, K., Kannurpatti, S. S., and Grinvald, A. (1997) Vascular imprints of neuronal activity: relationsships between the dynamics of cortical blood flow, oxygenation, and volume changes following sensory stimulation, *Proc. Nat. Acad. Sci. USA*, **94**, 14826–14831.

16. Schummers, J., Yu, H., and Sur, M. (2008) Tuned responses of astrocytes and their influence on hemodynamic signals in the visual cortex, *Science*, **20**(320), 1638–1643.

17. Logothetis, N. K., Pauls, J., Augath, M., Trinath, T., and Oeltermann, A. (2001) Neurophysiological investigation of the basis of the fMRI signal, *Nature*, **412**, 150–157.

18. Lehericy, S., Biondi, A., Sourour, N., Vlaicu, M., duMontcel, S. T., Cohen, L., Vivas, E., Capelle, L., Faillot, T., Casasco, A., Le Bihan, D., Marsault, C. (2002) Arteriovenous brain malformations: is functional MR imaging reliable for studying language reorganization in patients? Initial observations, *Radiology*, **223**, 672–682.

19. Turner, R. (2002) How much cortex can a vein drain? Downstream dilution of activation-related cerebral blood oxygenation changes, *NeuroImage*, **16**, 1062–1067.

20. Lu, H., Soltysik, D. A., Ward, B. D., and Hyde, J. S. (2005) Temporal evolution of the CBV-fMRI signal to rat whisker stimulation of variable duration and intensity: a linearity analysis, *NeuroImage*, **26**, 432–440.

21. Huppert, T. J., Hoge, R. D., Diamond, S. G., Franceschini, M. A., and Boas, D. A. (2006) A temporal comparison of BOLD, ASL, and NIRS hemodynamic responses to motor stimuli in adult humans, *NeuroImage*, **29**, 368–382.

22. Darquie, A., Poline, J. B., Poupon, C., Saint-Jalmes, H., and Le Bihan, D. (2001) Transient decrease in water diffusion observed in human occipital cortex during visual stimulation, *Proc. Natl. Acad. Sci. USA*, **98**, 9391–9395.

23. Le Bihan, D., Urayama, Si, Aso, T., Hanakawa, T., and Fukuyama, H. (2006) Direct and fast detection of neuronal activation in the human brain with diffusion MRI, *Proc. Natl. Acad. Sci. USA*, **103**, 8263–8268.

24. Kohno, S., Sawamoto, N., Urayama, S., Aso, T., Aso, K., Seiyama, A., Fukuyama, H., and Le Bihan, D. (2009) Water-diffusion slowdown in the human visual cortex on visual stimulation precedes vascular responses, *J. Cereb. Blood Flow Metab.*, **29**, 1197–1207.

25. Einstein, A. (1956) *Investigations on the Theory of Brownian Motion* (ed. Furthe, R., and Cowper, A. D.), Dover, New York. [Collection of papers translated from the German.]

26. Le Bihan, D., and Breton, E. (1985) Imagerie de diffusion *in vivo* par résonance magnétique nucléaire, *C. R. Acad. Sci., Ser. II*, **301**, 1109–1112.

27. Taylor, D. G., and Bushell, M. C. (1985) The spatial mapping of translational diffusion coefficients by the NMR imaging technique, *Phys. Med. Biol.*, **30**, 345–349.

28. Le Bihan, D., Breton, E., Lallemand, D., Grenier, P., Cabanis, E., and Laval Jeantet, M. (1986) MR Imaging of intravoxel incoherent motions : application to diffusion and perfusion in neurologic disorders, *Radiology*, **161**, 401–407.

29. Le Bihan, D. (2003) Looking into the functional architecture of the brain with diffusion MRI, *Nat. Rev. Neurosci.*, **4**, 469–480.

30. Tanner, J. E. (1979) Self diffusion of water in frog muscle, *Biophys. J.*, **28**, 107–116.

31. Tanner, J. E. (1978) Transient diffusion in a system partitioned by permeable barriers. Application to NMR measurements with a pulsed field gradient, *J. Chem. Phys.*, **69**, 1748–1754.

32. Cooper, R. L., Chang, D. B., Young, A. C., Martin, J., and Ancker-Johnson, B. (1974) Restricted Diffusion in Biophysical Systems, *Biophys. J.*, **14**, 161–177.

33. Le Bihan, D. (1995) Molecular diffusion, Tissue microdynamics and microstructure, *NMR Biomed.*, **8**, 375–386.

34. Moseley, M. E., Cohen, Y., and Mintorovitch, J. (1990) Early detection of regional cerebral ischemic injury in cats : evaluation of diffusion and T2-weighted MRI and spectroscopy, *Magn. Resonance Med.*, **14**, 330–346.

35. Warach, S., Baron, J. C. (2004) Neuroimaging, *Stroke*, **35**, 351–353.

36. Sotak, C. H. (2002) The role of diffusion tensor imaging in the evaluation of ischemic brain injury—a review, *NMR Biomed.*, **15**, 561–569.

37. Takahara, T., Imai, Y., Yamashita, T., Yasuda, S., Nasu, S., and van Cauteren, M. (2004) Diffusion-weighted whole-body imaging with background body signal suppression (DWIBS): technical improvement using free breathing, STIR and high-resolution 3D display, *Radiat. Med.*, **22**, 275–282.

38. Ide, M. (2006) Cancer screening with FDG-PET, *Q. J. Nucl. Med. Mol. Imaging.*, **50**, 23–27.

39. Siegel, B. A., and Dehdashti, F. (2005) Oncologic PET/CT: current status and controversies, *Eur. Radiol.*, **15**(Suppl 4), D127–D132.

40. Moseley, M. E., Cohen, Y., and Kucharczyk, J. (1990) Diffusion-weighted MR imaging of anisotropic water diffusion in cat central nervous system, *Radiology*, **176**, 439–446.

41. Douek, P., Turner, R., Pekar, J., Patronas, N. J., and Le Bihan, D. (1991) MR color mapping of myelin fiber orientation, *J. Comput. Assist. Tomogr.*, **15**, 923–929.

42. Basser, P. J., Mattiello, J., and Le Bihan, D. (1994) MR diffusion *tensor* spectroscopy and imaging, *Biophys. J.*, **66**, 259–267.

43. Le Bihan, D., Mangin, J. F., Poupon, C., Clark, C. A., Pappata, S., and Molko, N. (2001) Diffusion tensor imaging: concepts and applications, *JMRI*, **13**, 534–546.

44. Verhoeven, J. S., De Cock, P., Lagae, L., and Sunaert, S. (2010) Neuroimaging of autism, *Neuroradiology*, **52**, 3–14.

45. Kyriakopoulos, M., Bargiotas, T., Barker, G. J., and Frangou, S. (2008) Diffusion tensor imaging in schizophrenia, *Eur. Psychiatry*, **23**, 255–273.

46. Stebbins, G. T., and Murphy, C. M. (2009) Diffusion tensor imaging in Alzheimer's disease and mild cognitive impairment, *Behav. Neurol.*, **21**, 39–49.

47. Yacoub, E., Uludag, K., Ugurbil, K., and Harel, N. (2008) Decreases in ADC observed in tissue areas during activation in the cat visual cortex at 9.4T, *Magn. Reson. Imaging*, **26**, 889–896.

48. Jin, T., Zhao, F., and Kim, S. G. (2006) Sources of functional apparent diffusion coefficient changes investigated by diffusion-weighted spin-echo fMRI, *Magn. Reson. Med.*, **56**, 1283–1292.

49. Jin, T., and Kim, S. G. (2008) Functional changes of apparent diffusion coefficient during visual stimulation investigated by diffusion-weighted gradient-echo fMRI, *NeuroImage*, **41**, 801–812.

50. Miller, K. L., Bulte, D. P., Devlin, H., Robson, M. D., Wise, R. G., Woolrich, M. W., Jezzard, P., and Behrens, T. E. (2007) Evidence for a vascular contribution to diffusion FMRI at high b value, *Proc. Natl. Acad. Sci. USA*, **104**, 20967–20972.

51. Bratton, C. B., Hopkins, A. L., and Weinberg, J. W. (1965) Nuclear magnetic resonance studies of living muscle, *Science*, **147**, 738–739.

52. Cope, F. W. (1969) Nuclear magnetic resonance evidence using D2O for structured water in muscle and brain, *Biophys. J.*, **9**, 303–319.

53. Hazlewood, C. F., Rorschach, H. E., and Lin, C. (1991) Diffusion of water in tissues and MRI, *Magn. Reson. Med.*, **19**, 214–216.

54. Kasturi, S. R., Chang, D. C., and Hazlewood, C. F. (1980) Study of anisotropy in nuclear magnetic resonance relaxation times of water protons in skeletal muscle, *Biophys. J.*, **30**, 369–381.

55. Trantham, E. C., Rorschach, H. E., Clegg, J. S., Hazlewood, C. F., Nicklow, R. M., and Wakabayashi, N. (1984) Diffusive properties of water in Artemia cysts as determined from quasi-elastic neutron scattering spectra, *Biophys. J.*, **45**, 927–938.

56. Nicholson, C., and Philipps, J. M. (1981) Ion diffusion modified by tortuosity and volume fraction in the extracellular micorenvrionment of the rat cerebellum, *J. Physiol.*, **321**, 225–257.

57. Chen, K. C., and Nicholson, C. (2000) Changes in brain cell shape create residual extracellular space volume and explain tortuosity behavior during osmotic challenge, *Proc. Natl. Acad. Sci. USA*, **97**, 8306–8311.

58. Latour, L. L., Svoboda, K., Mitra, P. P., and Sotak, C. H. (1994) Time-dependent diffusion of water in a biological model system, *Proc. Natl. Acad. Sci. USA*, **91**, 1229–1233.

59. Norris, D. G., Niendorf, T., and Leibfritz, D. (1994) Healthy and infarcted brain tissues studied at short diffusion times: the origins of apparent restriction and the reduction in apparent diffusion coefficient, *NMR Biomed.*, **7**, 304–310.

60. Assaf, Y., and Cohen, Y. (1998) Non-mono-exponential attenuation of water and N-acetyl aspartate signals due to diffusion in brain tissue, *J. Magn. Reson.*, **131**, 69–85.

61. Le Bihan, D., Turner, R., and Douek, P. (1993) Is water diffusion restricted in human brain white matter? An echo-planar NMR imaging study, *NeuroReport*, **4**, 887–890.

62. Moonen, C. T. W., Pekar, J., De Vleeschouwer, M. H. M., Van Gelderen, P., Van Zijl, P. C. M., and Des Pres, D. (1991) Restricted and anisotropic displacement of water in healthy cat brain and in stroke studied by NMR diffusion imaging, *Magn. Reson. Med.*, **19**, 327–332.

63. Garcia-Perez, A. I., Lopez-Beltran, E. A., Kluner, P., Luque, J., Ballesteros, P., and Cerdan, S. (1999) Molecular crowding and viscosity as determinants of translational diffusion of metabolites in subcellular organelles, *Arch. Biochem. Biophys.*, **362**, 329–338.

64. Rorschach, H. E., Chang, D. C., Hazlewood, C. F., and Nichols, B. L. (1973) The diffusion of water in striated muscle, *Ann. N. Y. Acad. Sci.*, **204:445–52.**, 445–452.

65. Chang, D. C., Rorschach, H. E., Nichols, B. L., and Hazlewood, C. F. (1973) Implications of diffusion coefficient measurements for the structure of cellular water, *Ann. N. Y. Acad. Sci.*, **204**, 434–443.

66. Colsenet, R., Mariette, F., and Cambert, M. (2005) NMR relaxation and water self-diffusion studies in whey protein solutions and gels, *J. Agric. Food. Chem.*, **53**, 6784–6790.

67. Amiry-Moghaddam, M., and Ottersen, O. P. (2003) The molecular basis of water transport in the brain, *Nat. Rev. Neurosci.*, **4**, 991–1001.

68. Cohen, Y., Assaf, Y. (2002) High b-value q-space analyzed diffusion-weighted MRS and MRI in neuronal tissues—a technical review, *NMR Biomed.*, **15**, 516–542.

69. Niendorf, T., Dijkhuizen, R. M., Norris, D. G., Van Lookeren Campagne, M., and Nicolay, K. (1996) Biexponential diffusion attenuation in various states of brain tissue : implications for diffusion-weighted imaging, *Magn. Reson. Med.*, **36**, 847–857.

70. Karger, J., Pfeifer, H., and Heink, W. (1988) Principles and application of self-diffusion measurements by nuclear magnetic resonance, *Adv. Magn. Reson.*, **12**, 1–89.

71. Mulkern, R. V., Gudbjartsson, H., Westin, C. F., Zengingonul, H. P., Gartner, W., Guttmann, C. R. G., Robertson, R. L., Kyriakos, W., Schwartz, R., Holtzman, D., Jolesz, F. A., and Maier, S. E. (1999) Multi-component apparent diffusion coefficients in human brain, *NMR Biomed.*, **12**, 51–62.

72. Maier, S. E., Bogner, P., Bajzik, G., Mamata, H., Mamata, Y., Repa, I., Jolesz, F. A., and Mulkern, R. V. (2001) Normal brain and brain tumor: multicomponent apparent diffusion coefficient line scan imaging, *Radiology*, **219**, 842–849.

73. Clark, C. A., and Le Bihan, D. (2000) Water diffusion compartmentation and anisotropy at high b values in human brain, *Proc. 8th Int. Soc. Magn. Reson. Med.*, p. 759.

74. Nicholson, C., and Sykova, E. (1998) Extracellular space structure revealed by diffusion analysis, *Tins*, **21**, 207–215.

75. LeBihan, D., and vanZijl, P. (2002) From the diffusion coefficient to the diffusion tensor, *NMR Biomed.*, **15**, 431–434.

76. Sehy, J. V., Ackerman, J. J. H., and Neil, J. J. (2002) Apparent diffusion of water, ions, and small molecules in the Xenopus oocyte is consistent with Brownian displacement, *Magn. Reson. Med.*, **48**, 42–51.

77. Sehy, J. V., Ackerman, J. J. H., and Neil, J. J. (2002) Evidence that both fast and slow water ADC components arise from intracellular space, *Magn. Reson. Med.*, **48**, 765–770.

78. Stanisz, G. J., Szafer, A., Wright, G. A., and Henkelman, R. M. (1997) An analytical model of restricted diffusion in bovine optic nerve, *Magn. Reson. Med.*, **37**, 103–111.

79. Assaf, Y., Freidlin, R. Z., Rohde, G. K., and Basser, P. J. (2004) New modeling and experimental framework to characterize hindered and restricted water diffusion in brain white matter, *Magn. Reson. Med.*, **52**, 965–978.

80. Chin, C. L., Wehrli, F. W., Fan, Y. L., Hwang, S. N., Schwartz, E. D., Nissanov, J., and Hackney, D. B. (2004) Assessment of axonal fiber tract architecture in excised rat spinal cord by localized NMR q-space imaging: simulations and experimental studies, *Magn. Reson. Med.*, **52**, 733–740.

81. Novikov, E. G., Van Dusschoten, D., and VanAs, H. (1998) Modeling of self-diffusion and relaxation time NMR in multi-compartment systems, *J. Magn. Reson.*, **135**, 522–528.

82. Yablonskiy, D. A., Bretthorst, G. L., and Ackerman, J. J. H. (2003) Statistical model for diffusion attenuated MR signal, *Magn. Reson. Med.*, **50**, 664–669.

83. vanderWeerd, L., Melnikov, S. M., Vergeldt, F. J., Novikov, E. G., and VanAs, H. (2002) Modelling of self-diffusion and relaxation time NMR in multicompartment systems with cylindrical geometry, *J. Magn. Reson.*, **156**, 213–221.

84. Kroenke, C. D., Ackerman, J. J. H., and Yablonskiy, D. A. (2004) On the nature of the NAA diffusion attenuated MR signal in the central nervous system, *Magn. Reson. Med.*, **52**, 1052–1059.

85. Sukstanskii, A. L., Yablonskiy, D. A., and Ackerman, J. J. H. (2004) Effects of permeable boundaries on the diffusion-attenuated MR signal: insights from a one-dimensional model, *J. Magn. Reson.*, **170**, 56–66.

86. Sotak, C. H. (2004) Nuclear magnetic resonance (NMR) measurement of the apparent diffusion coefficient (ADC) of tissue water and its relationship to cell volume changes in pathological states, *Neurochem. Int.*, **45**, 569–582.

87. Van Der Toorn, A., Sykova, E., Dijkhuizen, R. M., Vorisek, I., Vargova, L., Skobisova, E., Van Lookeren Campagne, M., Reese, T., and Nicolay, K. (1996) Dynamic changes in water adc, energy metabolism, extracellular space volume, and tortuosity in neonatal rat brain during global ischemia, *Magn. Reson. Med.*, **36**, 52–60.

88. Thorne, R. G., and Nicholson, C. (2006) In vivo diffusion analysis with quantum dots and dextrans predicts the width of brain extracellular space, *Proc. Natl. Acad. Sci. USA*, **103**, 5567–5572.

89. Benveniste, H., Hedlund, L. W., and Johnson, G. A. (1992) Mechanism of detection of acute cerebral ischemia in rats by diffusion-weighted magnetic resonance microscopy, *Stroke*, **23**, 746–754.

90. OShea, J. M., Williams, S. R., vanBruggen, N., and GardnerMedwin, A. R. (2000) Apparent diffusion coefficient and MR relaxation during osmotic manipulation in isolated turtle cerebellum, *Magn. Reson. Med.*, **44**, 427–432.

91. Hasegawa, Y., Formato, J. E., Latour, L. L., Gutierrez, J. A., Liu, K. F., Garcia, J. H., Sotak, C. H., and Fisher, M. (1996) Severe transient hypoglycemia causes reversible change in the apparent diffusion coefficient of water, *Stroke*, **27**, 1648–1655.

92. Dijkhuizen, R. M., deGraaf, R. A., Tulleken, K. A. F., and Nicolay, K. (1999) Changes in the diffusion of water and intracellular metabolites after excitotoxic injury and global ischemia in neonatal rat brain, *J. Cereb. Blood Flow Metab.*, **19**, 341–349.

93. Buckley, D. L., Bui, J. D., Phillips, M. I., Zelles, T., Inglis BA, Plant, H. D., and Blackband, S. J. (1999) The Effect of Ouabain on Water Diffusion in the Rat Hippocampal Slice Measured by high Resolution NMR Imaging, *Magn. Reson. Med.*, **41**, 137–142.

94. Nicholson, C., and Tao, L. (1993) Hindered diffusion of high molecular weight compounds in brain extracellular microenvironment measured with integrative optical imaging, *Biophys. J.*, **65**, 2277–2290.

95. Zhong, J., Petroff, O. A. C., Prichard, J. W., and Gore, J. C. (1993) Changes in water diffusion and relaxation properties of rat cerebrum during status epilepticus, *Magn. Reson. Med.*, **30**, 241–246.

96. Zhong, J., Petroff, O. A. C., Pleban, L. A., Gore, J. C., and Prichard, J. W. (1997) Reversible, reproducible reduction of brain water apparent diffusion coefficient by cortical electroshocks, *Magn. Reson. Med.*, **37**, 1–6.

97. Busch, E., Gyngell, M. L., Eis, M., Hoehn Berlage, M., and Hossmann, K. A. (1996) Potassium-induced cortical spreading depressions during focal cerebral ischemia in rats: contribution to lesion growth assessed by diffusion-weighted NMR and biochemical imaging, *J. Cereb. Blood Flow Metab.*, **16**, 1090–1099.

98. Hasegawa, Y., Latour, L. L., Formato, J. E., Sotak, C. H., and Fisher, M. (1995) Spreading Waves of a reduced diffusion coefficient of water in normal and ischemic rat brain, *J. Cereb. Blood Flow Metab.*, **15**, 179–187.

99. Latour, L. L., Hasegawa, Y., Formato, J. E., Fisher, M., and Sotak, C. H. (1994) Spreading waves of decreased diffusion coefficient after cortical stimulation in the rat brain, *Magn. Reson. Med.*, **32**, 189–198.

100. Mancuso, A., Derugin, N., Ono, Y., Hara, K., Sharp, F. R., and Weinstein, P. R. (1999) Transient MRI-detected water apparent diffusion coefficient reduction correlates with c-fos mRNA but not hsp70 mRNA induction during focal cerebral ischemia in rats, *Brain. Res.*, **839**, 7–22.

101. Röther, J., De Crespigny, A. J., D'Arcueil, H., and Moseley, M. E. (1996) MR detection of cortical spreading depression immediately after focal ischemia in the rat, *J. Cereb. Blood Flow Metab.*, **16**, 214–221.

102. Dietzel, I., Heinemann, U., Hofmeijer, G., and Lux, H. D. (1980) Transient changes in the size of the extracellular space in the sensorimotor cortex of cats in relation to stimulus-induced changes in potassium concentration, *Exp. Brain Res.*, **40**, 432–439.

103. Phillips, J. M., and Nicholson, C. (1979) Anion permeability in spreading depression investigated with ion-sensitive microelectrodes, *Brain Res.*, **173**, 567–571.

104. Hansen, A. J., and Olsen, C. E. (1980) Brain extracellular space during spreading depression and ischemia, *Acta Physiol. Scand.*, **108**, 355–365.

105. Boyle, P. J., and Conway, E. J. (1941) Potassium accumulation in muscle and associated changes, *J. Cell Biol.*, **20**, 529–538.

106. Agre, P. (2005) Membrane water transport and aquaporins: looking back, *Biol. Cell*, **97**, 355–356.

107. Agre, P., Nielsen, S., and Ottersen, O. P. (2004) Towards a molecular understanding of water homeostasis in the brain, *Neuroscience*, **129**, 849–850.

108. Glynn, I. M. (2002) A hundred years of sodium pumping, *Annu. Rev. Physiol.*, **64**, 1–18.

109. Ling, G. N., Ochsenfeld, M. M., and Karreman, G. (1967) Is the cell membrane a universal rate-limiting barrier to the movement of water between the living cell and its surrounding medium? *J. Gen. Physiol.*, **50**, 1807–1820.

110. Pollack, G. H. (2001) Is the cell a gel—and why does it matter? *Jpn. J. Physiol.*, **51**, 649–660.

111. Whittam, R. (1961) Active cation transport as a pace-maker of respiration, *Nature*, **191**, 603–604.

112. Ling, G. N. (1988) A physical theory of the living state: application to water and solute distribution, *Scanning Microsc.*, **2**, 899–913.

113. Pollack, G. H. (2003) The role of aqueous interfaces in the cell, *Adv. Colloid Interface Sci.*, **103**, 173–196.

114. Ling, G. N. (1997) Debunking the alleged resurrection of the sodium pump hypotheses, *Physiol. Chem. Phys. Med. NMR*, **29**, 123–198.

115. Kellermayer, M., Ludany, A., Jobst, K., Szucs, G., Trombitas, K., and Hazlewood, C. F. (1986) Cocompartmentation of proteins and K+ within the living cell, *Proc. Natl. Acad. Sci. USA*, **83**, 1011–1015.

116. Cameron, I. L., Hardman, W. E., Fullerton, G. D., Miseta, A., Koszegi, T., Ludany, A., and Kellermayer, M. (1996) Maintenance of ions, proteins and water in lens fiber cells before and after treatment with non-ionic detergents, *Cell Biol. Int.*, **20**, 127–137.

117. Fullerton, G. D., Kanal, K. M., and Cameron, I. L. (2006) On the osmotically unresponsive water compartment in cells, *Cell Biol. Int.*, **30**, 74–77.

118. Arrio-Dupont, M., Foucault, G., Vacher, M., Devaux, P. F., and Cribier, S. (2000) Translational diffusion of globular proteins in the cytoplasm of cultured muscle cells, *Biophys. J.*, **78**, 901–907.

119. Luby-Phelps, K., Taylor, D. L., and Lanni, F. (1986) Probing the structure of cytoplasm, *J. Cell Biol.*, **102**, 2015–2022.

120. Seksek, O., Biwersi, J., and Verkman, A. S. (1997) Translational diffusion of macromolecule-sized solutes in cytoplasm and nucleus, *J. Cell Biol.*, **138**, 131–142.

121. Ling, G. N., and Walton, C. L. (1976) What retains water in living cells? *Science*, **191**, 293–295.

122. Mazur, P. (1970) Cryobiology: the freezing of biological systems, *Science*, **168**, 939–949.

123. Tanghe, A., Van Dijck, P., and Thevelein, J. M. (2006) Why do microorganisms have aquaporins? *Trends Microbiol.*, **14**, 78–85.

124. Bausch, A. R., Moller, W., and Sackmann, E. (1999) Measurement of local viscoelasticity and forces in living cells by magnetic tweezers, *Biophys. J.*, **76**, 573–579.

125. Wang, N., Butler, J. P., and Ingber, D. E. (1993) Mechanotransduction across the cell surface and through the cytoskeleton, *Science*, **260**, 1124–1127.

126. Vuilleumier, R., and Borgis, D. (2006) Molecular dynamics of an excess proton in water using a non-additive valence bond force yield, *J. Mol. Struc.*, **436**, 555–565.

127. Sciortino, F., Geiger, A., and Stanley, H. E. (1991) Effect of defects on molecular mobility in liquid water, *Nature*, **354**, 218–221.

128. Sciortino, F., Geiger, A., and Stanley, H. E. (1992) Network defects and molecular mobility in liquid water, *J. Chem. Phys.*, **96**, 3857–3865.

129. Toney, M. F., Howard, J. R., Richer, J., Borges, G. L., Gordon, J. G., Melroy, O. R., Wiesler, D. G., Yee, D., and Sorensen, L. B. (1994) Voltage-dependent ordering of water molecules at an electrode-electrolye interface, *Nature*, **368**, 444–446.

130. Pauling, L. (1945) The adsoprtion of water by proteins, *J. Am. Chem. Soc.*, **67**, 555–557.

131. Ling, G. N. (1965) The physical state of water in living cell and model systems, *Ann. N. Y. Acad. Sci.*, **125**, 401–417.

132. Ling, G. N., Miller, C., and Ochsenfeld, M. M. (1973) The physical state of solutes and water in living cells according to the association-induction hypothesis, *Ann. N. Y. Acad. Sci.*, **204:6–50.**, 6–50.

133. Ling, G. N. (1977) The functions of polarized water and membrane lipids: a rebuttal, *Physiol Chem. Phys.*, **9**, 301–311.

134. Clegg, J. S. (1984) Intracellular water and the cytomatrix: some methods of study and current views, *J. Cell Biol.*, **99**, 167s–171s.

135. Cameron, I. L., Kanal, K. M., Keener, C. R., and Fullerton, G. D. (1997) A mechanistic view of the non-ideal osmotic and motional behavior of intracellular water, *Cell. Biol. Int.*, **21**, 99–113.

136. Ling, G. N. (2003) A new theoretical foundation for the polarized-oriented multilayer theory of cell water and for inanimate systems demonstrating long-range dynamic structuring of water molecules, *Physiol. Chem. Phys. Med. NMR*, **35**, 91–130.

137. Wiggins, P. M. (1990) Role of water in some biological processes, *Microbiol. Rev.*, **54**, 432–449.

138. Rorschach, H. E., Lin, C., and Hazlewood, C. F. (1991) Diffusion of water in biological tissues, *Scanning Microsc. Suppl.*, **5**, S1–S9.

139. Clegg (1984) Properties and metabolism of the aqueous cytoplasm and its boundaries, *Am. J. Physiol. Regul. Integr. Comp. Physiol.*, **246**, 133–151.

140. Fullerton, G. D., and Amurao, M. R. (2006) Evdence that collagen and tendon have monolayer water coverage in the native state, *Cell Biol. Int.*, **30**, 56–65.

141. Steinhoff, H. J., Kramm, B., Hess, G., Owerdieck, C., and Redhardt, A. (1993) Rotational and translational water diffusion in the hemoglobin hydration shell: dielectric and proton nuclear relaxation measurements, *Biophys. J.*, **65**, 1486–1495.

142. Pissis, P., Anagnostopoulou-Konsta, A., and Apekis, L. (1987) A dielectric study of the state of water in plant stems, *J. Exptl. Bot.*, **38**, 1528–1540.

143. Gershon, N. D., Porter, K. R., and Trus, B. L. (1985) The cytoplasmic matrix: its volume and surface area and the diffusion of molecules through it, *Proc. Natl. Acad. Sci. USA*, **82** , 5030–5034.

144. Yoshizaki, K., Seo, Y., Nishikawa, H., and Morimoto, T. (1982) Application of pulsed-gradient ^{31}P NMR on frog muscle to measure the diffusion rates of phosphorus compounds in cells, *Biophys. J.*, **38**, 209–211.

145. Hubley, M. J., Rosanske, R. C., and Moerland, T. S. (1995) Diffusion coefficients of ATP and creatine phosphate in isolated muscle: pulsed gradient 31P NMR of small biological samples, *NMR Biomed.*, **8**, 72–78.

146. Wojcieszyn, J. W., Schlegel, R. A., Wu, E. S., and Jacobson, K. A. (1981) Diffusion of injected macromolecules within the cytoplasm of living cells, *Proc. Natl. Acad. Sci. USA*, **78**, 4407–4410.

147. Israelachvili, J., and Wennerstrom, H. (1996) Role of hydration and water structure in biological and colloidal interactions, *Nature*, **379**, 219–225.

148. Israelachvili, J. N., and McGuiggan, P. M. (1988) Forces between surfaces in liquids, *Science*, **241**, 795–800.

149. Horn, R. G., and Israelachvili, J. (1981) Direct measurement of structural forces between two surfaces in a nonpolar liquid, *J. Chem. Phys.*, **75**, 1400–1411.

150. Grannick, S. (1991) Motions and relexations of confined liquids, *Science*, **253**, 1374–1379.

151. Pashley, R. M., and Kitchener, J. A. (1979) Surface forces in adsorbed multilayers of water on quartz, *J. Colloid Interface Sci.*, **71**, 491–500.

152. Fisher, I. R., Gamble, R. A., and Middlehurst, J. (1981) The Kelvin equation and the condensation of water, *Nature*, **290**, 575–576.

153. Xu, X. H. N., and Yeung, E. S. (1998) Long-range electrostatic trapping of single-protein molecules at liquid-solid interface, *Science*, **281**, 1650–1653.

154. Shelton, D. P. (2000) Collective molecular rotation in water and other simple liquids, *Chem. Phys. Lett.*, **325**, 513–516.

155. Chou, K. C. (1992) Energy-optimized structure of antifreeze proteins and its binding mechanism, *J. Mol. Biol.*, **223**, 509–517.

156. Fitter, J., Lechner, R. E., and Dencher, N. A. (1999) Interactions of hydratation water and biological membranes studied by neutron scattering, *J. Phys. Chem. B*, **103**, 8036–8050.

157. Wassall, S. R. (1996) Pulsed field-gradient-spin echo NMR studies of water diffusion in a phospholipid model membrane, *Biophys. J.*, **71**, 2724–2732.

158. Duong, T. Q., Ackerman, J. J., Ying, H. S., and Neil, J. J. (1998) Evaluation of extra- and intracellular apparent diffusion in normal and globally ischemic rat brain via 19F NMR, *Magn. Reson. Med.*, **40**, 1–13.

159. Duong, T. Q., Sehy, J. V., Yablonskiy, D. A., Snider, B. J., Ackerman, J. J. H., and Neil, J. J. (2001) Extracellular apparent diffusion in the rat brain, *Magn. Reson. Med.*, **45**, 801–810.

160. Le Bihan, D. (2007) The "wet mind": water and functional neuroimaging, *Phys. Med. Biol.*, **52**, R57–R90.

161. Clark, C. A., and Le Bihan, D. (2000) Water diffusion compartmentation and anisotropy at high b values in the human brain, *Magn. Reson. Med.*, **44**, 852–859.

162. Flint, J. (1999) The genetic basis of cognition, *Brain*, **122**, 2015–2031.

163. Flint, J., and Hansen, B., Vestergaard-Poulsen, P., Blackband, S. J. (2009) Diffusion weighted magnetic resonance imaging of neuronal activity in the hippocampal slice model, *NeuroImage*, **46**, 411–418.

164. Inoue, H., and Okada, Y. (2007) Roles of volume-sensitive chloride channel in excitotoxic neuronal injury, *J. Neurosci.*, **27**, 1445–1455.

165. Inoue, I., Kobatake, Y., and Tasaki, I. (1973) Excitability, instability and phase transitions in squid axon membrane under internal perfusion with dilute salt solutions, *Biochim. Biophys. Acta*, **307**, 471–477.

166. Hagiwara, S., Chichibu, S., and Naka, K. I. (1964) The effects of various ions on resting and spike potentials of barnacke muscle fibers, *J. Gen. Physiol.*, **48**, 163–179.

167. Tasaki, I. (1999) Rapid structural changes in nerve fibers and cells associated with their excitation processes, *Jpn. J. Physiol.*, **49**, 125–138.

168. Naundorf, B., Wolf, F., and Volgushev, M. (2006) Unique features of action potential initiation in cortical neurons, *Nature*, **20**, 1060–1063.

169. Cohen, L. B., and Keynes, R. D. (1968) Evidence for structural changes during the action potential in nerves from the walking legs of Maia squinado, *J. Physiol.*, **194**, 85–6P.

170. Cohen, L. B., Keynes, R. D., and Hille, B. (1968) Light scattering and birefringence changes during nerve activity, *Nature*, **218**, 438–441.

171. Tasaki, I., and Byrne, P. M. (1993) The origin of rapid changes in birefringence, light scattering and dye absorbance associated with excitation of nerve fibers, *Jpn. J. Physiol.*, **43**(Suppl. 1), S67–S75.

172. Iwasa, K., and Tasaki, I. (1980) Mechanical changes in squid giant axons associated with production of action potentials, *Biochem. Biophys. Res. Commun.*, **95**, 1328–1331.

173. Iwasa, K., Tasaki, I., and Gibbons, R. C. (1980) Swelling of nerve fibers associated with action potentials, *Science*, **210**, 338–339.

174. Andrew, R. D., and Macvicar, B. A. (1994) Imaging cell volume changes and neuronal excitation in the hippocampal slice, *Neuroscience*, **62**, 371–383.

175. Schwartzkroin, P. A., Baraban, S. C., and Hochman, D. W. (1998) Osmolarity, ionic flux, and changes in brain excitability, *Epilepsy Res.*, **32**, 275–285.

176. Aitken, P. G., Fayuk, D., Somjen, G. G., and Turner, D. A. (1999) Use of intrinsic optical signals to monitor physiological changes in brain tissue slices, *Methods*, **18**, 91–103.

177. Rothman, S. M. (1985) The neurotoxicity of excitatory amino acids is produced by passive chloride influx, *J. Neurosci.*, **5**, 1483–1489.

178. Lux, H. D., Heinemann, U., and Dietzel, I. (1986) Ionic changes and alterations in the size of the extracellular space during epileptic activity, *Adv. Neurol.*, **44**, 619–639.

179. Meyer, F. B. (1989) Calcium, neuronal hyperexcitability and ischemic injury, *Brain Res. Rev.*, **14**, 227–243.

180. Holthoff, K., and Witte, O. W. (1998) Intrinsic optical signals in vitro: a tool to measure alterations in extracellular space with two-dimensional resolution, *Brain Res. Bull.*, **47**, 649–655.

181. Holthoff, K., and Witte, O. W. (1996) Intrinsic optical signals in rat neocortical slices measured with near-infrared dark-field microscopy reveal changes in extracellular space, *J. Neurosci.*, **16**, 2740–2749.

182. Takagi, S., Obata, K., and Tsubokawa, H. (2002) GABAergic input contributes to activity-dependent change in cell volume in the hippocampal CA1 region, *Neurosci. Res.*, **44**, 315–324.

183. Andrew, R. D., Fagan, M., Ballyk, B. A., and Rosen, B. (1989) Seizure susceptibility and the osmotic state, *Brain Res.*, **498**, 175–180.

184. Jefferys, J. G. R. (1995) Nonsynaptic modulation of neuronal activity in the brain: electrical currents and extracellular ions, *Physiol. Rev.*, **75**, 689–715.

185. Dudek, F. E., Yasumura, T., and Rash, J. E. (1998) 'Non-synaptic' mechanisms in seizures and epileptogenesis, *Cell Biol. Int.*, **22**, 793–805.

186. Muller, V., Birbaumer, N., Preissl, H., Braun, C., and Lang, F. (2002) Effects of water on cortical excitability in humans, *Eur. J. Neurosci.*, **15**, 528–538.

187. Manz, B., Chow, P. S., and Gladden, L. F. (1999) Echo-planar imaging of porous media with spatial resolution below 100 μm, *J. Magn. Reson.*, **136**, 226–230.

188. Inoue, H., Mori Si, Morishima, S., and Okada, Y. (2005) Volume-sensitive chloride channels in mouse cortical neurons: characterization and role in volume regulation, *Eur. J. Neurosci.*, **21**, 1648–1658.

189. Macvicar, B. A., and Hochman, D. (1991) Imaging of synaptically evoked intrinsic optical signals in hippocampal slices, *J. Neurosci.*, **11**, 1458–1469.

190. Macvicar, B. A., Feighan, D., Brown, A., and Ransom, B. (2002) Intrinsic optical signals in the rat optic nerve: role for K+ uptake via NKCC1 and swelling of astrocytes, *Glia*, **37**, 114–123.

191. Murase, K., Saka, T., Terao, S., Ideka, H., and Asai, T. (1998) Slow intrinsic optical signals in the rat spinal cord dorsal horn slice, *NeuroReport*, **9**, 3663–3667.

192. Holthoff, K., and Witte, O. W. (2000) Directed spatial potassium redistribution in rat neocortex, *Glia*, **29**, 288–292.

193. Ransom, B. R., Yamate, C. L., and Connors, B. W. (1985) Activity-dependent shrinkage of extracellular space in rat optic nerve: a developmental study, *J. Neurosci.*, **5**, 532–535.

194. Churchwell, K. B., Wright, S. H., Emma, F., Rosenberg, P. A., and Strange, K. (1996) NMDA receptor activation inhibits neuronal volule regulation after swelling induced by veratridine-stimulated NA+ influx in rat cortical cultures, *J. Neurosci.*, **16**, 7447–7457.

195. Tasaki, I., and Iwasa, K. (1982) Further studies of rapid mechanical changes in squid giant axon associated with action potential production, *Jpn. J. Physiol.*, **32**, 505–518.

196. Tasaki, I., Kusano, K., and Byrne, P. M. (1989) Rapid mechanical and thermal changes in the garfish olfactory nerve associated with a propagated impulse, *Biophys. J.*, **55**, 1033–1040.

197. Tasaki I, Kusano K, Byrne PM (1989) Rapid mechanical and thermal changes in the garfish olfactory nerve associated with a propagated impulse, *Biophys J.*, **55**, 1033–1040.

198. Aso, T., Urayama, S., Poupon, C., Sawamoto, N., Fukuyama, H., and Le Bihan, D. (2009) An intrinsic diffusion response function for analyzing diffusion functional MRI time series, *NeuroImage*, **47**, 1487–1495.

199. Abbott, B. C. (1958) The positive and negative heat production associated with a single impulse, *Proc. R. Soc. London B*, **148**, 149–187.

200. Howarth, J. V., Ritchie, J. M., and Stagg, D. (1979) The initial heat production in garfish olfactory nerve fibres, *Proc. R. Soc. London B*, **205**, 347–367.

201. Howarth, J. V., Keynes, R. D., and Ritchie, J. M. (1968) The origin of the initial heat associated with a single impulse in mammalian non-myelinated nerve fibres, *J. Physiol.*, **194**, 745–793.

202. Tasaki, I., and Byrne, P. M. (1991) Demonstration of heat production associated with spreading depression in the amphibian retina, *Biochem. Biophys. Res. Commun.*, **174**, 293–297.

203. Kusano, K., and Tasaki, I. (1990) Heat generation associated with synaptic transmission in the mammalian superior cervical ganglion, *J. Neurosci. Res.*, **25**, 249–255.

204. Tasaki, I., and Byrne, P. M. (1987) Heat production associated with synaptic transmission in the bullfrog spinal cord, *Brain Res.*, **407**, 386–389.

205. Tasaki, I., and Iwasa, K. (1981) Temperature changes associated with nerve excitation: detection by using polyvinylidene fluoride film, *Biochem. Biophys. Res. Commun.*, **101**, 172–176.

206. Attwell, D., and Gibb, A. (2005) Neuroenergetics and the kinetic design of excitatory synapses, *Nat. Rev. Neurosci.*, **6**, 841–849.

207. Williams, R. J. P. (1970) The biochemistry of sodium, potassium, magnsium and calcium, *Chem. Soc. Q. Rev.*, **24**, 331–365.

208. Tasaki, I., and Byrne, P. M. (1990) Volume expansion of nonmyelinated nerve fibers during impulse conduction, *Biophys. J.*, **57**, 633–635.

209. Kusano, K., and Tasaki, I. (1990) Mechanical changes associated with synaptic transmission in the mammalian superior cervical ganglion, *J. Neurosci. Res.*, **25**, 243–248.

210. Sato, H., Tasaki, I., Carbone, E., and Hallett, M. (1973) Changes in the axon birefringence associated with excitation: implications for the structure of the axon membrane, *J. Mechanochem. Cell Motil.*, **2**, 209–217.

211. Tsukita, S., Tsukita, S., Kobayashi, T., and Matsumoto, G. (1986) Subaxolemmal cytoskeleton in squid giant axon. II. Morphological identification of microtubule- and microfilament-associated domains of axolemma, *J. Cell Biol.*, **102**, 1710–1725.

212. Metuzals, J., and Tasaki, I. (1978) Subaxolemmal filamentous network in the giant nerve fiber of the squid (Loligo pealei L.) and its possible role in excitability, *J. Cell Biol.*, **78**, 597–621.

213. Endo, S., Sakai, H., and Matsumoto, G. (1979) Microtubules in squid giant axon, *Cell Struct. Funct.*, **4**, 285–293.

214. Tasaki, I., Singer, I., and Takenaka, T. (1965) Effects of internal and external ionic environment on excitability of squid giant axon. A macromolecular approach, *J. Gen. Physiol.*, **48**, 1095–1123.

215. Matsumoto, G., Kobayashi, T., and Sakai, H. (1979) Restoration of the excitability of squid giant axon by tubulin-tyrosine ligase and microtubule proteins, *J. Biochem. (Tokyo)*, **86**, 1155–1158.

216. Tasaki, I., and Byrne, P. M. (1992) Rapid structural changes in nerve fibers evoked by electric current pulses, *Biochem. Biophys. Res. Commun.*, **188**, 559–564.

217. Tasaki, I., and Byrne, P. M. (1994) Optical changes during nerve excitation: interpretation on the basis of rapid structural changes in the superficial gel layer of nerve fibers, *Physiol. Chem. Phys. Med. NMR*, **26**, 101–110.

218. Tasaki, I., and Byrne, P. M. (1992) Discontinuous volume transitions in ionic gels and their possible involvement in the nerve excitation process, *Biopolymers*, **32**, 1019–1023.

219. Tasaki, I. (2005) Repetitive abrupt structural changes in polyanionic gels: a comparison with analogous processes in nerve fibers, *J. Theor. Biol.*, **236**, 2–11.

220. Collins, C. M., Smith, M. B., and Turner, R. (2004) Model of local temperature changes in brain upon functional activation, *J. Appl. Physiol.*, **97**, 2051–2055.

221. Spyropolous, C. S. (1961) Initiation and abolition of electric response by thermal and chemical means., *Am. J. Physiol.*, **200**, 203–208.

222. Yablonskiy, D. A., Ackerman, J. J. H., and Raichle, M. E. (2000) Coupling between changes in human brain temperature and oxidative metabolism during prolonged visual stimulation, *PNAS*, **97**, 7603–7608.

223. Van Leeuwen, G. M., Hand, J. W., Kagendijk, J. J., Azzopardi, D. V., and Edwards, A. D. (2000) Numerical modeling of temperature distributions within the neonatal head, *Pediatr. Res.*, **48**, 351–356.

224. Gorbach, A. M., Heiss, J., Kufta, C., Sato, S., Fedio, P., Kammerer, W. A., Solomon, J., and Oldfield, E. H. (2003) Intraoperative infrared functional imaging of human brain, *Ann. Neurol.*, **54**, 297–309.

225. Stroman, P. W., Lee, A. S., Pitchers, A., and Andrew, R. D. (2008) Magnetic resonance imaging of neuronal and glial swelling as an indicator of function in cerebral tissue slices, *Magn. Reson. Med.*, **59**, 700–706.

226. Crick, F. (1982) Do dentritic spines twitch? *Tins*, 44–47.

227. Tasaki, I. (1999) Rapid structural changes in nerve fibers and cells associated with their excitation processes, *Jpn. J. Physiol.*, **49**, 125–138.

Chapter 17

WATER AND BRAIN

Hidenao Fukuyama

Human Brain Research Center, Kyoto University Graduate School of Medicine,
54 Shogoin Kawahara-cho, Sakyo-ku, Kyoto 606-8507, Japan
fukuyama@kuhp.kyoto-u.ac.jp

17.1 INTRODUCTION

About 70% of the human body consists of water, and it contains about one-third of the concentration of seawater minerals. This fact means that the internal milieu of the human body, as well as of many other animals, has an internal sea within the body, derived from the long history of evolution of life on this earth. This fact is important to medical doctors caring for patients, because patients lose water through the kidney, combined with metabolic waste products, in order to maintain homeostasis.

17.2 CONTROL OF WATER CONCENTRATION IN THE BODY BY THE PITUITARY GLAND

One of the very important functions of the brain is to maintain a constant concentration of body water by appropriate secretion of antidiuretic hormone (ADH) from the posterior lobe of the pituitary gland. ADH is measured by immunological or biological tests, but these tests cannot determine ADH accumulation in the pituitary gland. Dr. Fujisawa, Department of Radiology, Kyoto University [1], found a high intensity in the pituitary gland on T1-weighted images (Fig. 17.1) on the saggital plane, suggesting visualization of the secreted hormone by magnetic resonance imaging (MRI). Using this finding, we can identify ADH secretion from the posterior lobe of the pituitary gland appropriately to evaluate function, organic conditions (tumor, bleeding) [2, 3], and other findings.

Water: The Forgotten Biological Molecule
Edited by Denis Le Bihan and Hidenao Fukuyama
Copyright © 2011 by Pan Stanford Publishing Pte. Ltd.
www.panstanford.com

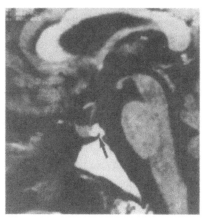
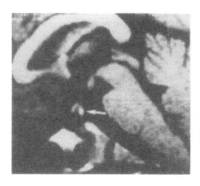

a   b

Figure 17.1 Pituitary gland. (a) Normal pituitary gland. The arrow indicates the posterior lobe showing high intensity. (b) A 10-year-old boy with growth hormone deficiency and normal ADH secretion. Transsection of the stalk at the level of the diaphragma sellae and the characteristic high signal intensity of the enlarged proximal stump (white arrow), as well as disappearance of the posterior gland. Figure courtesy of Prof. Yukio Miki, Department of Radiology, Osaka City University Graduate School of Medicine.

It has not yet completely been established why these are high signals in the posterior lobe due to ADH. There have been several explanations: magnetization transfer, high hormonal concentration in the secreting vesicle, and so on.

In cases of bleeding within the pituitary gland [4] or after surgery in the subarachnoid space, ADH secretion reduces, followed by diabetes insipidus. It was difficult to treat this condition, but ADH has recently. Recently identified disease involves syndrome of inappropriate secretion of ADH (SIADH) despite sufficient water volume in the body [5], causing hyponatremia. Despite severe hyponatremia, urinary sodium excretion was not suppressed and serum osmolality was lower than urine osmolality and arginine vasopressin (AVP) remained within the normal range. It is necessary to rule out pseudo SIADH, caused by a high volume of water intake, adrenal insufficiency, or renal dysfunction. Under this condition, it is important to check the level of sodium concentration, such as hyponatremia, especially when the condition is less than 110 mEq/L. Central pontine or extrapontine myelinolysis was regarded as an idiopathic disease, but rapid correction of hyponatremia has been identified as a cause of such localized demyelination [6–8]. These clinical abnormalities originate from dysfunction of the posterior lobe of the pituitary gland, which maintains homeostasis throughout the whole body.

17.3 CEREBROSPINAL FLUID CIRCULATION

The brain is the only organ that does not use lymph to maintain cell functions. Instead, cerebrospinal fluid (CSF) circulates from the choroid plexus of the lateral and third ventricle through the aqueduct of the midbrain, the fourth ventricle, and foramen Lushka or Magendie to the suarachnoid space. CSF circulates to the spinal cord and returns to the ventral arachnoid space of the brainstem. Finally, CSF circulates to the brainstem, cerebellum, and the cerebral cortex and is absorbed into Paccioni bodies (Fig. 17.2).

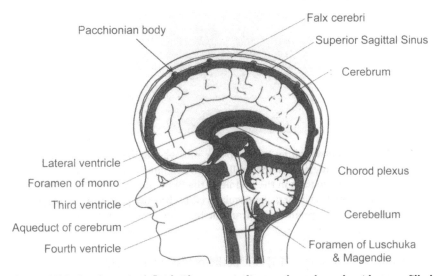

Figure 17.2 Cerebrospinal fluid. Blue area indicates the subarachnoid space filled with CSF. Lines indicate the flow of CSF. For color reference, see p. 392.

In addition, CSF circulates through the brain parenchyma through Virchow Robin spaces, which are located around the small penetrating arteries, and maintains the condition of the brain cells, just like lymph does in other organs.

This circulating system has several problems. In the developmental stage, the aqueduct of the midbrain has a thin membrane that prevents smooth CSF flow from the third ventricle to the fourth ventricle, resulting in flow disturbance and hydrocephalus, which because of the obstruction by a tumor or other pathology is known as obstructive hydrocephalus. The former type of hydrocephalus is called communicating hydrocephalus, and aqueduct obstruction can been seen on MRI as a flow void phenomenon. This condition does not necessarily produce symptoms, but advances in MRI have facilitated diagnosis.

An important issue related to CSF is disturbance of absorption from Paccioni bodies. Accordingly, communicating hydrocephalus will occur and produce cognitive decline combined with urinary incontinence and gait disturbance. The disease entity was proposed by Adams *et al.*, who also indicated that affected persons can be successfully managed by shunting of CSF from the ventricle to the abdomen. In those days, there was no way to diagnose such changes easily. In the 1970s, radioisotope examination, RI-cisternography [9], was developed, and absorption abnormalities could then be objectively diagnosed by showing the radioisotope accumulation in the lateral ventricle (Fig. 17.3).

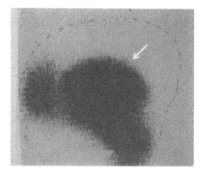

Figure 17.3 RI-cisternography. In-111-DTPA injected into the subarachnoid space of the lumbar region going up into the skull. The dotted line indicates the lateral view of the skull. High accumulation in the middle of the brain indicates the lateral ventricle (white arrow), which is called ventricular reflux. Normal-pressure hydrocephalus involves deficient absorption of CSF from the high convexity and then reflux into the lateral ventricle.

Recent MRI studies [10] of pathological abnormalities in CSF flow demonstrated high convexity of the cerebrum attached to the skull due to various etiologies, reflecting disturbance of CSF absorption into Paccioni bodies. This finding is easily recognized on a coronal section of the brain (Fig. 17.4).

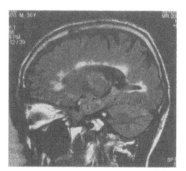

Figure 17.4 Coronal view of MRI in idiopathic normal-pressure hydrocephalus. The upper part of the brain parenchyma (above the level of the Sylvian fissure; arrow) contacts to the skull tightly.

17.4 NEUROMYELITIS OPTICA, OPTIC-SPINAL FORM OF MS

Since the introduction of computed tomography (CT), multiple sclerosis (MS) showed abnormal intensities around the lateral ventricle. There was no definite evidence to prove that such abnormal findings indicated a pathological process. Recent advances in immunology have demonstrated a specific type of MS, which shows a long spinal cord lesion with optic myelitis, and the antibody for aquapoline 4 was detected in 2006 [11]. The findings of neuromyelitis optica on MRI and clinical findings are specifically different from those in the common type of MS [12, 13].

However, this finding leads to diagnostic evidence for MS on MRI or CT, which show abnormal intensities around the lateral ventricle, especially the corpus callosum showing firelike high intensities on sagittal T2-weighted images. According to this finding, antibodies circulating through CSF cause pathological lesions around the ventricle (Fig. 17.5).

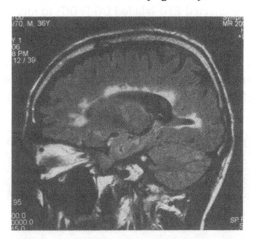

Figure 17.5 MRI of MS. Sagittal T2-weighted image indicates flairlike high intensities on the corpus callosum, as well as other periventricular area.

17.5 NEUROVASCULAR COUPLING

Brain circulation is thought to be controlled by local neural activities. This hypothesis was proposed by Roy and Sherrington in 1890 [14]. Since then, neuroscientists have measured cerebral blood flow (CBF) instead of neural activities, which cannot be clearly recorded by the noises from various origins. Roy and Sherrington showed evidence of coupling neuronal activities with CBF, but there has not been any biochemical explanation until recently.

PET studies of ischemic brain lesions indicated the uncoupling of CBF with neuronal activity, such as luxury perfusion syndrome [15] and misery perfusion syndrome [16].

We have performed research using PET and MRI to investigate this hypothesis in an animal model. We found initially that an administration of an anticholinergic drug, atropine, abolished the blood flow control [17]. More precise examination using bromo-pyruvate, which blocks the synthesis of acetylcholine on the membrane of mitochondria [18], was then performed. Microinjection of bromo-pyruvate into the local brain area, where acetylcholine will not be synthesized, can demonstrate the importance of acetylcholine in coupling CBF and neuronal activities. We stimulated the forepaw of a cat and activated the sensorimotor cortex. After measuring local CBF increase using animal PET with O15-labeled water, we injected bromo-pyruvate into the activated area and measured CBF again after bromo-pyruvate injection. We found that there is no increase of CBF in the same area (Fig. 17.6). After measuring CBF, we injected F18-labeled DG (FDG) to measure the neuronal activities. FDG showed increased glucose metabolism, while there was no increase in CBF [19].

According to recent evidence, cholinergic projection from the nucleus basalis of Meynert, or small intrinsic cholinergic neurons of the cortex, may function to regulate CBF appropriately in response to the demand of the nervous system [20].

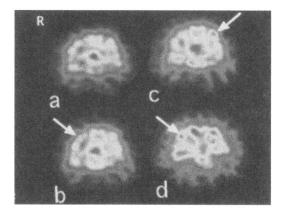

Figure 17.6 CBF on animal PET images. (a) Coronal image of the sensory cortex at rest. (b) Stimulating the left forepaw. There was no activation on the right side of the brain after injection of bromo-pyruvate into the sensory cortex. (c) Contralateral side stimulation (right forepaw stimulation) induced CBF increase on the left side of the cortex, which had not received any treatment. (d) Right as well as left sensory cortex showed high accumulation of FDG after right-side injection of bromo-pyruvate. R: right side of the cat brain. For color reference, see p. 392.

Another important hypothesis has been proposed. Astrocytes regulate CBF to maintain the function of the nervous system [21]. This hypothesis will also explain diffusion hyperintensity in acute ischemic stroke. Astrocytes absorb water from blood vessels, as well as other nutrients, to reserve the energy requirements of the neurons. This water increase explains the hyperintensity on diffusion MRI in acute stroke immediately after onset (Fig. 17.7).

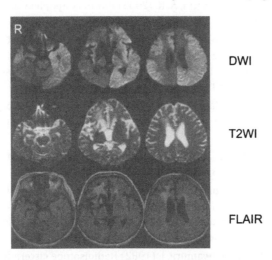

Figure 17.7 MRI in acute stroke. Old infarction of the right frontal lobe was shown on T2-WI and FLAIR, but DWI showed high intensities on the left, which indicate that acute stroke occurred on the left hemisphere. *Abbreviations*: DWI, diffusion weighted image; T2-WI, T2-weighted image; FLAIR, fluid-attenuated inversion recovery; R, right hemisphere.

17.6 CONCLUSION

Water has various functions in the brain as well as in body organs. Water molecules are the most basic component of living cells, not only for animals, but also for plants. There are many things that remain to be clarified in the future.

References

1. Fujisawa, I., Asato, R., Kawata, M., Sano, Y., Nakao, K., Yamada, T., Imura, H., Naito, Y., Hoshino, K., and Noma, S. (1989) Hyperintense signal of the posterior pituitary on T1-weighted MR images: an experimental study, *J. Comput. Assist. Tomogr.*, **13**, 371–377.

2. Iida, M., Takamoto, S., Masuo, M., Makita, K., and Saito, T. (2003) Transient lymphocytic panhypophysitis associated with SIADH leading to diabetes insipidus after glucocorticoid replacement, *Intern. Med.*, **42**, 991–995.

3. Sato, N., Endo, K., Ishizaka, H., and Matsumoto, M. (1993) Serial MR intensity changes of the posterior pituitary in a patient with anorexia nervosa, high serum ADH, and oliguria, *J. Comput. Assist. Tomogr.*, **17**, 648–650.

4. Agrawal, D., and Mahapatra, A. K. (2003) Pituitary apoplexy and inappropriate ADH secretion, *J. Clin. Neurosci.*, **10**, 260–261.

5. Iwai, H., Ohno, Y., Hoshiro, M., Fujimoto, M., Nishimura, A., Kishitani, Y., and Aoki, N. (2000) Syndrome of inappropriate secretion of antidiuretic hormone (SIADH) and adrenal insufficiency induced by rathke's cleft cyst: a case report, *Endocr. J.*, **47**, 393–399.

6. Bibl, D., Lampl, C., Gabriel, C., Jungling, G., Brock, H., and Kostler, G. (1999) Treatment of central pontine myelinolysis with therapeutic plasmapheresis, *Lancet*, **353**, 1155.

7. Menger, H., and Jorg, J. (1999) Outcome of central pontine and extrapontine myelinolysis (*n* = 44), *J. Neurol.*, **246**, 700–705.

8. Nagamitsu, S., Matsuishi, T., Yamashita, Y., Yamada, S., and Kato, H. (1999) Extrapontine myelinolysis with parkinsonism after rapid correction of hyponatremia: high cerebrospinal fluid level of homovanillic acid and successful dopaminergic treatment, *J. Neural. Transm.*, **106**, 949–953.

9. Fukuyama, H., and Kawamura, J. (1982) Radioisotope cisternography in acute viral encephalitis. A reappraisal, *Arch. Neurol.*, **39**, 293–297.

10. Kitagaki, H., Mori, E., Ishii, K., Yamaji, S., Hirono, N., and Imamura T. (1998) CSF spaces in idiopathic normal pressure hydrocephalus: morphology and volumetry, *Am. J. Neuroradiol.*, **19**, 1277–1284.

11. Misu, T., Fujihara, K., Nakamura, M., Murakami, K., Endo, M., Konno, H., and Itoyama, Y. (2006) Loss of aquaporin-4 in active perivascular lesions in neuromyelitis optica: a case report, *Tohoku. J. Exp. Med.*, **209**, 269–275.

12. Nakamura, M., Miyazawa, I., Fujihara, K., Nakashima, I., Misu, T., Watanabe, S., Takahashi, T., and Itoyama, Y. (2008) Preferential spinal central gray matter involvement in neuromyelitis optica. An MRI study, *J. Neurol.*, **255**, 163–170.

13. Wingerchuk, D. M. (2007) Diagnosis and treatment of neuromyelitis optica, *Neurologist*, **13**, 2–11.

14. Roy, C. S., and Sherrington, C. S. (1890) On the Regulation of the Blood-supply of the Brain, *J. Physiol.*, **11**, 17, 85–158.

15. Lassen, N. A. (1966) The luxury-perfusion syndrome and its possible relation to acute metabolic acidosis localised within the brain, *Lancet*, **2**, 1113–1115.

16. Baron, J. C., Bousser, M. G., Rey, A., Guillard, A., Comar, D., and Castaigne, P. (1981) Reversal of focal "misery-perfusion syndrome" by extra-intracranial arterial bypass in hemodynamic cerebral ischemia. A case study with 150 positron emission tomography, *Stroke*, **12**, 454–459.

17. Ogawa, M., Magata, Y., Ouchi, Y., Fukuyama, H., Yamauchi, H., Kimura, J., Yonekura, Y., and Konishi, J. (1994) Scopolamine abolishes cerebral blood flow response to somatosensory stimulation in anesthetized cats: PET study, *Brain. Res.*, **650**, 249–252.

18. Gibson, G. E., Jope, R., and Blass, J. P. (1975) Decreased synthesis of acetylcholine accompanying impaired oxidation of pyruvic acid in rat brain minces, *Biochem. J.*, **148**, 17–23.

19. Fukuyama, H., Ouchi, Y., Matsuzaki, S., Ogawa, M., Yamauchi, H., Nagahama, Y., Kimura, J., Yonekura, Y., Shibasaki, H., and Tsukada, H. (1996) Focal cortical blood flow activation is regulated by intrinsic cortical cholinergic neurons, *NeuroImage*, **3**, 195–201.

20. Peruzzi, P., Von Euw, D., Correze, J. L., and Lacombe, P. (2007) Attenuation of the blood flow response to physostigmine in the rat cortex deafferented from the basal forebrain, *Brain Res. Bull.*, **72**, 66–73.

21. Parri, R., and Crunelli, V. (2003) An astrocyte bridge from synapse to blood flow, *Nat. Neurosci.*, **6**, 5–6.

Chapter 18

WATER, A KEY MOLECULE FOR LIVING, WATER IN METABOLISM AND BIODIVERSITY

Gilles Boeuf

Université Pierre et Marie Curie, Paris6/CNRS, Laboratoire Arago, Avenue du Fontaulé,
66650 Banyuls-sur-mer and Muséum national d'Histoire naturelle,
57 rue Cuvier, 75005, Paris, France
gboeuf@obs-banyuls.fr, boeuf@mnhn.fr

18.1 INTRODUCTION

Planet Earth is unique in our solar system, its size and the distance from the sun allowing a crucial significant range of temperature and the existence of both liquid water and water vapor. Solid water, as ice, also exists in cold conditions, mainly near the poles and at altitudes, but Earth is dominated by land masses between oceans and water vapor above them. Water, as we already saw before in previous chapters of this book, is actually the key molecule for the living. Life appeared and developed in water. Living beings are constituted of water. Without water, a land rapidly becomes a desert with very poor and specialized life, or without life. Water is life: take a look at the periphery of a Chilean desert, which receives rainfall only every 10–12 years. In a matter of days it is covered by flowers (with the requisite cortege of insects!) that last for a few weeks, the temporary ecosystem returning to its extreme arid state for the next years. This is a natural phenomenon, but humans can also cause an explosion of life by irrigating the desert. So, life emerged from water and is constituted by water! We propose to successively present here the relationships between water and life origin and emergence from the ocean, water in biodiversity, and the major facts in terms of water and ion metabolism before concluding.

Water: The Forgotten Biological Molecule
Edited by Denis Le Bihan and Hidenao Fukuyama
Copyright © 2011 by Pan Stanford Publishing Pte. Ltd.
www.panstanford.com

18.2 WATER AND ANCIENT LIFE

The age of the earth has been estimated at 4.6 billion years. Life appeared in the oceans relatively rapidly after the initial cooling and condensation of water masses, which occurred approximately 3.9 billion years ago. Christian de Duve, the 1974 Nobel laureate in physiology and medicine, wrote in his 1995 book *Vital Dust: Life as a Cosmic Imperative* that the earth was in such an ideal position with regard to the sun that the emergence of life was inevitable [1]. The most ancient sedimentary rocks known to man (on Akilia Island in southern Greenland) contain originated organic carbon dating back 3.85 billion years. Primitive life must be conceived as being very simple at the start, emerging from a world of RNA and protocellular structures [2]. The current deposits of stromatolites in Australia are very precious because their silicified segments contain the oldest fossils of microorganisms, cyanobacteria. A small piece of stromatolite encodes biological activity perhaps representing thousands of years. In broad terms, stromatolites are fossil evidence of the prokaryotic life that remains today, as it has always been, the major source of biomass in the biosphere. Cyanobacteria belong to the first ocean colonies of life that emerged approximately between 3.7 and 3.4 billion years ago, when there was no oxygen in the atmosphere. With the presence of specific cell pigments, photosynthesis developed, producing in the water environment oxygen and sugars from light and carbon dioxide (CO_2). This process appeared on earth around 3.5 billion years ago. Oxygen began to diffuse outside the aquatic environment around 3.2 billion years ago. So, our planet has the only oxygen-rich atmosphere in the solar system, though this is primarily a consequence of the presence of life rather than a prerequisite for life to have evolved. The biosphere has been able to develop from the lithosphere, the hydrosphere, and the atmosphere.

In these ancient seas, some events were crucial for the outcome of life and biodiversity: (i) the development of the nuclear membrane and the individual nucleus (transition from prokaryote to eukaryote status) around 2.2 billion years ago, (ii) the emergence of multicellular organisms and metazoans around 2.1 million years ago, and (iii) the capture of surrounding cyanobacteria that would become symbionts and cell organelles such as mitochondria and plastids with their own small DNA, around 2 billion years and 1.4 billion years ago, respectively. In addition, something exceptional happened in this ancestral sea: the emergence of sexual reproduction (about 1.5 billion years ago), which would turn out to be extremely important, the "best engine," for the development of biodiversity.

Organized metazoan life emerged from the oceans after the explosion of oceanic Cambrian life 550 million years ago (Fig. 18.1).

Period	Age (Myr)	Event
Pliocene	5.3–1.7	Earliest humans
Paleocene	66–58	Radiation of mammals and birds, flowering plants, pollinating insects
Cretaceous-Paleocene	66	Fifth great crisis: extinction of dinosaurs
Triassic	245–208	First dinosaurs, first mammals
Permian-Triassic	245	Third great crisis of extinction
Carboniferous	360–300	Radiation of land life
Devonian	408–360	First amphibians
Silurian	438–408	First arthropods on lands
Ordovician	489–438	First land plants
Precambrian	550	Precambrian radiation in the sea
	800	First cyanobacteria on lands
	1900, 1400	Cell organelles (mitochondria, plastids)
	2200	Cell nucleus, eukaryotic life
	3200–2500	Release of O2 from the ancestral ocean
	3900	Emergence of life, only prokaryotic life
	4600	Age of the Earth

Figure 18.1 Geological time scale and life major events.

The first terrestrial plant life (first vascular plants, as bryophytes during the Ordovician, 475 million years ago) developed on ancient paleosoils, and land animals (including arthropods and vertebrates) left their mark on the continents approximately 430 million years ago at the end of the Silurian period. Initially, the first terrestrial plants remained near the coast, protected from and preventing dehydration. At the end of the Devonian period, 360 million years ago, the plant cover was organized in the same way as today, of course with different groups, but all strata were present. Vertebrates went through an amphibious stage (*dipnoi*, lungfish, *rhipidistia*, *Ichthyostega*, etc.) oscillation between arid and rainy periods on the *old red sandstone continent*. The settlement of a terrestrial plant cover obviously played a crucial role for the possibility of land conquests by faunas. The first myriapods appeared during the Silurian period and *chelicerata* during the Devonian period (Fig. 18.1) and were able to resist dehydration and high ultraviolet (UV) radiation, hiring and probably developing nocturnal activity. Like plants and tetrapods, arthropods were able to emerge several times from water during the evolution [3]. Huge expansion of the terrestrial living so occurred during the Devonian period and mostly during the Carboniferous period (360–300 million years).

An abundant entomofauna was already present in the deep wet forests, and huge forms existed such as dragonflies with a 75 cm wingspan, scorpions and centipedes several meters long, and so forth!

Numerous new adaptations developed in plants as well as in animals because the shift to terrestrial life and air breathing was an exceptional development in the history of life. After lichens and mosses, new vegetal forms developed roots able to search and absorb water and nutritive elements from the soil, cuticles against dehydration, lignin for vertical carriage, and stomates for transpiration and gas exchanges. The distinction between aquatic and aerial organisms is fundamental for the physiologist. Due to the specific density and viscosity of water, the energy required for ventilation is much higher for aquatic animals: it represents only 1–2% of the energetic budget of humans, whereas it is 5% to more than 30% for fish, depending on species and conditions. Very oxyphilic species like huge pelagic fish, for example, are very high demanders. They are able to reach extremely rapid speeds in swimming (>130 km/h). Small animal species draw oxygen from the water by diffusion toward their deepest parts, while larger species (or stages) use gills. Seawater, equilibrated with air, contains approximately 30 times less oxygen than the same volume of air. Aquatic animals maintaining a constant osmolarity cannot develop a very large surface area for exchange (gills) because of the inherent dangers of physical osmotic flow (water and electrolytes): the animal has the risk of losing its water in the salty ocean or of being "drowned" in freshwater. In fact, fish are constantly subjected to a delicate compromise between developing a maximum surface area for oxygen capture in a poorly oxygenated and changing environment or a minimum surface area to avoid a severe loss or gain (depending on the environment) of water and mineral equilibrium. Aquatic animals secrete ammonia, and the vast majority, as for terrestrial species, cannot regulate their body heat. Terrestrial animals on the other hand face UV rays, dehydration, and different weight-bearing requirements (requiring a heavier and more resistant skeleton and a consequently heavier muscle mass) and have to develop a different type of excreta with little or no toxicity (uric acid, urea). Much later, during the Triassic period, approximately 210 million years ago and following the third large extinction, the first models of thermoregulation emerged and were optimally efficient in large dinosaurs and then especially in birds and mammals. On the opposite, a very good example of returning to the ocean is illustrated by the Cetaceans, who began this reacclimatization to marine life, from primitive artiodactyls closely related to *Diacodexis* that were first terrestrial and then amphibian (such as *Ambulocetus* and *Pakicetus*) around 50–55 million years ago. The current gigantic forms, which are the largest animals that ever inhabited the planet since the origins of life (and which humans have cheerfully hunted down and killed for the past 160 years), have a recent origin.

Today, scientists are searching for traces of extraterrestrial life by focusing their efforts on RNA, DNA, amino acids, adenosine triphosphate (ATP), and so on, keeping in mind that the first key molecule of life is water. Every living organism is made of water—a low percentage of water in the "driest" organisms, such as plant seeds and animal stages in cryptobiosis (anhydrobiosis, tardigrades, nematods, insect eggs), and up to 98% for certain aquatic species such as algae, jellyfish, and tunicates. The human body is made up of two-thirds of water. Water was therefore a truly determining factor in the history of life. The ocean is salted, primarily with sodium chloride, and was so for a very long time. Today, we understand that this salinity is very stable: billion of tons of cations (calcium, potassium, magnesium, sodium) have always flowed into the oceans from estuaries. The calcium income is caught by marine sediments and the formation of limestone; the potassium is compensated by an adsorption in clay deposits. Magnesium and sodium are taken up by oceanic ridges (serpentinization and clay enrichment of pyroxene and olivine). Bicarbonates are constantly being exchanged in the atmosphere and biosphere. For chlorides, which are not taken up by any large biogeochemistry cycles, it is now believed that chlorine was one of the original volatile elements that dissolved into the first ocean and remained there (no fluvial contributions take place today).

18.3 WATER IN BIODIVERSITY

Today, water constitutes more than 60% of animal bodies. The distribution of body water varies from 60% to more than 90%, depending on the phyla [4, 5]. The total water content varies between 79% and 92 % (extracellular 6–65%, intracellular 27–76%) for mollusks, and 60–81% (extracellular 6–65%, intracellular tissues 50–57% with also extracellular space) for vertebrates. The concentration of urine is also a good indicator of the ability to escape the aquatic environment, changing from 520 mOsm/L to 9,400 from the beaver to the kangaroo rat in the desert. Urine/plasma ratios change from 1.7 to 31.3 [5]. Integumentary evaporation also varies a lot according to the size, the environment, and the species: from 6 mg H_2O per h/cm^2 (for a vapor pressure of 1 mmHg) to 870 in the active snail (only 39 in the rested snail) and 400 for the earthworm [6]. There is a very interesting case in amphibians: they all live in freshwater and are not able to support salinity since their skin is very permeable. Only one species, *Rana cancrivora* (*R. cancrivora*), living in mangroves and eating crabs in South Africa, is able to cope with salty water.

Such species retain urea just like *Elasmobranches* (sharks) do. It remains hyperosmotic in seawater. *R. cancrivora* tadpoles are also very tolerant to salinity even if they need freshwater for their metamorphosis [6]. Marine tetrapods have developed different strategies and can be divided into two groups, reptiles and birds on the one hand and mammals on the other hand. Like teleost fish, the first ones are unable to produce a significantly hyperosmotic urine to eliminate excess salts. So, they have differentiated salt glands allowing them to actively extrude salts: up to 1,170 mM Na^+ and 1,330 mM Cl^- in marine iguana and 1,100 mM NaCl in Leach's petrel [7]. Glands are localized in the nasal cavity (iguanas), eye sockets (turtles), mouth (sea snakes), and tongue (marine crocodile). In birds they are on the top of the skull. Marine mammals are able to produce a very hypertonic urine through a very efficient kidney (elongated loop of Henlé). They avoid to drink seawater and are taking advantage of the osmoregulatory abilities of their prey by choosing only their preregulated fluids (e.g., flesh of the prey) [7].

Today, 14 animal phyla (of the 35 recorded) are exclusively marine and have never left the ocean (echinoderms, brachiopods, *Chaetognatha*, etc., Table 18.1). There are only two groups (and no complete phyla) that are exclusively terrestrial: myriapods and amphibians (apart from one species). In addition, the ocean biomass is considerable: bacteria from the subsurface of the ocean alone represent 10% of the entire carbon-based biomass on the earth [8]. The marine environment has therefore played a determinant role in the history of life, and the current ocean preserves its primordial role in biological and climate evolution.

Table 18.1 Animal phyla exclusively marine

Phylum	Pelagic	Benthic
Placozoa		X
Ctenophora	X	
Kinorhynca		X
Priapulida		X
Loricifera		X
Pogonophora		X
Echiurians		X
Phoronidians		X
Brachiopoda		X
Echinodermata	X	X
Chaetognatha	X	X
Hemichordata		X
Cephalochordata		X
Tunicata		X

Marine and terrestrial diversity are quite different from each other, including freshwater systems. Among the 1.9 million species described and deposited in museums, less than 300,000 live in the oceanic environment. This is low and may be due to two reasons. The first one is that our knowledge of ocean life—particularly with regard to ocean depths and microorganisms, bacteria, and, especially microalgae—is still incomplete: we therefore considerably underestimate the overall ocean biodiversity. New equipments and technologies, such as the combination of flow cytometry and molecular probes used to detect organisms with high specificity, have uncovered an extraordinary and completely unanticipated degree of biological diversity.

"Ocean sequencing," which is the sequencing of all DNA in a volume of filtered oceanic water, tends to the same conclusion as obtained data appear to be largely unknown. For all prokaryotes and very small eukaryotes, recent molecular tools (ribosomal RNA sequencing, including 16S and 18S) are constantly providing an astonishing array of knowledge. As for the second reason, it is clear that marine ecosystems present a continuous and pretty homogenous environment. Species that populate them—through gamete and larval stage dispersion—are less predisposed to strict endemicity than terrestrial biotopes are. There are many more barriers and isolates on lands that can undergo speciation more easily. This leads to significant differences in specific diversity, wherein marine ecological niches do not reach the wealth of diversity of land niches, which are much more compartmentalized and thus encourage the development of new species. On the other hand, the marine biomass is huge. Nowadays we still describe and classify 16,000–18,000 species a year, 10% of which are marine [9].

Today, the erosion of biodiversity increases dramatically [9, 10–13, 20], and diversity has to be protected: oceanic life is very affected by pollution, dissemination of alien invasive species, overexploitation, and climate change [9, 14].

18.4 WATER AND ION MOVEMENTS: OSMOTIC AND IONIC STRATEGIES

All the aspects we developed before have to be kept in mind. In metazoans, all the living equilibriums that always existed between the external environment, the *milieu intérieur* (extracellular fluids), and the intracellular medium keep this life origin story. Why are ions such as Na^+, Cl^-, K^+, Ca^{2+}, or Mg^{2+} so important for the living? Iodine is a typical marine element and is crucial for vertebrate physiology and endocrinology!

But it remains to extrapolate these data to the original composition of seawater. The ocean we know today is very stable and has the same composition since at least the Cretaceous period, 100 million years ago. It is so stable that it has been deposited in Copenhagen an etalon of open seawater, representing the world ocean. Of course, the more we approach the coasts, the more the seawater composition varies according to the local sites (mixed with freshwater to form brackish water). So it is difficult or impossible to specify what "representative" freshwater is (all are different). For open seawater, it is much easier and reliable. The composition of sea water is given in Table 18.2.

Table 18.2 Seawater composition

Anions	g/kg	Cations	g/kg
Cl^-	18.98	Na^+	10.56
$SO4^{2-}$	2.85	Mg^{2+}	1.27
$HCO3^-$	0.14	Ca^{2+}	0.40
Br^-	0.06	K^+	0.38
H_3BO^{3-}	0.03	Sr^{2+}	0.01
F^-	0.001		

Freshwater mainly contains CO_3^-, CO^{2+}, SiO_2, and SO_4^{2-}. For a 35.5 practical salinity unit (psu), 35.5 g NaCl/L seawater, generally admitted as representing the open ocean salinity, the associated osmolarity (for osmotic pressure) is 1,050–1,100 mOsm/L, while it is only 10–20 for freshwater.

Life emerged from the ancestral ocean, probably in shallow mild coastal waters, and was then able to conquer the lands. If we try to compare animal life in water and air, differences are drastic (Table 18.3).

Table 18.3 Major differences between life in water and air (adapted from Ref. [7])

Property	Water/Aquatic	Air/Terrestrial	Approx. Ratio Water/Air
Density	1.00 g/mL	0.0012	850
Viscosity	1.00 kg/m/s	0.02	50
Thermal capacity	1.00 J/mL/°C	0.0003	3,300
Velocity of sound	1,485 m/s	343	4.33
Refractive index	1.33	1.00	1.33
Oxygen content	4-7 mL/L	210	1/30

Contd.

Property	Water/Aquatic	Air/Terrestrial	Approx Ratio Water/Air
O_2 diffusion ratio			1/300,000
Carbon dioxide	0.4	46	1/115
Salts	Freely available	No directly available	
Water	Abundant but may be osmotically unavailable	Rare, always hard to find and keep	

Between animal life in water and air:

- Aquatic animals live and breathe in water.
- Stabilization and movements are very different due to the Archimedes principle.
- Conditions for respiration and excretion are very different due to fluid density, viscosity, and oxygen content.
- Osmoregulation and ionoregulation are very different due to salinity.
- Light influence is very different; irradiance and spectrum are very affected by water.
- New factors as hydrostatic pressure and pH create new specific conditions.
- Development and growth characteristics are different (often ectothermy; fish maintain cell hyperplasia during the adult stage).

By checking the current ionic and osmotic values of the extracellular fluids (*milieu intérieur*) of all the animal phyla, we get very interesting and useful data [4–7, 15]. This is summarized in two tables, Table 18.4 (for invertebrates) and Table 18.5 (for vertebrates). In fact, from the beginning of the metazoans organized animal life, at that time only living in seawater, values of osmolarities of both extra- and intracellular fluids reflect the seawater osmotic pressure: the internal body extracellular fluids match seawater for major ions. However, it is different for Mg^{2+} and SO_4^{2-}. Of course, if the osmotic pressure is the same in both extra- and intracellular fluids, the composition is different. Sodium and chloride are the major ions responsible for the osmolarity in the *milieu intérieur* (extracellular fluids), potassium, free amino acids, or specific molecules like trimethylamine oxide (TMAO), doing so into the cell. The marine ancestral environment influenced for a very long time capabilities of changes and migration. The physical and chemical characteristics of water played a critical role in the establishment of life on earth. As we already said, all the crucial steps of life occurred in water. Species first developed some simple answers, where organisms had the same composition as saltwater. Then, some more elaborated strategies emerged in order to survive in a more diverse set of osmotic conditions. Osmotic parameters constitute in the ocean the main environmental factor that limits the geographic distribution of animal species. Geographic dispersion, along with genetic isolation, is a key component involved in speciation. If vertebrates and arthropods hadn't competed—developing strategies to regulate their extracellular compartments—to colonize new and hostile terrestrial ecosystems while getting out of the oceans, a totally different set of organisms would have succeeded and invaded some empty niches: the current profile of living organisms would be completely different [15].

Table 18.4 Extracellular fluid compositions in a range of invertebrates from different habitats

	Habitat	Osmolarity (mOsm/L)	Na$^+$ (mM/L)	K$^+$ (mM/L)	Mg^{2+} (mM/L)	Ca^{2+} (mM/L)	Cl$^-$ (mM/L)
Jellyfish *Aurelia*	SW	1,000	460	10	51	10	554
Annelid *Arenicola*	SW	1,000	459	10	52	10	537
Mollusk *Anodonta*	FW	40	16	0.5	0.2	8.5	12
Mollusk *Aplysia*	SW	1,000	492	10	49	13	543
Crustacea crayfish	FW	310	146	4	4	8	139
Crustacea lobster	SW	1,000	472	4	47	16	470
Norwegian lobster	SW	1,100	541	8	9	12	552
Echinoderm *Holothuria*	SW	1,063	484	10.5			563
Starfish *Asterias*	SW	1,050	428	9.5	49.2	11.7	487
Insect cockroach	T	325	162	8	6	4	144
Insect *Locusta*	T	325	160	12	25	17	144

Abbreviations: FW, freshwater; SW, seawater; T, terrestrial.

Table 18.5 Extracellular fluid compositions in a range of vertebrates from different habitats

	Habitat	Osmolarity (mOsm/L)	Na⁺ (mM/L)	K⁺ (mM/L)	Mg²⁺ (mM/L)	Ca²⁺ (mM/L)	Cl⁻ (mM/L)
Cyclostome Myxine fish	SW	1,000	554	7	23	9	532
Cyclostome FW lamprey	FW	248	120	3	2	2	96
Chondrychthian shark	SW	1,075	269	4	1	3	258 376 urea
FW teleost goldfish	FW	293	142	2	3	6	107
SW teleost Japanese flounder	SW	337	180	4	1	3	160
Amphibian frog	FW	210	92	3	2	2	70
Reptile alligator	FW	278	140	4	3	5	111
Bird duck	T	294	138	3	2	2	103
Mammal rat	T	295	145	6	2	3	116
Mammal human	T	302	142	4	2	5	104

Abbreviations: FW, freshwater; SW, seawater; T, terrestrial.

So, as already specified, we may observe from Tables 18.4 and 18.5 that marine life has always depended on this osmolarity and developed a universal strategy of intracellular isosmotic regulation, in which the vast majority of invertebrates and some vertebrates have the same osmotic pressure (inner environment and cells) as seawater. Another strategy, which first appeared in certain crustaceans, is called extracellular anisosmotic regulation: it enables a strong migration capacity and adaptation capability to new environments by maintaining an osmotic pressure in cells and bodily fluids within strict limits in any degree of exterior salinity (between 250 and 350 mOsm/L; humans are at 302). In fact, with this strategy, death from dehydration is possible in seawater, with the salty environment causing water to flow from the organism toward the exterior through surfaces of exchange in intimate contact with saltwater (blood-water) and epithelia inside the mouth and gills (the salts from seawater migrating in the opposite direction). This is highlighted in Figs. 18.2 and 18.3. Marine osmoregulators (such as bony fish) had to adopt strategies to constantly drink seawater, while excreting salts via the gills, with the kidney also undertaking this role. One of the main problems of land-based life is water conservation and the struggle to avoid dehydration. The role of the kidney is therefore essential: a good example is the small kangaroo rat of the desert, which never has available water to drink and which secretes urine that is nine times "saltier" than seawater!

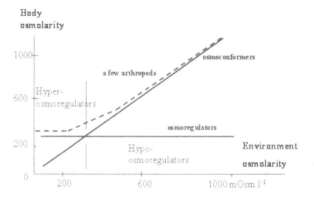

Salinity factor: aquatic organisms response

Figure 18.2 Ionic and osmotic strategies in aquatic animals. Osmoconformers have the same osmolarity as the environment; osmoregulators are independent; a few arthropods (dotted line), e.g., crabs, are intermediary, able to maintain an independent osmolarity only in a limited salinity range.

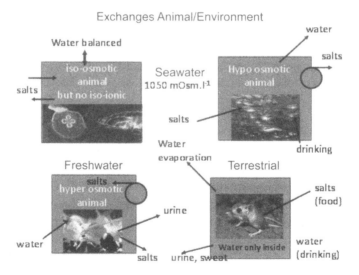

Exchanges Animal/Environment

Figure 18.3 Major situations and strategies developed by animals for water and ionic exchanges.

Euryhaline species, able to migrate from freshwater to seawater, or vice versa, constitute very interesting models to better understand all these questions concerning osmo- and ionoregulation in aquatic environments. The few invertebrates that regulate their intracellular osmolarity and are able to support differences in salinity are generally marine species. They operate from full seawater (35–40 psu) to mid-salinity brackish water (15–20 psu). They are mollusks, both bivalves and gastropods. And they manage to do so in using specific volumes, intervalvular (iv) in bivalves, or intracapping (ic) in gastropods. In fact, they act on four compartments: external water, these two specific volumes (iv or ic), extracellular fluids, and intracellular fluids. They adjust the osmolarity between intracellular and external environments through the balancing or buffering of iv and ic spaces. Among mollusks, cephalopods are a very ancient group without species able to survive in brackish water or freshwater. Several crustaceans were already for a long time able to adapt to salinity changes through anisosmotic extracellular regulation. The Chinese crab, *Eriocheir sinensis*, is so able to adapt from seawater to freshwater, even if it needs seawater for reproduction.

In vertebrates, we find the best models of euryhalinity with very long and achieved migrations: among them, eels and salmons are exceptional. The eels hatch in deep seawater and cross the oceans as *leptocephalus*; then glass eels, after metamorphosis, when they approach the coasts, grow in freshwater. They return to the origin deep seawaters for spawning after decades. Salmon and migrating trout hatch in freshwater and spend years in rivers or lakes before getting ready for migration and seawater adaptation. Parr-smolt transformation, or smolting, [16] corresponds to a profound metamorphosis, occurring and developing in freshwater, transforming the freshwater juvenile (parr) into a premarine fish (smolt). Parr-smolt transformation truly corresponds to a preadaptation to the marine environment, fish in freshwater having already developed most of the mechanisms they will need to adapt and grow in hyperosmotic surroundings. The different cytological, anatomical, morphological, biochemical, endocrinological, and physiological changes studied in freshwater demonstrate that at the end of the freshwater stage, the smolt is closer to a marine fish than to a freshwater salmon [16]. All the essential hormones involved in both freshwater and seawater osmoregulation are concerned and change during smolting achievement in freshwater: the blood plasma insulin level increases and then decreases; prolactin (PRL) decreases; thyroid hormones T_3 and T_4, growth hormone (GH), and cortisol and IGF1 levels increase. In fish, the main hormone involved in osmoregulation in freshwater is PRL, while cortisol and GH are the major hormones in seawater. Fish in freshwater do not drink, elaborate a very abundant and diluted urine (up to 1 mL/h/100 g body weight!) and actively uptake ions from freshwater through the gills. Fish in seawater permanently drink, produce a little urine, concentrated, and actively extrude ions through the gills. Gill (Na^+-K^+)-ATPase and cystic fibrosis transmembrane conductance regulator (CFTR, chloride channel) are very active also in the kidney and in salt glands in species having them. In fish, many indications suggest that better growth in brackish water (not systematic but very frequent) would depend on controlled food intake because many species adapt food ingestion to external water salinity. Calculations have been made to estimate the energetic cost of osmoregulation; it averages between 10% and 50 % of the total energetic available budget, according to species and physiological stage [17].

Far from water, terrestrial fauna has three main options to access water: drinking, taking water out of the food, and using metabolic water produced as a by-product of metabolism. Additional avenues for water gain by a few animals include integumental osmotic absorption (e.g., amphibians absorb water through their skin) and water vapor absorption from the air (some arthropods like mites and ticks). The main avenues for water loss are urine, feces, and integumental osmotic loss. These losses occur only through the body surface for hypoosmotic aquatic animals and both through evaporation from the body surface (cutaneous water loss) and the respiratory surface (respiratory water loss) for terrestrial animals [4]. A few animals may temporarily buffer a disequilibrium between gains and losses by storing water and solutes. All these considerations are summarized in Table 18.6 and Fig. 18.3.

Table 18.6 Avenues for water and ion exchange in terrestrial species (adapted from Ref. 4)

In	Storage	Out
Water		
Drinking	Body fluids	Feces
Food	Urine	Urine
Metabolism		Evaporation from skin
Transcutaneous diffusion		Evaporation through respiration
Water vapor		Transcutaneous diffusion
Ions		
Drinking	Body fluids ions	Feces
Food	Skeleton	Urine
Transcutaneous diffusion		Transcutaneous diffusion
Active transport		Extrarenal secretion

In summary, in aquatic environments, animals regulate water and ion fluxes in response to passive fluxes of ions and water across their integument. An osmotic loss of water is readily replenished by drinking the ambient medium, although this may increase the ionic efflux. An osmotic gain of water can be eliminated by copious urine production, although this may involve a depletion of body ions. However, integumentory ion pumps may compensate for such influxes or effluxes of water. Freshwater invertebrates must osmo- and ionoregulate because it is impossible to be isosmotic or isoionic with such a diluted medium. In seawater, animals can be isosmotic and so, not influenced by an osmotic gradient, or hypoosmotic, having to drink seawater permanently and actively excreting electrolytes (Fig. 18.3). In freshwater, they always are hyperosmotic and do not drink, able to actively absorb ions. For terrestrial animals, they have to drink and keep water (Fig. 18.3). The first migrations from seawater to freshwater were not so easy! And then, imagine what the difficulties were to leave water and to conquer air for the first times!

18.5 CONCLUSION

In an entire life, a human being needs about 75 m³ of water only for his or her physiological needs. According to sex and age, the human water content varies between 60% and 75% of body weight, females and old people having less water. A newborn is constituted of 75% water with extremely fast fluxes between different compartments. Humans do not support more than 10–15% dehydration; it rapidly leads to death. We could remain for three weeks without eating, but it is impossible to remain for more than two to three days without water. We need 2.5 L of water per day, obtained by either drinking (1.5 L) or eating (1 L). Daily losses correspond to 1 L of urine, 0.1 L of stools, 0.5 to 1 L of sweat, and 0.5 L when breathing. Like other mammals, humans keep the memory of the ancestral oceanic life, our *milieu intérieur* being about 3.5 times less concentrated than seawater (see Table 18.7).

Table 18.7 Human extracellular fluids compared to SW concentration

Human *milieu intérieur* (blood)	SW, open ocean
Osmolarity 302 mOsm/L	Osmolarity 1,050 mOsm/L
Cl⁻ 100–105 mM/L	Cl⁻ 560 mM/L
Na⁺ 138–142 mM/L	Na⁺ 450 mM/L
K⁺ 3–5 mM/L	K⁺ 11 mM/L
Kidney cell and fluid 3,000 mOsm/L	"Extreme," Dead Sea 2,500 mOsm/L

Most phyla of animals live in the sea and have done so throughout their evolution. Indeed, most of the evolutionary history of the planet has been marine (90% in time), and most of the really important evolutionary innovations in design occurred there. In estuaries and shorelines, littoral communities are among the most finely patterned and perturbation sensitive that can be found. Due to high and rapid external changes (tide), including temperature, oxygen, salinity, emersion from and exposition to the air, hydrostatic pressure, and pH, every organism has its own very tightly defined niche, very often within a vertically zoned habitat. Life in freshwater is also well diversified, and it is very difficult to generalize, as all freshwaters are different compared with the sea. All is a matter of skin permeability and ionic water content. Important marine groups have not been able to colonize freshwater systems as cephalopods, echinoderms, tunicates, etc. (Table 18.1). Terrestrial life became a marvelous success that began more than 400 million years ago, but the real success stories are restricted to just a few *taxa* whose impact on lands has been more striking. Arthropods (cuticule) took the small size niches and vertebrates (dermis) the larger ones. Mollusks do well in all biomes. In extreme terrestrial environments, the convergence of adaptation mechanisms and behaviors allowed to acquire the same kind of solutions, in polar, desert, or high mountain areas [7].

Another interesting point corresponds to the different services offered by marine organisms in term of extraction of specific and original molecules, like pharmaceuticals and cosmetics. Over 50% of the medicines sold in pharmacies correspond to natural products (or are produced from natural products), and more than 15,000 of these molecules are obtained from marine organisms [18]. Thirty percent of marine substances have been found in spongelike organisms. Today, systematic screening of biological activity is carried out in various countries and in large pharmaceutical laboratories. They mainly are anticancer, antibiotics, antiviral, antifungi, immunostimulators, immunosuppressives, growth factors, bone regenerators, molecular tools (polymerases, fluorescent proteins, etc.).

In terms of supplying original and pertinent research models from the sea, in 1865, physiologist Claude Bernard stated:

> Some experiments are impossible to carry out on certain animal species and the intelligent selection of an animal with the right disposition is often the essential condition for success and to solve very important physiological problems ... comparative physiology is a very fertile field for general physiology.

Closer to our time, the words of August Krogh, the Nobel Laureate in 1920, are stated as a principle: "For each problem in physiology, there is an ideal living model." Finally, in 1997, Francois Jacob, the 1965 Nobel Laureate, added that "to attack an important problem, to have a reasonable chance of finding a solution, the biologist must use the appropriate materials." The world's oceans may supply many possible species with the *right disposition* or the *appropriate materials*! Some marine models have been the source of essential discoveries in life sciences and several Nobel Laureates have obtained them thanks to marine models: the discovery of phagocytosis, the postfertilization processes, the anaphylactic shock reaction, the transmission of nerve impulses, the molecular basis for memory, the discovery of cyclins in cancer research, the green fluorescent protein from jellyfish, the development of the eyes, the role of membrane neurotransmitter receptors, the fundamentals of specific immunity, and many more. Marine models are essential to understanding the ancestral roles and mechanisms of many human systems and sometimes in deducing the applications for effective treatments. The ocean provides humankind with renewable resources, which are very threatened today and which must be more adequately managed in preserving ocean ecosystems, stocks, and biodiversity [9].

One of the most worrying problems humans are facing now is the lack of good-quality water enhanced by the global climatic change [10, 11, 19]. As with all terrestrial species, humans need water and cannot live without it. Millions of people die every year for lack of water or by drinking polluted water. Water development underpins food security, people's livelihoods, industrial growth, and environmental sustainability throughout the world. In 1995 the world withdrew 3,906 km³ of water for these purposes. By 2025 water withdrawal for most uses (domestic, industrial, livestock) is projected to increase by at least 50%. This will severely limit irrigation water withdrawal, which will increase by only 4%, constraining food production in turn [19].

All the chapters of this book highlight how life is totally linked with this apparently so simple and so abundant molecule, from the general physiology to the intimate functioning of the brain. Tomorrow, wars could be triggered for water. It has been often in the past the most protected resource for many peoples. It is extremely unequally shared and wasted today. Hasn't water, so vital, became the forgotten molecule?

References

1. De Duve, C. (1995). *Vital Dust: Life as a Cosmic Imperative*, Basic Books, New York.

2. Maurel, M. C. (2003) *La naissance de la vie. De l'évolution prébiotique à l'évolution biologique* [The Birth of Life: From Prebiotic to Biological Evolution], Dunod, UniverScience.

3. Steyer, S. (2009) *La Terre avant les dinosaures*, Belin, Pour la Science.

4. Withers, P. C. (1992) *Comparative Animal Physiology*, Thomson Learning, Brooks/Cole.

5. Gilles, R., Charmantier, G., and Charmantier, M. (2006) Relations ioniques, hydriques et osmotiques, in *Physiologie animale* (ed. Gilles, R.), de Boeck, pp. 5–118.

6. Schmidt-Nielsen, K. (1998) *Physiologie animale, adaptation et milieu de vie*, 5th ed., Dunod.

7. Willmer, P., Stone, G., and Johnston, I. (2000) *Environmental Physiology of Animals*, Blackwell Science.

8. Parkes, R. J., Cragg, B. A., Bale, S. J., Getliff, J. M., Goodman, K., and Rochelle, P. A. (1994) Deep bacterial biosphere in Pacific Ocean sediments, *Nature*, **371**, 410–413.

9. Boeuf, G. (2008) What does the future hold for biodiversity? in *A Better World for Everybody*, directed by J. P. Changeux and J. Reisse, Collège de France, Odile Jacob, Paris, pp. 47–98.

10. Vitousek, P. M., Mooney, H. A., Lubchenco, J., and Melillo, J. M. (1997) Human domination of Earth's ecosystems, *Science*, **277**, 494–499.

11. Palumbi, S. R. (2001) Humans as the world's greatest evolutionary force, *Science*, **293**, 1786–1790.

12. Millennium Ecosystem Assessment (MEA). (2005) *Ecosystems and Human Well-Being: Synthesis*, Island Press, Washington, DC.

13. Thuillier, W. (2007) Climate change and the ecologist, *Nature*, **448**, 550–553.

14. Blondel, J., Aronsson, J., Bodiou, J. Y., and Boeuf, G. (2010) *The Mediterranean Basin: Biodiversity in Space and Time*, Oxford University Press, New York and London.

15. Eckert, R., Randall, D., Burggren, W., and French, K. (1999) *Physiologie animale*, 4th edn. (ed. Eckert, R.), de Boeck.

16. Boeuf, G. (1993) Salmonid smolting, a preadaptation to the oceanic environment, in *Fish Ecophysiology* (ed. Rankin, J. C., and Jensen, F. B.), Chapman & Hall, London, pp. 105–135.

17. Boeuf, G., and Payan, P. (2001) How should salinity influence fish growth? *Comp. Biochem. Physiol., C: Comp. Pharmacol. Toxicol.*, **130**, 411–423.

18. Boeuf, G., and Kornprobst, J. M. (2009) Marine bio- and chemical diversities, *Biofutur*, **301**, 28–32.

19. Rosegrant, M. W., Cai, X., and Cline, S. A. (2002) Global Water Outlook to 2025, Averting an Impending Crisis. IFPRI and IWMI Institutes Report, Columbo, Sri Lanka.

20. Butchart, S. H. M., *et al.* (2010) Global biodiversity: indicators of recent declines, *Science*, **328**, 1164–1168.

Color Inserts

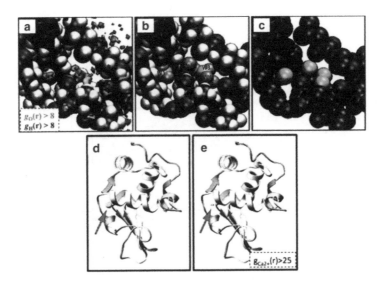

Figure 4.1 Comparison of distribution of water and Ca²⁺ ion evaluated by 3D-RISM and X-ray. The isosurfaces of water oxygen (green) and hydrogen (pink) for the three-dimensional distributions larger than 8 (a), the most probable model of the hydration structure reconstructed from the isosurface plots (b), and the crystallographic water sites (c). The position of Ca²⁺ ion measured by X-ray (d) and the isosurfaces of Ca²⁺ ion for the three-dimensional distributions larger than 25 (e) [17, 26, 27].

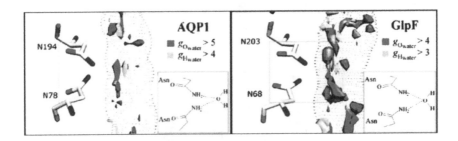

Figure 4.3 The three-dimensional distribution functions of water in the NPA region of AQP1 and G1pF: pale red and light blue surface represent the distribution function of oxygen and hydrogen of water, respectively, and the dot surface denotes the pore area.

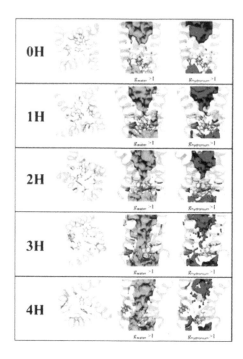

Figure 4.6 Structure of Trp41 gating (orange) of different protonated histidine tetrad, and 3D-DFs of water in the channel (cyan) and hydronium ion (red), with $g > 1$. For color reference, see p. 362.

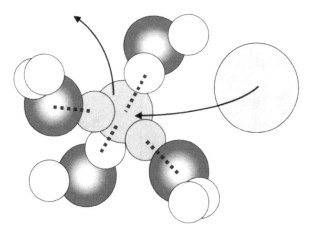

Figure 5.1 To create a cavity inside liquid water, at least one water molecule must be extracted, at the cost of breaking four H bonds.

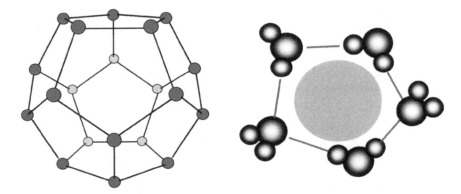

Figure 5.3 The structure of a cavity in a chlorine hydrate crystal. Water molecules are at the corners of a pentagonal dodecahedron. H bonds are formed along edges of the dodecahedron and also between the dodecahedron and adjacent dodecahedra and between the dodecahedron and interstitial water molecules [4].

1 **High solubility** **High permeability**	2 **Low solubility** **High permeability**
3 **High solubility** **Low permeability**	4 **Low solubility** **Low permeability**

Figure 5.5 General classification scheme for drugs administered through the oral route [6].

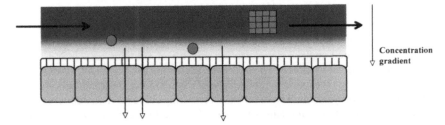

Concentration
gradient

Figure 5.7 Transport of sparingly soluble solids through the gut. The layer of epithelial cells is represented in grey, supporting the mucus layer (in vertical lines). Large crystals (solubility 10^{-6} g/g to 10^{-5} g/g) go through the digestive tract with almost no dissolution. Amorphous particles have much higher solubility. Nanoparticles have a large specific area (10–100 m²/g) and can supply solute molecules at a high rate if these molecules are taken up efficiently by the epithelial cells.

1	SPPGKPQGPPQQE
14	GNKPQGPPPP
24	GKPQGPPPA
33	GGNPQQPQAPPA
45	GKPQGPPPPPQ
56	GGRPPRPAQGQQPPQ

1	FLISGKPVGRRPQGGNQPQRPPPPP
26	GKPQGPPPQGGNQSQGPPPPP
47	GKPEGRPPQGRNQSQGPPPHP
68	GKPERPPPQGGNQSQGTPPPP
89	GKPERPPPQGGNQSHRPPPPP
110	GKPERPPPQGGNQSRGPPPHR
131	GKPEGPPPQEGNKSRSAR

"IB-5"
Prolin rich (41 %)
Not glycosylated

"II-1"
Prolin rich (35 %)
Glycosylated

Figure 5.11 Amino acid sequences of the human salivary proteins IB5 and II-1. Both have repeated proline sequences that prevent the formation of alpha helices and that are involved in hydrophobic bonding to other solutes, such as plant tannins. IB5 is not glycosylated; in the presence of tannins, such as EGCG, it precipitates. II-1 is glycosylated on one or a few of the N amino acids; in the presence of tannins, such as EGCG, it remains in solution.

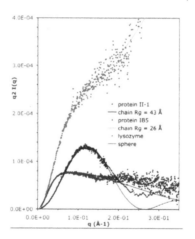

Figure 5.12 Small-angle X-ray-scattering spectra of proteins in solution, in the Kratky-Porod representation. Horizontal scale: Scattering vector q, in Å^{-1}. Vertical scale: Scattered intensity multiplied by the square of the scattering vector, $q^2I(q)$. Symbols and lines: Data for lysozyme and theoretical scattering curve for dense spheres (black dots and black line); data for salivary protein II-1 and theoretical curve for a random chain (blue dots and pink curve); data for salivary protein IB-5 and scattering curve for an elongated chain (pink dots and blue curve).

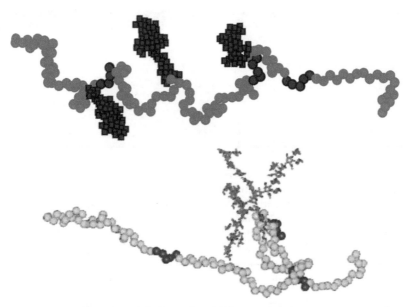

Figure 5.13 Conformations of glycosylated salivary proteins II-1, constructed using a mathematical algorithm that looks for structures that provide the best fit to the small-angle X-ray scattering spectrum [11]. The blue balls are amino acids in the protein chain, and the orange side groups are the branched sugar chains. The top panel presents a protein with three sugar side groups, and the bottom panel a protein with a single sugar side group.

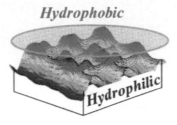

Figure 6.7 Schematic illustration of the capillary waves of water meeting the hydrophobic surface with a flickering vapor phase in between.

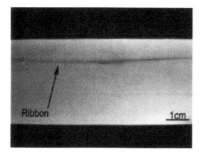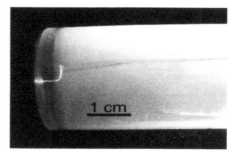

Figure 7.2 (a) Section of exclusion "ribbon" running along the length of a cylindrical tube, nucleated at the surface of the polyacrylic acid gel beyond the left side of the panel. The tube was filled with a solution containing distilled water and 15 drops of stock solution of 0.45 μm–diameter carboxylate-coated microspheres, yielding an estimated concentration of 6×10^8 microspheres per mL. (b) Top view of the chamber showing the wedge-shaped origin of the exclusion ribbon. The wedge grows out of the original disk-shaped exclusion zone adjacent to a polyacrylic acid gel surface. The prominent white object on the left is a pin used to hold the gel against the stopper.

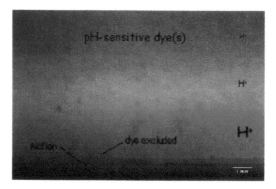

Figure 7.3 Chamber containing a Nafion tube (*bottom*) filled with water containing a pH-sensitive dye. View is normal to the wide face of a narrow chamber. Image obtained 5 min after dye solution poured into chamber.

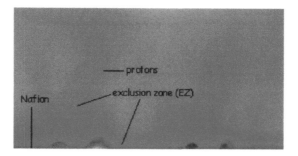

Figure 7.4 Similar to Fig. 7.3 but taken 20 min after the water with pH indicator dye is poured into the chamber. Note the exclusion zone growth perpendicular to the Nafion surface, surrounded by areas of low pH.

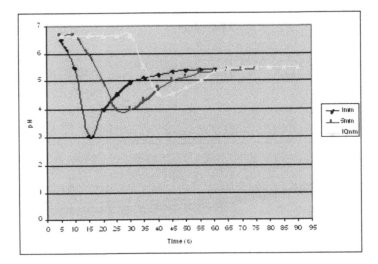

Figure 7.5 Time course of pH following addition of water to a sheet of Nafion. The pH was measured at three distances, indicated to the right, from the Nafion surface. A wave of protons appears to be generated.

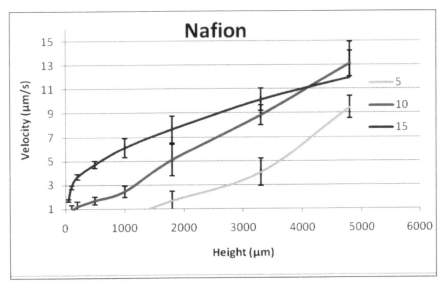

Figure. 7.6. Velocity as a function of height above the Nafion surface, for 5, 10, and 15 μm–diameter polystyrene microspheres.

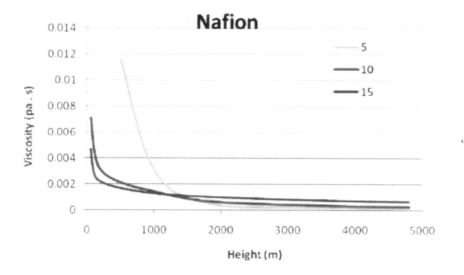

Figure. 7.7. Computed viscosity as a function of distance from the Nafion surface for 5, 10, and 15 μm–diameter polystyrene microspheres.

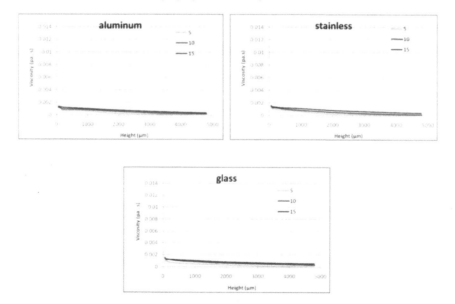

Figure 7.8 Computed viscosity as a function of distance from the surface for 5, 10, and 15 μm–diameter polystyrene microspheres for aluminum, glass, and stainless steel.

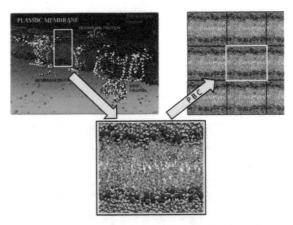

Figure 8.1 Modeling membrane bilayers. *Left*: Sketch of a plasmic membrane with incorporated membrane proteins. *Bottom*: Configuration of a palmitoyl-oleyl-phosphatidyl-choline (POPC) hydrated bilayer system from a well-equilibrated constant-pressure MD simulation performed at 300 K. Water molecules (gray), phosphate (magenta), and nitrogen atoms (blue) are represented as solid van der Waals spheres, while the acyl chains (cyan) are displayed as sticks. *Right*: During the simulation, 3D periodic boundary conditions (PBCs) are used, that is, the system is replicated infinitely in the three directions of space.

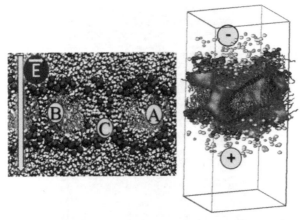

Figure 8.3. *Left*: Configuration taken from an MD simulation of a POPC bilayer subject to an electric field generating a TM potential of ~3 V, after a 2 ns run. Note that the simultaneous presence of (A) water (grey van der Waals spheres) fingers, (B) water wires, and (C) large water pores stabilized by lipid head groups (black spheres). *Right*: Configuration taken from an MD simulation of a POPC bilayer embedded in a 1 M NaCl electrolyte solution and subject to TM voltage of 2 V generated from a net charge imbalance between both sides of the membranes (unpublished data). Note here that the water baths are terminated by water–air interfaces. Na⁺ and Cl⁻ ions are represented as yellow and cyan spheres, respectively, and water is not shown for clarity. An isosurface of the potential is drawn in magenta to highlight the contour of the created hydrophilic pore.

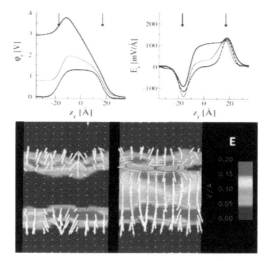

Figure 8.4 Electrostatic potential (*top left*) and electric field profiles (*top right*) along the normal to lipid membranes subject to increasing TM voltages Δ*V*, estimated at the initial stage of the MD simulations. The arrows indicate the approximate position of the water–membrane interface. *Bottom*: Cross-sectional view of the 3D maps of the electric field derived as the gradient of the electrostatic potential (*left*: Δ*V* = 0 V; *right*: Δ*V* = 3 V). The arrows indicate the direction and strength of the field.

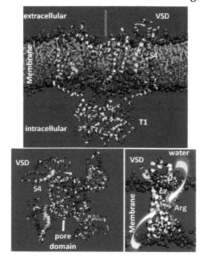

Figure 8.5 *Top*: Configuration of the Kv1.2 channel (light pink ribbons) embedded in a lipid bilayer. Only two monomers are represented for clarity. The central helices define the hydrophilic pore (direction indicated by the orange arrow) and the selectivity filter can accommodate two K⁺ ions (orange spheres). The helix S4 of the voltage sensor domain (VSD) containing highly conserved positively charged residues is highlighted in yellow. *Bottom left*: Corresponding top view. *Bottom right*: A close-up view of the VSD. MD simulations of the structure show that water (light blue) penetrates from both sides of the membrane deeply in the VSD. Atoms of the Arginine 303 depicted as yellow spheres come in contact with both water crevices.

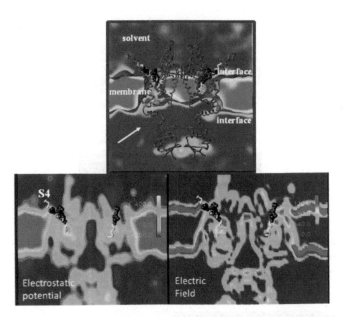

Figure 8.6 3D Electrostatic environment of the Kv1.2 channel embedded in a lipid bilayer. *Top*: An isosurface of the potential is drawn in magenta to highlight the contour of hydrophilic pore created by the water protruding inside the VSD. *Bottom*: Close-up view of the electrostatic environment of the top gating charges (Arg; black) showing the abrupt local variation of the electrostatic potential in mV (*left*) and of the electric field in mV/Å (*right*), due to the penetration of water within the TM region of the K_v channel.

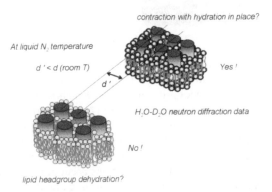

Figure 9.1. A schematic illustration of the results of Zaccai (1987). The decrease in crystallographic cell dimension is similar for dehydration of the membranes at room temperature and for when the membranes are cooled to liquid-nitrogen temperature. The neutron diffraction results, from experiments using H_2O/D_2O exchange, showed unambiguously that cooled membranes were not dehydrated but had contracted with the head group hydration still in place.

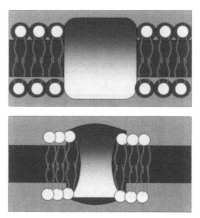

Figure 9.2 A side view of BR in the membrane, showing how dehydration of lipid head groups could lead to tightening of the protein environment, which could interfere with its internal dynamics.

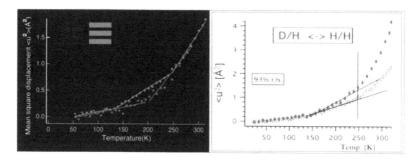

Figure 9.3 (a) Elastic temperature scans of the intermembrane water layer (red) and the membrane (green) obtained by specific H/D labeling of the components. Note that dynamical transitions occur at different temperatures for the red and green lines: ~200 K for the water and ~250 K for the membrane (the inflection in the membrane data at ~150 K is due to methyl group rotations) [7]. (b) Elastic temperature scans of the lipid (red) and membrane (black) fractions in PM, obtained by H/D labeling and membrane reconstitution [16]. The inflection in both plots at ~150 K is due to methyl group rotations (the density of methyl groups is higher in the lipid than in the membrane fraction) Note that dynamical transitions occur at the same temperature, ~250 K, for the membrane and lipid fractions, suggesting that they are coupled.

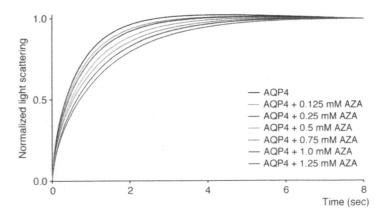

Figure 10.2 Dose dependence of the inhibition of AQP4 water permeability by AZA. The graph shows averaged time courses: 0.125 mM AZA, $n = 16$; 0.25 mM AZA, $n = 15$; 0.5 mM AZA, $n = 17$; 0.75 mM AZA, $n = 17$; 1.0 mM AZA, $n = 19$.

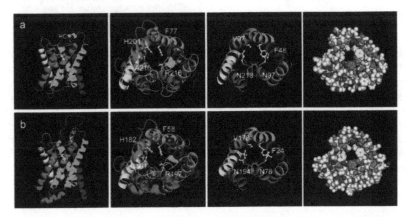

Figure 10.3 Structures of (a) rAQP4M23 and (b) bAQP1 showing the typical AQP fold. AQP4 features a short 3_{10} helix, named, HC, which is important for the adhesive function, but AQP1 has no such HC (first and second figures from the left in both [a] and [b]). The narrowest region in the AQP4 and AQP1 pores, previously termed ar/R [40], is located close to the extracellular entrance of the pore (figures in the second row). The pore diameter at NPA motifs in AQP4 is similar to that in AQP1, although the main chain of Ala210 is in a different conformation and His201 is slightly shifted away from the pore centre. At the NPA sites in the centre of the pores of AQP4 and AQP1 (third figure), the pore diameters of both channels are about 3 Å of Van der Waals distances (fourth figure).

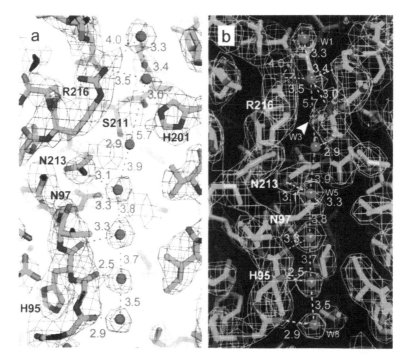

Figure 10.4 Water molecules in the AQP4 channel. (a) In the 2Fo-Fc map contoured at 1.2 σ (marine mesh), water molecules in the channel (red spheres) are clearly resolved as spherical densities. (b) In the Fo-Fc map contoured at 3.0 σ (orange-yellow mesh), a small density (marked by a white arrowhead and labeled W3) appeared in the channel, which we interpreted as the quasi-stable position of a water molecule. Distances (Å) between water molecules and their closest protein atoms are depicted as dotted lines. Distances (Å) between neighboring water molecules are depicted as dotted lines if in hydrogen-bonding distance or as dotted cyan lines if the distance is too far for hydrogen bonding.

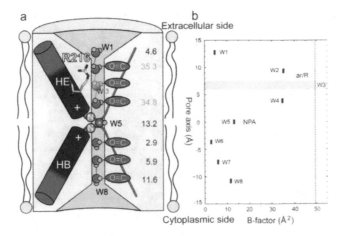

Figure 10.5 Schematic diagram of the water arrangement in the AQP4 channel and the temperature factors of water molecules. (a) The two short pore helices, HB and HE, which are shown as rods with the blue and red colors indicating their electrostatic dipoles, form an electrostatic field that causes water molecules in the two half channels to adopt opposite orientations. All water molecules, except the one at the channel entrances, form two or three hydrogen bonds with the neighboring water, main chain carbonyl oxygen (C=O in the red oval), and/or nitrogen of the NPA motifs (N in the cyan circle). The residue Arg216, part of the ar/R constriction site, is shown in a ball-and-stick representation. The water molecules in the pore are designated W1–W8. The temperature factors of the water molecules are denoted to the right of the channel. (b) plot of the B-factors of the water molecules in the channel. All eight water molecules in the pore are represented. The labeling of the water molecules is the same as in (a). The channel region occupied by W3 is depicted as a wheat-colored band because the B-factor of W3 could not be determined. The red dashed line indicates the average B-factor of all the atoms in the AQP4 molecule.

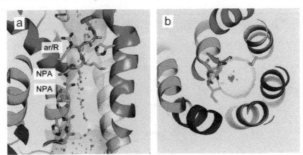

Figure 10.6 AQP4 channel with positions of all water molecules seen in AQP structures determined to date. The pore in AQP4 as calculated by "HOLE" [49] is shown as a transparent surface. The color indicates the pore radius from red (narrow; ar/R) to green (NPA) to purple (wide). Residues interacting with water molecules in the channel are represented as ball-and-stick models (carbon, oxygen, and nitrogen atoms are shown in green, red, and blue, respectively). Red crosses indicate the positions of water molecules seen in other AQP structures. Side view (a) and slice around (b) the NPA region of the AQP4 channel with water molecules seen in the structures of highly water-selective AQPs. Reproduced from Ref. [43] by permission.

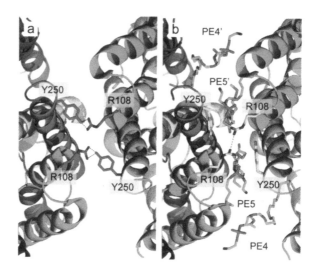

Figure 10.7 Interaction to stabilize the orthogonal arrays of AQP4. (a) In AQP4 wild-type crystals, AQP4 tetramers are stabilized by a specific interaction, such as tyrosine and arginine. (b) Structure of the AQP4S180D mutant analyzed at a resolution of 2.8 Å revealed the insertion of lipid molecules between tetramers. In this crystal, the arrays are mediated by hydrophobic interaction with lipid molecules.

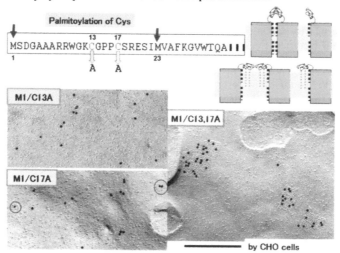

Figure 10.8 The array formation on CHO cell membranes was observed by deleting amino acids one by one from the N-terminus on the basis of the freeze-fracture method. When the 17th amino acid cysteine residue was deleted, suddenly large arrays were observed. Cysteine residues 13 and 17 were revealed to be essential for array disruption by palmitoylation. The lipid modification even on one cysteine was also confirmed to disrupt the arrays of AQP4. Without any lipid modification, interaction of intertetramers has no hindrance and forms orthogonal arrays, but lipid-modified amino terminus hinders the array-forming interaction. The scale bar indicates 4,000 Å.

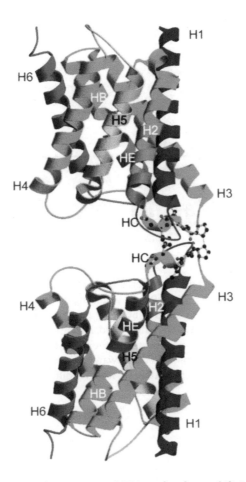

Figure 10.9 One pair of interacting AQP4 molecules stabilizing the membrane junction is shown in a ribbon representation. A ribbon diagram of a pair of interacting AQP4 molecules highlighting residues Pro139 and Val142 of 3_{10} helices responsible for junction interactions is shown.

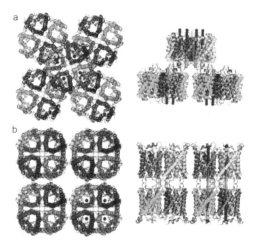

Figure 10.10 Two types of double-layered interactions, which were observed in the 2D crystals of (a) AQP4 and (b) AQP0. All adjoining molecules in AQP0 crystals can adhere even without array formation, but only one pair of molecules of AQP4 can adhere when orthogonal arrays were disrupted, because the tetramer of AQP4 at the upper layer was shifted from the tetramer at the lower layer. As shown by blue rods, adjoining channels of AQP4 are shifted and water permeation is blocked when these two membrane layers are strongly adhered, which is different from the case of AQP0.

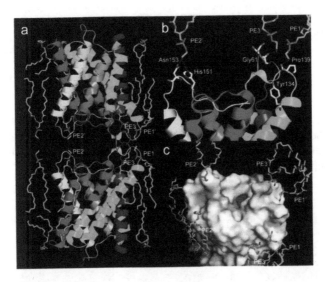

Figure 10.11 AQP4 and lipid interactions stabilizing the membrane junction. (a) Two interacting AQP4 molecules (ribbon diagram) with surrounding lipid molecules (ball-and-sticks representation) viewed parallel to the membrane plane. (b) Close-up view of the interactions between the extracellular AQP4 surfaces with lipids of the adjoining membrane. Lipids (cyan) and their interacting AQP4 residues (yellow) are shown in a ball-and-stick representation. Yellow dotted lines depict hydrogen bonds between lipid head groups and AQP4 residues. (c) Electrostatic potential of the AQP4 surface, ranging from −4kT (red) to +4kT (blue), and a ball-and-stick representation of the associated PE molecules, showing that the lipid head groups interact with neutral (grey) or negatively charged surface areas of AQP4. Reproduced from Ref. [43] by permission.

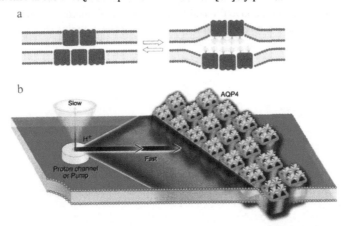

Figure 10.12 (a) Schematic for explaining a possible function of a cell-adhesive channel (adhennel [7]). Adhesive AQP4 molecules may separate membrane layers by pressure of gushing water, as shown in this figure, because the packing of AQP4 tetramers is not end-on but a staggered arrangement between the two adjoining channels. (b) Schematic for explaining proton signal transfer and regulation mechanism by the water channel as gushing water model 2.

Figure 11.2 In a simple binary mixture of two phospholipids, one with short acyl chains, in the liquid-crystalline phase, the other with longer acyl chains, in the gel phase, a hydrophobic mismatch occurs, exposing portions of the acyl chains in the long-chain lipid to water. As a result, the two lipid phases separate.

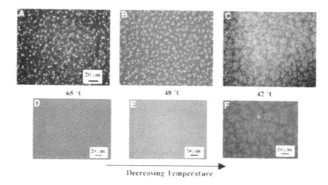

Figure 11.3 Epifluorescence images of DMPC/DSPC-supported single bilayers labeled with 0.5 mol% NBD-PE (di14:0). As the mole ratio of DMPC/DSPC is decreased (A–C) the fluorescent domain increases in size, indicating that this is the DSPC fraction. When the system was heated above the DSPC transition, the domain structure was lost (D). Cooling then resulted in reformation and growth of the domains (E, F). Modified from Ref. [27].

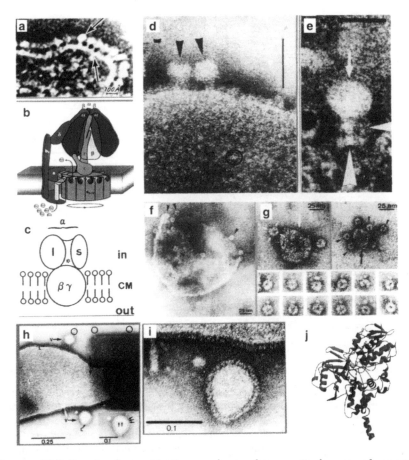

Figure 12.5 Structural organization and membrane attachment of enzyme complexes. Electron micrograph (a) and diagrammatic view (b) of ATPases of the FoF1and the AoA1 type (after Ref. [9], modified); (c) oxaloacetate decarboxylase from *Klebsiella pneumoniae*: size and shape of the enzyme and location of its prosthetic biotin group; (d,e) methano-reductosomes (arrowheads) attached to the cytoplasmic side of the (artificially reversed) cytoplasmic membrane of *Methanobacterium thermoautotrophicum*; (e) in addition to the "head," details of the stalk and the membrane integration of the basis of the stalk are visible; (f,g) F420-reducing hydrogenase complexes attached to the cytoplasmic side of the (artificially reversed) cytoplasmic membrane of *Methanobacteriumthermoautotrophicum*; (g) details of the stalks connecting the enzyme complex with the membrane are visible; (h,i) remnants of cells of *Thermoanaerobacterium thermosulfurogenes* EM1after removal of the cell envelope by specific growth conditions. The surface of the exposed cytoplasmic membrane and vesicles thereof is densely covered by α-amylase complexes attached to the membrane by stalks; (j) a ribbon diagram of human monoamine oxidase B (MAO B). The protein has a single transmembrane helix that anchors it to the outer membrane of the mitochondrion. Nevertheless, the protein is considered monotopic because the bulk of the 520 residues, including the active center, is outside the membrane. Reproduced with permission: c from Ref. [11], d,e from Refs. [12] and [13], g from Ref. [14], h,i from Ref. [15], and j from Ref. [16].

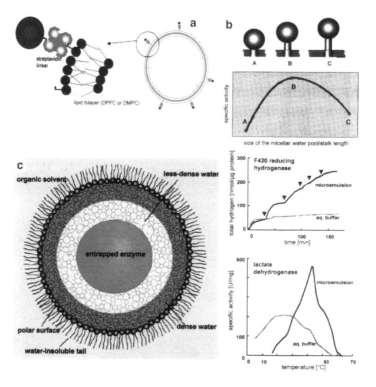

Figure 12.6 Experiments designed to simulate *in vivo* conditions of enzyme complexes. (a) Assembly of enzyme–liposome complexes. At the right side, a schematically drawn liposome with attached enzymes is depicted. Part of the diagram (circled) is drawn, at the left side, with more details: a Streptavidin linker connects an enzyme molecule with the surface of the liposome. The size of the linker determines the distance of the enzyme from the liposome. A biotin tag not occupied by a linker is also shown attached to the outside of the liposome. *Abbreviations*: DPPC, 1,2-palmitoyl-*sn*-glycero-3-phosphocholine; DMPC, 1,2-dimyristoyl-*sn*-glycero-3-phosphocholine. (b) Dependency of the measurable specific activity of an enzyme from the length of a linker as shown in Fig. 12.6a or from the inner surface of a reversed micelle (Fig. 12.6c) (original drawing by M. Hoppert). (c) (*left*) An enzyme molecule (size about 5–8 nm) placed inside a reversed micelle with a diameter of less than about 20 nm. The properties of the components making up the reversed micelle and the overall dimensions of the system were chosen in such a way that the system can be used for the simulation of the *in vivo* situation of an enzyme; (*right*) comparison of the specific enzyme activities of enzymes (F420-reducing hydrogenase, lactate hydrogenase), measured, in the conventional way, in aqueous buffer, or trapped inside a reversed micelle; (*top*) hydrogen production of the F420-hydrogenase at 60°C. The arrows indicate the points of time of methylviologen additions to both assays. The specific activity of the enzyme in the reversed micelle is very substantially higher compared to that in aqueous buffer; (*bottom*) thermal stability and specific activity of the enzyme are increased for enzymes in reversed micelles compared to in aqueous buffer. Reproduced with permission: a from Ref. [20] and c from Refs. [21] and [22].

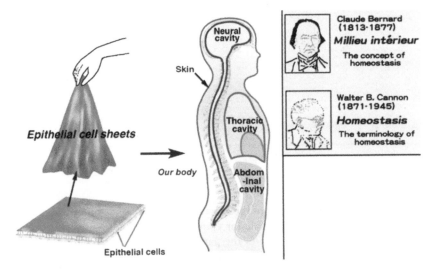

Figure 13.1 Illustration of the functions of epithelial cell sheets in higher organisms. Epithelial cells adhere to each other in a side-by-side fashion to form the sheetlike configuration shown. The epithelial cell sheet covers the surface of every compartment inside and outside the body of higher organisms to maintain the homeostasis of the separate fluid-filled compartments of multicellular organisms (*left*). The concept of homeostasis was defined by Claude Bernard, and the term, "homeostasis," by Walter Cannon (*right*).

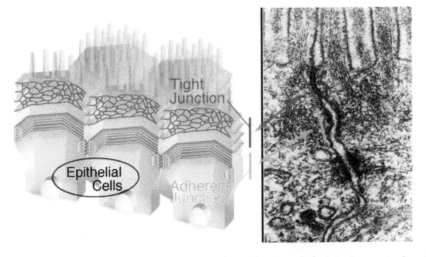

Figure 13.2 Schematic drawing of epithelial cell sheets and electron micrographs of adherens and tight junctions. Epithelial cells adhere to each other with specialized adhesion complexes, such as tight junctions, adherens junctions, and desmosomes. An electron microscopic image of tight junctions, adherens junctions in the intestinal epithelial cells is shown (*right*).

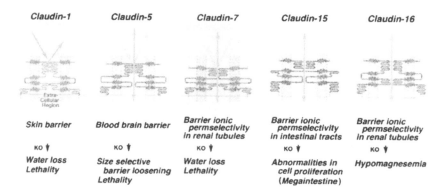

Figure 13.3 Schematic drawings of the function of particular claudins described in this review. Claudins are responsible for epithelial cell–cell adhesion and regulate the permselective barrier function of the paracellular space by providing selective permeability to water and other molecules according to their size and/or charge, as revealed by knockout mice analyses.

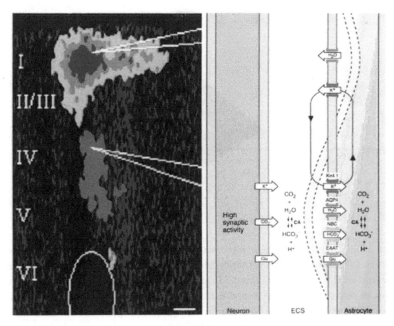

Figure 15.1 Hypothetical scheme of the coupling between activity-dependent changes of extracellular (ECS) volume, and water and ion transport across glial membranes. *Left*: Changes in intrinsic optical signals (IOS) following afferent stimulation of rat neocortical slice. Slices were electrically stimulated through an electrode in layer VI (indicated) with stimulus trains lasting 2 s. IOS were captured 6 s after stimulation was started. The bright signal (red) in layer IV (which is target of the afferent stimulation) corresponds to a reduction of the extracellular space. Darkening of the IOS ("black wave", blue)) corresponds to a widening of the extracellular space and is seen in the superficial cortical layers. Electrodes for measurement of potassium activity in layers I and IV are indicated. *Right* (scale bar 200 μm): Activity dependent volume changes and ion transport. Active neurons (orange) release K^+ and glutamate and produce increased amounts of CO_2. As indicated in left panel, activity is associated with a shrinkage of the extracellular space locally and with a widening of the extracellular space distal to the site of stimulation (i.e., in layer I). We assume that these volume changes reflect water influx and efflux, respectively, primarily across astrocyte plasma membranes (dashed lines indicate position of astrocyte membrane after neuronal activation; cf. left panel). Reproduced from Ref. [6] by permission.

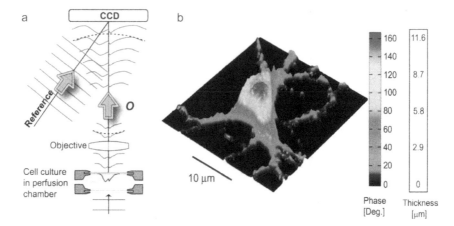

Figure 15.2 Digital holographic microscopy (DHM) of living mouse cortical neurons in culture. (a) Schematic representation of cultured cells mounted in a closed perfusion chamber and (b) trans-illuminated 3D perspective image in false colors of a living neuron in culture. Each pixel represents a quantitative measurement of the phase retardation or cellular optical path length (OPL) induces by the cell (with a sensitivity corresponding to a few tens of nanometers). By using the measured mean value of the neuronal cell body refractive index, resulting from the decoupling procedure, scales (at the right), which relate OPL (Deg) to morphology in the z axis (μm), can be constructed.

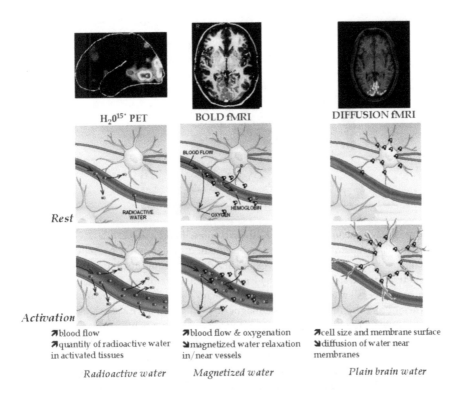

Figure 16.1 General principles of PET and MRI functional neuroimaging methods. *Left*: Brain activation mapping with PET. Radioactive water (H_2O^{15}) is used as a tracer to detect neuronal activation–induced increases in blood flow. *Middle*: Brain activation mapping with BOLD fMRI. Water magnetization in and around small vessels is modulated by the flow of red blood cells containing paramagnetic deoxyhemoglobin. (Adapted from Ref. [4]). *Right*: Brain activation mapping with diffusion fMRI. The reduction in water diffusion, which occurs during activation, is thought to originate from a membrane-bound water layer, which expands during activation-induced cell swelling. While with PET and BOLD fMRI water is only an indirect means to detect changes in blood flow through the imaging scanner, the changes in water properties seen with diffusion fMRI seem to be an intrinsic part of the activation process.

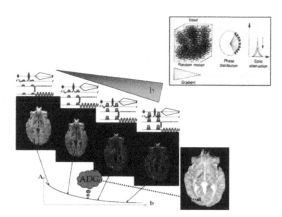

Figure 16.2 Diffusion-weighted MRI principles. *Right corner insert*: The overall, random displacement of water molecules due to diffusion in the presence of a space-varying magnetic field (gradient) results in a dispersion of the phases of the spinning magnetization carried by water hydrogen nuclei. This phase dispersion results, in turn, in an overall attenuation of the MRI ("echo") signal, depending on the diffusion coefficient and the intensity of the magnetic field gradient (*b* value). *Bottom*: Set of diffusion-weighted images are obtained using different *b* values, which are obtained by changing the intensity of the diffusion gradient pulses in the MRI sequence. In diffusion-weighted images the overall signal intensity decreases with the *b* value. Tissues with high diffusion (such as ventricles) get darker more rapidly when the *b* value is increased and become black. Tissues with low diffusion remain with a higher signal. As diffusion-weighted images also contain relaxation (T_1 and T_2) contrast, one may want to calculate pure diffusion (or ADC) images. To do so the variation of the signal intensity, $A(x, y, z)$, of each voxel with the *b* value is fitted using Eq. 16.1 to estimate the ADC for each voxel. In the resulting image the contrast is inverted: bright corresponds to fast diffusion and dark to low diffusion.

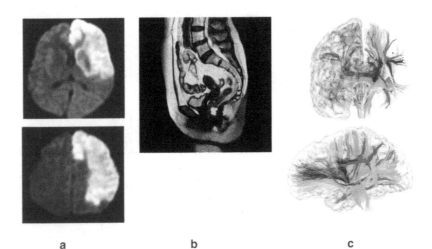

a   b   c

Figure 16.3 Major current applications of diffusion MRI. (a) Diffusion MRI in brain acute stroke. The region with a bright signal corresponds to brain regions where water diffusion is reduced, as a result of acute cerebral ischemia and associated cytotoxic edema. (Image courtesy of Dr. Openheim [Radiology Department, Hôpital Saint-Anne, France]). (b) Diffusion MRI in cancer. Colored areas correspond to regions where the water diffusion coefficient is decreased. Such regions have been shown to match areas where malignant cells are present (primary lesions or metastases). (Image courtesy of Dr. Koyama, Radiology Department, Kyoto University, Graduate School of Medicine, Kyoto, Japan). (c) Diffusion and brain white matter fiber tracking. Water diffusion in brain white matter is anisotropic. As a result it is possible to determine for each voxel of the image the direction in space of the fibers. Using postprocessing algorithms the voxels can be connected to produce color-coded images of the putative underlying white matter tracts. (Images courtesy of Y. Cointepas and M. Perrin, SHFJ/CEA, Orsay, France).

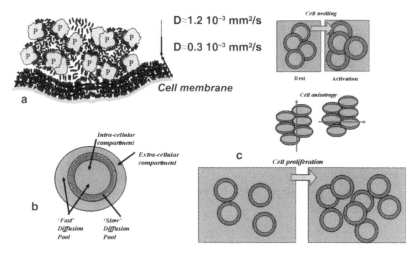

Figure 16.6 Membranes, water structure, and diffusion. (a) Schematic representation of the structuring effect of charged proteins (P) and membranes on water molecules. Bulk-water molecules are exchanging rapidly with the water molecules in the protein hydration shells. Other water molecules are trapped in a membrane-bound layer. Charges of the protein membranes and the underlying cytoskeleton strongly influence the water network in this layer, resulting in an order, a lower density, and a slower diffusion coefficient. (b) Conceptual biphasic water diffusion model. The slow diffusion pool is made of a water layer trapped by the electrostatic forces of the protein membranes and associated cytoskeleton, as indicated in (a). The remaining of the water molecules, whether in the intra- or in the extracellular compartment, constitutes the fast diffusion pool (which remains, however, slower that free water). (c) Diagram showing some predictions of the biphasic water diffusion model. Water diffusion would be reduced through an increase of the slow diffusion pool fraction associated with membrane expansion, as during cell swelling (brain activation, *top*) or cell proliferation (cancer, *bottom*). It should be noted that MRI diffusion measurements are always performed along one particular direction. In most cases, tissues are isotropic. However, in tissues with anisotropic cells (*middle*), such as brain white matter, the number of membrane surface intersections with the measurement direction will vary according the respective angle of the measurement direction and the long axis of the cells. The slow diffusion phase is, thus, expected to be the largest when those directions are perpendicular.

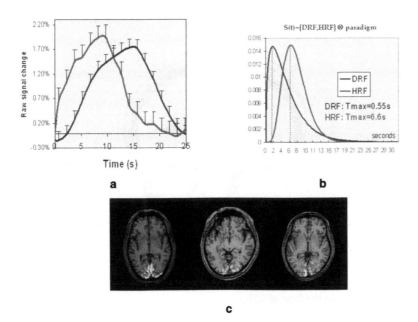

Figure 16.8 Diffusion fMRI. (a) Time course of the BOLD and diffusion fMRI responses in visual cortex (raw signal changes) for a 10 s stimulus and an image resolution of 2 s. The amplitudes of the responses are comparable, but the diffusion response is clearly ahead of the BOLD response by several seconds, both at onset and at offset. (b) Intrinsic response functions derived from the raw signal time courses of (a) after deconvolution of the signal from the stimulation time course. These intrinsic response functions, DRF for water diffusion and HRF for the hemodynamic response, represent the theoretical responses that should be observed from impulse stimuli. The DRF picks much earlier than the HRF (T_{max}) and presents a more asymmetric shape with a sharp increase and a slightly slower return to baseline, a pattern also observed in IOS responses (Fig. 16.7). (c) Activation maps obtained with diffusion fMRI during visual stimulation in three subjects, showing that the localization of the voxels where water diffusion is reduced are well localized within the brain cortex (Courtesy of D. Le Bihan, S. Urayama, T. Aso, T. Hanakawa, and H. Fukuyama, Human Brain Research Center, Kyoto University). *Abbreviations*: DRF, diffusion response function; HRF, hemodynamic response function.

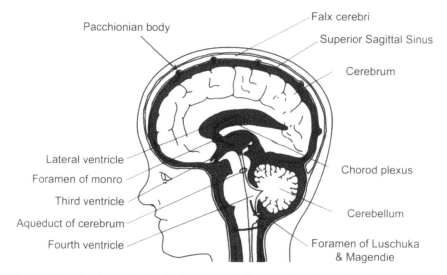

Figure 17.2 Cerebrospinal fluid. Blue area indicates the subarachnoid space filled with CSF. Lines indicate the flow of CSF.

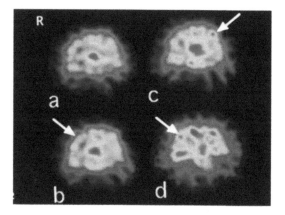

Figure 17.6 CBF on animal PET images. (a) Coronal image of the sensory cortex at rest. (b) Stimulating the left forepaw. There was no activation on the right side of the brain after injection of bromo-pyruvate into the sensory cortex. (c) Contralateral side stimulation (right forepaw stimulation) induced CBF increase on the left side of the cortex, which had not received any treatment. (d) Right as well as left sensory cortex showed high accumulation of FDG after right-side injection of bromo-pyruvate. R: right side of the cat brain.

Index

T - #0291 - 101024 - C0 - 235/191/22 [24] - CB - 9789814267526 - Gloss Lamination